ADVANCES IN MEDICINE AND BIOLOGY

ADVANCES IN MEDICINE AND BIOLOGY. VOLUME 14

ADVANCES IN MEDICINE AND BIOLOGY

Additional books in this series can be found on Nova's website under the Series tab.

Additional E-books in this series can be found on Nova's website under the E-books tab.

ADVANCES IN MEDICINE AND BIOLOGY

ADVANCES IN MEDICINE AND BIOLOGY. VOLUME 14

LEON V. BERHARDT
EDITOR

Nova Science Publishers, Inc.
New York

Library of Congress Cataloging-in-Publication Data

ISSN: 2157-5398

ISBN: 978-1-61761-930-4

Published by Nova Science Publishers, Inc. † New York

Contents

Preface

This continuing series gathers and presents original research results on the leading edge of medicine and biology. Each article has been carefully selected in an attempt to present substantial topical data across a broad spectrum. This compilation focuses on the symptoms, diagnosis and treatment of Pompe disease; insulin resistance and its relation to intra-uterine growth retardation; evidence-based medicine research; the therapeutic values of Vitamin E supplements; megalin-mediated endocytosis in renal proximal tubule cells; and biofilm development kinetics.

Chapter I - The study of the energy metabolism in children is a critical part of their clinical evaluation in cases of illness, malnutrition or obesity, because it is necessary to meet their energy requirements. This article reviews the characteristics and limitations of methods of assessment of energy expenditure. Predictive equations based in anthropometric data have been used for many years for the estimation of energy expenditure, but there are many both physiologic and pathological conditions that cause considerable discrepancies between real and estimated values so it's preferable to use direct measurements. Doubly labeled water method is considered as a gold standard but it's usually used only for research studies. And it informs about prolonged periods so it's not useful in very changing conditions. Indirect calorimetry, also considered as a gold standard, is the most used technique in clinical practice, but it can't be employed in some critically ill children because of technical restrictions and it can provide information about the energy expenditure in a laboratory, as well in basal conditions as during several types of exercise, but not of the energy expenditure among a day in free-living conditions. The reports of physical activities are not often reliable. Several methods have been designed to estimate the daily energy expenditure in free-living conditions. These methods are based in heart rate data or temperature or motion assessment (accelerometry) or are combination of several of these variables, but they also present several limitations.

Chapter II - Pompe disease is a rare neuromuscular glycogen storage disorder with a variable frequency of 1:138,000 in Caucasians, 1:40,000 in USA and higher in other ethnic groups. It is one of the 41 lysosomal storage disorders occurring in children and adults due to deficiency of lysosomal hydrolase acid-α-Glucosidase (acid maltase). This results in the accumulation of glycogen in numerous tissues with marked deposition in cardiac, skeletal and hepatic tissues of infants. The clinical presentation varies among infants and adults. Infants commonly present with muscle weakness, hypotonia and floppy baby with delayed motor milestones; whereas in adult onset, proximal muscle weakness with gait abnormalities is more distinct. Respiratory insufficiency with obvious cardiomyopathy is seen in infantile group but is less common in adults. The early presenting time of the disease in our centre is found to be 4.5 months to 3.5 years with an average frequency of 3.1% among five hundred and sixty six children with neuroregression and developmental delay. In older children and adults, diagnostic delay averages to 7 years. Initial diagnostic tests for cardiomyopathy include X-ray of the chest, EEG and cardiac echo for specific Pompe pattern and elevated serum Creatine Kinase (CPK). Muscle biopsy is often performed for glycogen accumulation. However, normal values in adults do not rule out this possibility. Hence, the confirmatory diagnosis is made by estimating acid-Glucosidase from skin fibroblasts or muscle tissue. Though collection of both tissues are invasive and time consuming estimating enzyme activity from leucocytes, using natural acid-Glucosidase inhibitor (acarbose) is found to be more appropriate. Infantile Pompe disease has virtually undetectable activity in fibroblasts and 30-40% in leucocytes. However, a newer method using ratio of acid-Glucosidase activity with and without acarbose of less than 0.2 is found to be as sensitive and specific as the fibroblast study. Mutation analysis of GAA gene is the recommended method for confirming diagnosis as it additionally helps in carrier identification and prenatal diagnosis in the subsequent pregnancy. The availability of precise diagnosis and awareness has made it possible to offer enzyme replacement therapy at an early age. The naturally occurring form of human GAA that acts like a native enzyme is found to be effective in degrading glycogen by catalysing the hydrolysis of lysosomal glycogen and compensates for the GAA deficiency seen in Pompe disease. Reversal of cardiomyopathy and measurable motor gains have been observed in 70-100% of the study subjected reported to date. With the availability of specific diagnosis and therapeutic measures, timely diagnosis and treatment as well as appropriate genetic counseling for the future, reproductive decisions can be provided.

Chapter III - Overweight (BMI > 25) and obesity (BMI > 30), in our modern world of food abundance has become epidemic and is associated with a whole range of health problems like hypertension, diabetes-2, and an excess of cardiovascular and renal diseases. We will describe intra-uterine growth retardation due to maternal undernutrition, life-style and genetic susceptibility as causes for insulin resistance and β-cell dysfunction. The 'fetal' or 'early' origins of adult disease hypothesis provided evidence for the association between the perturbation of the early nutritional environment and the major risk factors (hypertension, insulin resistanc , and obesity) for cardiovascular disease, diabetes, and the metabolic syndrome in adult life. In this review we give some epidemiological examples from the "Dutch Hunger Winter" a war induced famine between August 1944 and April 1946 in Amsterdam with less than 1,000 calories per day from government food rations. Epidemiological studies demonstrate that exposure to famine during gestation resulted in increases in impaired glucose tolerance, obesity, coronary heart diseases, atherogenic lipid profile, hypertension, microalbuminuria, schizophrenia, antisocial personality and affective disorders. Maternal undernutrition during gestation has important effects on health in later life, but it is hypothesized that the timing of the nutritional insult determines which organ system is affected. Maternal undernutrition can express itself in the second or third generation. Obesity due to diet and a sedentary lifestyle may result in excess adipose tissue which actively participates in the integration of whole-body energy and fuel metabolism by the secretion of many hormones like leptin, adiponectin and resistin. Life style features like a) disturbance in the autonomous nervous system (biological clock), b) gender, c) socioeconomic status, d) race and e) stress all may be contributing factors in the origin of metabolic syndrome and insulin resistance. Genetic factors influence the different components of the Metabolic Syndrome. Type 2 diabetes is a polygenic disorder. The "thrifthy gene theory" states that it can be inherited from our ancestors living in an environment with unstable food supplies and famine. For them it could increase their probability of survival if they could maximize their storage of food surpluses. With the term "thrifthy genes" one can think about underlying hormonal pathways and metabolic routes like growth hormone and insulin. Inflammation, a second characteristic of insulin resistance, can be caused by Interleukin-6 (IL-6) and Tumor necrosis factor-α (TNF-α). Peroxisome proliferator-activated receptor-gamma (PPAR-gamma) is a transcription factor involved in regulating genes in adipogenesis and, by implication, insulin action. Diet, exercise and preventing stress are the first-line strategies in the management of type 2 diabetes. Exciting results for medical glycaemic control are found in (thiazolidinediones (TZD eg. Pioglitazone). They are agonist for the PPARγ and also have an effect on the recently discovered hormone resistin which probably can link obesity to diabetes-2.

Chapter IV - Wastewater comprises of many different sources and contaminants, and the combinations of these require specific types of remediation strategies. Biological approaches based on the use of microbial biomass against physic-chemical processes are preferred as the low-cost alternative. Contaminated groundwater, urban and industrial wastewater may contain contaminants such as nitrates, heavy metals, crude oil derivatives or even high levels of nutrients. Microbial populations continually described for degradation of recalcitrant xenobiotics, accumulation or adhesion of heavy metals and for their contribution in geochemical cycles such as nitrogen, exceed expectations when presented with new water contaminants. Thus, their highly rate of adaptation make them ideal for wastewater treatments from different sources. Biofilm technology has been proved to be one of the most advanced methods to remove in a selective way, contaminants from water resources. Biofilms are very complex habitats where the microbial cells responsible for the treating process are embedded in a polymer matrix. Water treatment by submerged filter technology requires a biofilm formation around an inert substrate. The water to be treated passes through a biofilter consisting of the inert substrate and the biofilm established. The main components of the biofilm are different physiological and morphological bacterial cells and their composition depends on the environmental conditions. The relationships among the microbial populations established, whether mutualistic or competitive, along with the combined enzymatic activity, determine the efficiency of the bioremediation process, thus the importance in the exploration of the biodiversity. Different culture dependent and culture independent techniques have been used to describe the constant and shifting biodiversity present in wastewater biotechnologies. Trends in community profiling using the 16S ribosomal RNA as a taxonomic marker have allowed the establishment of a large taxonomic database available for microbiologists, but most importantly, it has created a general consensus to describe microbial populations to date. A review of the methods used for biofilm biodiversity determination and the corresponding microbial structures in submerged biofilm technologies is presented, along with the future trends in microbial population classification.

Chapter V - Project management (PM) may be used as a guideline for Evidence-Based Medicine (EBM) research. Such kind of research consists in the use of statistically sound estimates derived from scientific studies on population samples for supporting clinical decision making in diagnosis, investigation or management of individual patients. They are put in hierarchy in the following order: systematic review and meta-analysis randomized controlled trials with definitive results, randomized controlled trials with non-definitive results, cohort studies, case-controlled studies, cross-sectional surveys, and case reports. In such studies medical protocols are of major importance and a PM approach may be applied. For instance, a patient transplantation process starts with the acknowledgment of transplant necessity, where restrictions and hypothesis are determined defining medical procedures, psycho-social attendance strategies and performance criteria (initiation). Then, transplantation protocols and plans are formulated, that is, its schedules, activities, risk identification, staff assignment, medication, exams, evaluations and surgical intervention planning are elaborated (planning). Therefore, the preparation of the surgical transplant act requires that medical procedures are performed while the patient progression is evaluated constantly (executing). Moreover, to get the patient physiologically ready for it, several exams are performed and some parameters are monitored, because some specific limits or values must be kept or achieved (controlling). After the successful surgical intervention, the patient is still monitored for a while, usually one year, and then he returns from time to time for a follow up. Basically, after the surgery until a year his medical records should be reviewed and finalized to avert potential problems on him and on future transplants (closing). Consequently, as a methodology for scientific research, PM concepts and techniques may be used for treating a patient. Furthermore, regarding medical studies' hierarchy, PM may be simultaneously applied from bottom-up, that is, if it is used in every kind of research then it would be easier to perform systematic reviews and meta-analysis. In addition, a computational data-base could be created for EBM using PM, it could have functions to statistically support primary studies focused on treatment, diagnoses, tracking, prognoses and causality. Therefore, a PM approach allied to a computational data-base could speed up medical research while facilitating the physicians work, because there would be more and new kind of ready to use statistical data for supporting EBM researches. Also, due to that standardized approach, information retrieval techniques would be more effective in discovering new knowledge.

Chapter VI - The complex nature of glycogen storage disease type II or Pompe disease requires a multi-disciplinary team approach. A genetic counselor trained in metabolic or neuromuscular disorders can be a valuable member of the health care team working with patients with Pompe disease. The roles and unique clinical skills of a genetic counselor working with Pompe disease are highlighted here. These roles include assisting with the confirmation of a diagnosis, educating the family about the disease, supporting the management of disease and disease progression, and facilitating the communication process between the patient and the health care team.

Chapter VII - Free radical generation, redox imbalance and oxidative stress have been implied as the key pathological mechanism in the development and process of many diseases. Vitamin E is a very common dietary compound that functions as a potent antioxidant to scavenge toxic free radicals, regulate redox status and resolve further oxidative damage. Vitamin E has been comprehensively investigated in cell line experiments in vitro and animal models in vivo. Recently it has been extensively evaluated its protection potentials in large scale epidemiological and population studies. However the therapeutic values of vitamin E supplements in promoting health conditions in general population and preventing or treating disease conditions in major randomized controlled trials are still open to question. The studied diseases include inflammation, ischemia and reperfusion, degeneration, aging and cancers. In addition, many other antioxidants including vitamin E do interfere with the efficacy of other antioxidants and pharmaceutics used in clinical management and result in unwanted toxic side effects. In this chapter, we will critically review the therapeutic values of vitamin E in clinical applications from its chemistry, pharmacology, metablism, clinical significance to medical recommendation. The prospective and significant implications of Vitamin E to the clinical practices of modern medicine are discussed.

Chapter VIII - The attachment of microorganisms to surfaces and the subsequent biofilm development are very complex processes affected by several variables. Among these variables, it has been shown that the nutrient/substrate concentration may influence the structure of biofilms growing in aqueous environments. The goal of this study was to assess the effect of the nutrient load on biofilm formation by Escherichia coli JM109(DE3) by using three substrate feed concentrations (glucose concentrations of 0.55, 0.15 and 0.050 g.L-1) and four dilution rates (0.0042, 0.0126, 0.0800 and 0.1650 h-1) on separate experiments. Biofilms were formed in a flow cell reactor under turbulent conditions, with a Reynolds number of approximately 5000, by using a recirculating system that enables independent control of the dilution rate and flow velocity. Three nutrient load ranges were obtained on the independent experiments and our results show that when high nutrient loads were used, planktonic cell concentration was reduced and glucose consumption was proportional to the nutrient load. Additionally, when moderate nutrient loads were tested, higher planktonic cell concentrations were observed when compared to the high nutrient load conditions. When the nutrient load was further reduced planktonic cell concentration was further increased. Unlike the glucose consumption values obtained for the high nutrient load conditions, glucose consumption for the moderate and low nutrient loads was not dependent on the nutrient load that was used. Furthermore, when the system was fed with the highest nutrient concentration (0.55 g.L-1), glucose consumption was the same even when the dilution rate was increased. However, for the intermediate nutrient concentration (0.15 g.L-1), glucose consumption was enhanced with the increase in dilution rate.

As a general trend, E. coli biofilm formation under flow conditions was favored at higher nutrient loads and this could be measured both in terms of biofilm thickness as well as wet weight. There were no significant effects on the amount of biofilm formed by moderate and low nutrient loads.

Chapter IX - Receptor-mediated endocytosis is a pivotal function of renal proximal tubule cells (PTC) to reabsorb and metabolize substantial amounts of proteins in glomerular filtrates. The function accounts for the essential conservation of nutrients, carrier-bound vitamins and trace elements filtered by glomeruli. Impairment of the process results in a loss of such substances and development of proteinuria, an important clinical sign of kidney disease and a risk marker for cardiovascular disease. Megalin is a multi-ligand endocytic receptor abundantly expressed primarily at clathrin-coated pits of PTC, playing a central role in the process. Megalin is a large (~600 kDa) glycoprotein member of the low-density lipoprotein receptor family. Megalin cooperates with other membrane molecules, such as the cubilin-amnionless complex, Na+/H+ exchanger isoform 3, type IIa sodium phosphate cotransporter and chloride channel-5 in PTC, in order to transport various ligands into the endocytic pathway to lysosomes. Many intracellular adaptor proteins have been identified to interact with the cytoplasmic tail of megalin for intracellular trafficking. Megalin is also known to participate in signal transduction in the cells. Megalin-mediated endocytic overload of glomerular-filtered proteins leads to damage of PTC. Advanced glycation end products, generated in the circulation of patients with diabetes, are also presumed to injure PTC via megalin-mediated endocytosis. Further studies are needed to elucidate the mechanism of megalin-mediated endocytosis and to promote future development of strategies for preventing damage of PTC.

Chapter X - CD8+ natural killer T (NKT) cells possess potent cytolytic abilities and may be implicated in immune tolerance; however, the phenotype and function of this minor cell population are poorly defined. We have recently developed a protocol to culture CD8+ NKT-like cells and demonstrated the immunosuppressive activity of these cells towards CD4+ responder T cells. Data in this study indicate that these CD8+ NKT-like cells are anergic to stimulation by anti-CD3 and antigen presenting cells (APC) and can also efficiently suppress the proliferation of CD8+ naïve T cells. Most of the suppressive activity is cell contact-dependent, although soluble factors may also play a role in the function of their suppressive activity. The CD8+ NKT-like cells can efficiently kill antigen presenting cells but not naive T cells. APC-free suppression assays further demonstrate that the CD8+ NKT-like cells suppress T cell proliferation using APC-independent mechanisms including direct blockade of IL-2 production by responder T cells. Gene expression profiling suggested that over 3000 genes differ between the cultured CD8+ NKT-like cells and the naïve CD8+ T cells by more than 2-fold with >100 genes upregulated by more than 10-fold in the CD8+ NKT-like cells. A large proportion of the highly upregulated genes are soluble factors or surface markers that have previously been implicated in immune suppression or are likely to possess immunosuppressive properties. Many of these genes are regulated by two key cytokines, IL-10 and IFN-γ, suggesting that these pathways play a critical role in the suppressive function of these cells.

Chapter XI - Pompe disease is a rare genetic metabolic myopathy caused by deficiency of lysosomal acid α-glucosidase (GAA). Age of onset and rate of progression cover a broad clinical spectrum. Presentation of disease ranges from rapidly progressive infantile form with cardiac, skeletal and respiratory muscles always fatal without specific treatment to juvenile forms with muscular, respiratory and sometimes cardiac impairment and late onset forms consisting in progressive limb girdle myopathy beginning in adulthood, usually around the third decade, involving skeletal muscles but sometimes more specifically respiratory muscles.

Severe forms of Late Onset Pompe Disease (LOPD) are characterized by a progressive respiratory involvement that can lead to severe respiratory failure, which can require ventilatory support by mechanical ventilation, and/or muscular involvement that can lead to loss of ambulation. The management of these patients usually requires a multidisciplinary approach involving the neurologist, the myology specialist, the pulmonologist and the physical medicine specialist.

In this book chapter, the authors describe the particularity of these severe forms: clinical presentation with the description and relationship of the muscular and respiratory impairments. Muscular and functional explorations, respiratory and sleep explorations and criteria for starting ventilation are also discussed focusing on Pompe disease particularities. Recent advances in therapeutics, like enzyme replacement therapy (ERT) have mainly focused on infantile forms; however, late onset forms due to their frequency and to the heterogeneous presentation and progression they are susceptible to present, might also benefit from those advances.

Chapter XII - The microbial biofilm development is theoretically analyzed in combination of Monod-based microbial kinetic equation with a biofilm architecture model. It is assumed that the development of biofilm is primarily dominated by the net decay rates of microorganisms while an increase in the population of microorganisms are expressed by the sum of the net growth and decay rates. The biofilms are modeled so that they consist of the stack of micro-colonies and a density function of micro-colonies within biofilms is proposed in order to express the biofilm architecture in the direction perpendicular to the surface of substratum. By modeling a micro-colony as disk-shaped, it has been demonstrated that the maximum thickness of biofilms is in a range of 20 - 55 μm at the mean density of 7x10-6 g/cm3 of microorganisms in biofilms, while it has been suggested that an increase in the population of microorganisms significantly causes an increase in the density of microorganisms in micro-colonies besides an increase in both the maximum thickness of biofilms and the number of the stacks of micro-colony in biofilms.

Chapter XIII - The puncture test is an empirical test that uses a texturometer. The maximum force on a plunger is analyzed as it penetrates a sample. In Japan, the physical properties of liquid, gel-like, and semi-solid foods for babies and people with dysphagia are measured using the puncture test. This study analyzed the forces on the base and the curved surfaces of the plunger during the puncture test. It was foundthat the contribution of the curved surface to the overall force on the plunger could not be ignored. It was concluded that a simple calculation involving the combined forces on both surfaces of the plunger and the area of its base is inappropriate. Based on these findings, the compressive forces reported in the literature should be recalculated if they were estimated without considering the contribution of the curved surface to the compressive force.

In: Advances in Medicine and Biology Research, Vol. 14
Editor: Leon V. Berhardt
ISBN: 978-1-61761-930-4

Chapter I

Methods of Study of Energy Expenditure

Ana Sancho Martínez[1]
Infanta Sofía Hospital, San Sebastián de los Reyes, Madrid (Spain)[1]

Abstract

The study of the energy metabolism in children is a critical part of their clinical evaluation in cases of illness, malnutrition or obesity, because it is necessary to meet their energy requirements. This article reviews the characteristics and limitations of methods of assessment of energy expenditure. Predictive equations based in anthropometric data have been used for many years for the estimation of energy expenditure, but there are many both physiologic and pathological conditions that cause considerable discrepancies between real and estimated values so it's preferable to use direct measurements. Doubly labeled water method is considered as a gold standard but it's usually used only for research studies. And it informs about prolonged periods so it's not useful in very changing conditions. Indirect calorimetry, also considered as a gold standard, is the most used technique in clinical practice, but it can't be employed in some critically ill children because of technical restrictions and it can provide information about the energy expenditure in a laboratory, as well in basal conditions as during several types of exercise, but not of the energy expenditure among a day in free-living conditions. The reports of physical activities are not often reliable. Several methods have been designed to estimate the daily energy expenditure in free-living conditions. These methods are based in heart rate data or temperature or motion assessment (accelerometry) or are combination of several of these variables, but they also present several limitations.

Introduction

Importance of the Study of the Energy Metabolism

Prevalence of overweight and obesity has raised in last years, and it is estimated than more 30% of children in USA and 20% in Europe has overweight or obesity [1]. Overweight during childhood is associated with a higher risk of cardiovascular disease in adulthood [2]. Risk factors for cardiovascular disease such as hypertension, dyslipemia, impaired glucose tolerance, and vascular abnormalities are already present in overweight children [3, 4].

Surprisingly and simultaneously with the high prevalence of overweight, the incidence and prevalence of undernutrition in children in developed countries is higher as what is commonly thought, although it almost always appears in illness conditions. Children have a lower energetic reserve [5, 6], so they are more vulnerable in this conditions. The risk of malnutrition raises as the illness reaches more severity, so it is maximum in critically ill children [7-16]. The incidence of proteic-caloric malnutrition in Intensive Care Units usually varies between 16-55% [17, 18] but it may even reach percentages as high as 72% [14]. During the first days of hospitalization at Intensive Care Units nutritional support usually don't cover the requeriments [9, 11, 19, 20] and it may even be of 20% or less than these [19]. Nevertheless, as the stay gets longer the nutritional regimen tends to exceed the requirements [9, 11]. Malnutrition is related to an increase in morbility [7-9, 17, 18, 21] and mortality [9, 19] but the excessive giving of nutrients, as it was recommended some years ago, has also harmful consequences, creating a respiratory compromise by the increase in CO2 production [22-24] and altering the liver function [24]. The knowledge of energy metabolism is necessary to the design of well-done nutritional strategies [25-28].

Energy Expenditure and its Components

The daily energy requirements (DER) are the result of the sum of the rest energy expenditure (REE), the energy spent as a consequence of the physical activity (PA), the thermogenic effect of food and the energy consumed for growth[h] [25-27, 29, 30]. REE represents the consume of basal metabolism (BM, basal metabolic rate) [31] plus the energy consumption of the maintenance of a vigil activity in a situation of muscle rest [25, 32, 33]. Although some authors consider REE and BM equivalent [24, 34, 35], REE is between 8-10% higher than BM [32, 33], because shorter fasting and rest periods are asked for the measurement of REE than for the measurement of BM (for the measurement of BM a fast during twelve hours and a total rest during 8 hours are required).

Interindividual REE variations are related to age [5, 16, 17, 25, 25, 36-42], sex [25,36-51] weight [19,25,32-55], body lean mass [34,39-42,52-53,56-63], genetic factors [42] including parental weight [57,64,66-67] and ethnic group [40,49,51,57,64], clinical and nutritional situation [9,18,32,34,36,39,40,49,60,68-75], sympathetic activity [75,76] and ambient factors like temperature [39,77]. Influence of height is controversial [25,36-38,43,59].

During the growth period the muscle mass increase is proportionally higher than the increase of the organic tissue (liver, kidneys…) [5,6,40]. As organs are metabolically more active than muscle, the MB per kg of weight decreases along the years [15,6,25,40,49,50,65.]

REE accounts for approximately 70% of DER26,53, but it can vary between 45-70% in healthy people REE, according with their age and life-style25. For critically ill patients, REE is often between 75 and 100% of TEE [76].

The thermogenic effect of food is the energy required for the ingestion and digestion of food [22,79], and its about 8-10% of the daily total energy expenditure (TEE). The thermogenic effect of food depends on its composition, the way of administration [79,80] and the basal situation of the child [81,82].

Physical activity is the most variable component of the total energy expenditure [25]. It seems to expend less than it used to be estimated in children [83]. In five years old children percentages of 16-17% (± 7-11%) of TEE due to physical activity have been found [41,83]. The energetic cost of an activity also varies in relation to age [46,65], weight [46,47,57], body composition [46,47,51,57,65] and ethnic group [49,51,57.64.65].

Growth represents the 35% of the DER during the first 3 months of life [25], and then it progressively decreases along the two first years of life (6% at six months, 5% at 12 months34, 3% in the second year [25,29]). Between the third year of life and the adolescence it keeps stable around 1-2% [25].

The Effect of Nutritional Situation in Energy Metabolism

TEE of obese people is higher as their weight is higher [39,51,84]. Nevertheless, these people have less metabolic active tissue (fat free mass) per kg of body weight, so they can have a lower TEE per kg of body weight [38,39,56]. Obese people's physical activity is usually lower than thin people's physical activity [46,83,84].

Malnutrition causes important changes in energy expenditure [9,18,68,87-89]. REE [89] and thermogenic effect of food [90] decreases. But even though global TEE is lower than in not malnourished children. TEE per kg of weight was been found higher than in well-nourished kids [18,88], as in adults [87]. And BM usually increases at the first stages of nutritional rehabilitation [69,70,91].

The weight at birth, a reflect of fetal nutrition, conditions a difference in REE at long term, thus, REE has been seen lower in prepuberal children born at term and small for their gestational age in comparison with children born at term with adequate weight for their gestational age [92]. Conversely, REE has been seen higher in prepuberal children born preterm and small for their gestational age in comparison with children born prematurely but with adequate weight for their gestational age [92].

The Effect of Illness in Energy Metabolism

There is a strong variability in the clinical and metabolic responses to illness, to the different types of illnesses, to the different grades of severity [25,27,71-73] and to the time evolution. Illnesses such as fever [25] or heart failure [74], often increases the REE, but simultaneously the physical activity often decreases [74], so the DER may be decreased [26].

It's necessary considerer the increase in energetic losses with the depositions during diarrhea [25].

The body response are similar in critical illness or severe traumatism, it only differs in duration and severity [93-98]. It consist of the creation of a hyperdinamic situation with an increase in body temperature [16,.93,95-98], CO2 production, O2 and glucose consumption [16.93-98], glucogenolysis, proteolysis and futile cycle of substrates [16,93,95-98]. This general complex of metabolic processes is produced by the interaction of tissue mediators and the vascular endotelium, inflammatory cells and neuroendocrine system [93-95,99]. Its goal is the mobilization of substrates and energy to cope with the inflammation and tissue repairing, even though deleterious effect on body lean mass [93,96-98]. These metabolic alterations can keep on during weeks and even months [94], but their duration of is highly variable, as in adults [10] as in children [9,10,12,14,88,96,96,97,100].

But even though there always is hypercatabolism [95,98], hypermetabolism as stress response is not as frequent as it was usually thought [9,10,101-107]. In children stress response is not as hypermetabolic and not as stronger as it is in adults [9,10,103,104,107] and hypometabolism is found in many situations [9,10,14,15,100,103-107]. Nevertheless, some critically children show remarkable rises in energy metabolism [9,10,16,88,100,104,108-110]. In malnourished adults REE increase in response to the injure seems to be limited, but it doesn't seem limited in children [9,10,88]. EE varies from day to day and it's influenced by a lot of factors. Sedation decreases EE [9,10], the effect of muscle relaxation is not so clear [9,10,88,94,111,112]. Fever increases EE [10,113,114,115] hypothermia decreases it [114-115]. Mechanical ventilation decreases EE [12,109,116], while inotropic drugs raises it [9,75,101]. The administration of nutrition increases EE [9,22,111], as common care does transiently [116] (if they are appropriate, the global effect is a reduction [83]). So energy expenditure in critically ill children is poorly predictable.

REE doesn't serve as biologic marker of severity [10,16,15], even though some studies have found some correlation between illness severity [10,16,15] and rise of REE [21,17,97,100,110]. Some authors found different energy expenditure for every diagnosis [10,14,15,111], but others didn't [9,20].

The grade of hypermetabolism or hypometabolism in critically ill patients is assessed by the relation between the estimated energy expenditure and the measured energy expenditure, and this relation is defined as metabolic index [117]. The normal values for metabolic index are 90-110 [117]. Other authors use the relation between preoperative energy expenditure and post-operative energy expenditure [87].

Methods of Study of Energy Expediture

By the use of methods of direct measurement of EE (indirect calorimetry, doubly labeled water method, heart rate monitoring, accelerometry...) 24 hours measurements includes the thermogenic effect of food and the energetic cost of synthesis of tissues [25,31], but not the energy deposited at them [25]. Thus, the result of these measurements is equivalent to DER, but children and adolescents require more energy, this energy can be estimated by calculations of growth rate [25]. After the age of 2 years, the growth represents less than 1 or

2% of the DER, so it can be enough multiply the value of the physical activity level (it is the quotient TEE/BMR) by 1.01 (making TEE a 1% higher) [25].

When measurement of TEE is not disposable, it can be estimated by factorial calculations based in the time spent in daily activities and the energy cost of these activities [25]. During growth is necessary to sum the component of EE caused by the growth, as set out above [25,26].

Methods of Estimation of Energy Expenditure

Methods for estimate the individual energy expenditure have been used for years because of the lack of direct measurements of TEE in daily clinical practice, although these methods are disposable at research studies yet [25,26].

Holliday made first approach to the estimation of energy expenditure [118]. He established the supports to intravenous rehydration as a function of the weight, assuming that energy expenditure were equivalent to water needs, so 100 kcal/kg/day = 100 ml/kg/day [118]. But mammalian need really 140 ml/kg/day every 100 kcal/kg/day [119]. Darrow employed this physiologic equivalence to make the table showed in table 1 [119]. This table is still used, although it isn't as extended as Holliday's approximation. At the same time, Darrow created the concept of metabolized calorie, which is considered today as equivalent to the actual REE [119].

Table 1. Hollyday and Darrow's approximations.

	Body weight (kg)	Energy requirements (Kcal/kg/day)
Holliday	3-10	100
	11-20	50
	> 20	20
Darrow	< 3 kg	47 ± 4
	3 - 10	70 ± 5
	11-15	57 ± 4
	16-25	45 ± 4
	26-35	37 ± 3

There have been a lot of other approaches to the estimation of REE later. More than 200 equations have been published39. All these equations are based in anthropometric features [36-38,43-45,52-54,56,81]. They have been validated in several studies.

Since its publication in 1919, Harris Benedict's equation [36] has often been used for the prediction of REE in healthy adults with normal weight. However, the evidence is that frequency of error in REE estimations is high with this equation [39,54,68,81,120], most times by overestimating it [53,54,68,120]. It's not appropriate to/for REE estimation in malnourished patients [68]. It's not adapted to pediatric population, although it's can be used in 10 years old and older children. In 1981, Caldell and Kennedy validated a equation from Harris Benedict's one for its use in children younger than 3 years (table 2)[43]. Some other authors developed other methods to estimation of REE in children: tables, as Talbot [44] and

Lewis [45] done, or equations as Fleisch did, according to age, sex and body surface [37]. In 1985, Schofield evaluated the data of 114 studies on energy expenditure, and he published other equations, for children and adults [38]. Schoefield's equations are considered today as the more adequate [47,121,122]. WHO published its own equations in 1985 [81].

Table 2. Equations for the estimation of REE.

Infants (<2 years)	
Caldwell-Kennedy [43]	22 + (31,05 · W + (1,16 · H)
Children	
Males	
Schofield [38] <3 years	(0,0007 : W) + (6,349 · W) – 2,584
Schofield [38] 3-10 years	(0,082 · W) + (0,545 ·H) · 1,736
Schofield [38] 11-18 years	(0,068 · W) + (0,574 · H) + 2,157
Fleisch [37] 1-12 years	24 · BS · [54 – (0,885 · A)]
Fleisch [37] 13-19 years	24 · BS · (42,5 – [0,643 · (A – 13)])
Females	
Schofield [38]< 3 years	(0,068 · W) + (4,281 · H) – 1,730
Schofield [38] 3-10 years	(0,071 · W) + (0,677 · H) · 1,553
Schofield [38] 11-18 years	(0,035 · W) + (1,9484 · H) + 0,837
Fleisch [37] 1-12 years	24 · SC · (54 – [1,045 · A])
Fleisch [37] 13-19 years	24 · SC · (42,5 – [0,778 · (A – 11)])
Adults	
Males	
Harris and Benedict [36]	66 + 13,8 · W + 5 · H – 6,8 · A
Females	
Harris and Benedict [36]	655 + 9,5 · W + 1,9 · H – 4,7 · A

Schoefield equations give results in MJ. 1 MJ = 1.000 kJ;
1 kJ = 4,184 kcal.
H: height (m); A: age (year)s; W: weigh (kg); BS: body surface.

In general, these equations overestimate EE in children [41,55,60,121]. A lot of physiological situations can cause important discrepancies between estimated and real values [8,21,26,34,42-112], as set out above.

All the mentioned predictive equations are usually employed in ill children in clinical practice. But they were designed to the estimation of needs of healthy kids with normal life, with an usual level of physical activity [172], so they are inadequate for the calculation of energy requirements in illness [8-10,12,14,20,81,89,96-100]. As many physiological conditions, pathological situations can cause important variations between estimated and real values [2,5,7,30,39-42,46-51,55-77,79-80,82-84,87-91,93-117,124,125]. In ill children these equations usually overestimate EE [9,10,14,88,103,106,107], although some researches have found underestimations [10,15,16,104]. However, the number of children included in studies about the adequateness of this equations have been small, except for some ones with more than 40 children [12,14,15,88,111].

Other approaches have been designed with the purpose of improving the EE estimations in critical illness. For example, by the use of stress quotients [9,21,88,94,101,104,108,110], usually multiplying the predictions by Harris Benedict's or Schoefield's equations in children (quotients of 1.3 [108], 1.5 [104] and both [94]). With the use of such quotients overestimations of EE have been still found, either in adults [101,110] than in children [88,108]. Specific equations for children under mechanic ventilation were developed [15], but then they haven't shown usefulness [14]. The troubles in weighting the critically ill patient introduce even more uncertainty in anthropometric-based estimations [6,7,10,26,117], and the use of the ideal weight is controversial [14].

Methods of Estimation of Physical Activity

The knowledge of the children's physical activity is important to known TEE [25,41,46,47,49,51,57,64,65,83,86,126] but also to promote changes either at individual as at community levels [86,126]. Increased physical activity in children has shown to prevent obesity at older ages [86].

There have been tables which assign a physical activity ratio to each activity [25]. This ratio is based in the quotient energy spent in an activity/RBM by time unit [25] (EE is divided by RMR or BMR to eliminate the influence of individual weight [65]). Thus, knowing the time expended in each activity is possible to calculate the energy expenditure due to physical activity.

For long time, researchers have tried to quantify physical activity on the basis of self-reports and self-report questionnaires [126-132] (like Modifiable Activity Questionnaire or the Activitygram software [131]. But the general consensus is that self report tends to overestimate the intensity and duration of physical activity [126,133] (correlation between questionnaires and direct measures has been found low-to-moderate) [126] and that children themselves aren't able to provide contrasted information [128,130,132-134]. Self-report information is not considered reliable in children younger than 10 years [132]. Questionnaires show bias in children, because they are not designed to register non-organized physical activity [132-134]. With children, best way to get better data seems to ask about the last day by several calls [132-134]. Information from parents or relatives is not valid/reliable [134-135].

Nevertheless, direct observation can provide reliable information about the pattern of children's activities [126,128,129,133,134]. Some specific scales have been designed to categorize the intensity of the children's physical activity by direct observation, for example CARS (Children's Activity Ranging Scale) [136,137] and its modification OSRAP [85]. This task is held up by the highly changing and little routine pattern of children's physical activities [128,129]. And direct observation can generate itself a change in the physical activity if the child feel watched [134]. Besides, it requires a lot of time and ability from the observer and it's very expensive [128,129,134-136].

Self reports' intrinsic validity and reliability problems supports the use of direct measurement of physical activity [130,132-135]. EE during exercise can be measured by indirect calorimetry at laboratory [78,84,127,128,138-148]. Doubly labeled water meted allow the study in ordinary life conditions providing information of the patterns of physical activity during periods of 7-14 days, if the REE is known (for example, by indirect calorimetry) [25,46,47,51,58,59,64], but it doesn't inform about the pattern of physical activity along a 24 hours period [25,146,149]. Other methods to measure physical activity are heart rate monitors [126,133,134,139,147,149-157], body temperature measurements [158,159] and motion sensors [46,47,51,78,86,126-134,138-146,151-153,160-168].

Indirect Calorimetry

This method is considered as a gold standard. It provides direct measurements of EE through the quantification of oxygen and dioxide of carbon consumption. $O2$ and $CO2$ consumption are measured by the analysis of inspired and expired gases [7,26-28,170]. Using equations as Weir's [171], indirect calorimetry allows the calculation of the quantity of energy produced by the body (Weir equation: $EE = 3.94\ VO2 + 1.106\ VCO2 – 2{,}17$ ureic nitrogen). It's necessary to assume that the gaseous exchange is only done for the oxidative degradation of substrates [7,26-28,170,172]. Indirect calorimetry doesn't reflect the metabolic ways that don't use oxygen, as anaerobic glucolysis [173]. It's necessary to assume, too, than all the nitrogen produced in the protein oxidation is recovered in the urine [174], or the nitrogen eliminated with the insensible losses (5-8 mg/kg in children) must be added [175]. However, the importance of ureic nitrogen at the total amount of EE by the Weir's equation is small [27]. Without the measurement of ureic nitrogen, EE would be modified only in less than 1%175 - 2% [33,177]. Then it would be calculated [171] as $\text{Kcal/min} = O2\ (\text{l/min}) \cdot [3.9 + (1.1\ RQ)]$, being RQ (respiratory quotient) = $VCO2/VO2$.

There are several limitations to the use of indirect calorimetry in clinical practice. The measurement are not valid in compensating periods from metabolic acidosis or alcalosis, hypoventilation, hemodynamic instability, hemodialysis (there is elimination of $CO2$ through the dialized, so it is missed for the measurement) [178]. $FiO2$ mustn't be higher than 0.6 [9,10,15,105,179,180] and gases apart from $O2$, $CO2$ or nitrogen can't be used during measurements [9,14,15].

Indirect calorimetry is the most extended technique to measure EE in clinical practice [26,27], even though equipments are expensive, and their use requires validation and expert staff [15,172]. It can provide repeated EE measurements as it's not invasive [26,27], but it can't measure EE in free-living conditions [143], and the assessment of TEE finally consists on estimations based in the sum of REE measured by indirect calorimetry and the EE estimated for the physical activity of the child. Although there are mobile instruments [139,140], they are expensive, cumbersome and less precise than fixed metabolic carts [140]. At laboratory we can measure REE and EE during different activities, but it's necessary estimate the time spent in these activities or similar activities for the calculation of TEE. So TEE obtained by this method is only approximately. And they can change the patterns of exercise because they can limit the movements. Whereas their validity is well established at laboratory conditions, where indirect calorimetry is often used as a gold standard, there aren't well-informed studies using these machines in free-living situations.

Usual assessments in clinical practice are REE measurements. These must be done in a dark room [33], after a rest at home of 8 hours [25,31-33,39,40,51-54] and a rest at laboratory of 15 [35] or 30 minutes [25,31-33,39,40,51-54]. The child must lie face up, because simple activity of staying sitting can increase EE around 8% [33].. A fasting period of 12 hours is required to eliminate the thermogenic effect of food [25,31-33,39-40,52-54,60], although some authors consider sufficient a period of 2 hours [155,156].

Indirect calorimetry can estimated calories consumed during specific activities by the measurement of the oxygen consumption [154,155] (using Fick's equation: kcal = VO2máx (l/min) · 4.825 (kcal/lO2)[170, see table 1]). The physical activities performed usually consist in walking or running at a treadmill or at an cycle ergometer. Intensity of aerobic exercise is related with the maximum oxygen consumption [64]. For the estimation of physical activities in children, see subject before calorimetry.

Indirect calorimetry allows measurement of substrate utilization by the association of ureic nitrogen [26,27,121,174]. For that ureic nitrogen has to be measured in 24-hours urine, and there has to add non-ureic nitrogen [175,181], nitrogen from insensible losses [175] and, most of all if there is renal failure, the changes in the pool of ureic nitrogen [181]. Non-ureic nitrogen represents 10% or urinary nitrogen, and in children it's estimated as 55 mg/kg/day175. Changes in the pool of ureic nitrogen can calculated as Urinary Nitrogen not eliminated (mg) = (BUN2 – BUN 1) ·10 (0.6 · weight)[181].

Indirect calorimetry also can help to evaluate the adequateness of nutrition and its respiratory effect [28]. If RQ is higher than 1, carbohidrates oxidation predominates, and there is a high risk or lipogenesis and ventilatory problems because of accumulation of CO2[31]. If RQ is around 0.7, lipolysis and neoglucogenesis predominate [31]. But at clinical practice the accuracy of these assumptions is lower than EE measurement [31].

Indirect calorimetry allows as well the measurement of cardiac expenditure and the measurement of the respiratory death space when the data of gases in arterial and venous blood are disposable [182].

For some authors, indirect calorimetry is a useful tool for the prediction of the success of the weaning from mechanical ventilation, by the elevation in VO2 produced [183]. Some other authors have argued that it could be caused by a raise in catecolamines, and not only by a harder work of respiratory muscles [184].

Indirect calorimetry is the first-choice method for EE measurement in critically ill children. But transportable metabolic carts are not disposable at all centers [15], and they are quite expensive (20,000-100,000 dollars) [15]. For the moment, not clearly favorable effect in mortality, clinical situation173, weight or nitrogen balance has been demonstrated [173,185]. Besides, it can't measure EE in all critically children, because of technical reasons which will be commented later [7-9,14,15,17,106,112,179,186,187].

If the child isn't sedated, or the nutrition is not suspended, calorimetry reflects the EE at that moment, not the REE. Van Lanschot et al considered EE values measured by calorimetry in PICU as TEE [101], but some other authors have added activity factors, even in children with mechanical ventilation [21,104,108].

Patient must be at steady state to measurement represent the EE of 24 hours in a valid and reliable way [7,179]. This state is achieved when variability of VO2 and VCO2 is less than 5% [39,52, 126,185]-10% [52,110,126,179,185] (10% for some authors [188]) and RQ variability is less than 5% [110,126] at least during 30 minutes [126, 185]. For other researchers a duration of 5 minutes in this situation would be enough [45,72,126,132,185,186], as well in critically ill patients (in whom this duration has been validated [189]). Some other authors employ a stabilization period of 10 minutes [10,188].

The duration of the measurement is not definitively established. Along an hour approximately 15 liters of CO2 and oxygen are exchanged, so if the measurement last less than an hour the mistake may be of 10% for the VCO2 and 1% for the VO2 [190]. However, most studies in children employ measurements of 2049, [55,104] to 30 minutes [14,15,35,46,51,187,188]. Some researchers have used 45 minutes [10,40,50,60]. In a study with children measurements during 30 minutes at the PICU correlates well with the measurements along 24 hours [14]. The minimal duration which has been published in series of children is 15 minutes, and the reproducibility of two measurements was good ($r = 0.87$). This study was done with healthy children [59]. Some studies with children have employed the media of two measurements as definitive value of REE [57]. The correlation coefficient found between repeated measurements have been high [41,48,62]. In critically ill patients an intrasubject variability within the day less than 7-10% is accepted, varying according to the duration of the study [110,126,187,189].

There are two ways for indirect calorimetry: close circuit and open circuit. The most use type of indirect calorimetry is the open circuit. The close circuit has been employed at research studies [141,180]. In close circuit indirect calorimetry a mix of gases containing oxygen in a known concentration of O2 is breathed. Expired CO2 is extracted and analyzed, so air serves to be breathed again [180]. This technique has shown overtimations [25] and it is very sensible to leaks [146]. Its advantage is that it does not require measurement of concentration and flow of oxygen [180].

Open-circuit calorimetry requires an exact measurement of respiratory flows and volumes, and of inspiratory and expiratory fractions of oxygen and CO2 [27,28,170,185,178]. Every metabolic cart is equipped to measuring the O2 and CO2 concentrations, the expired flow air, temperature, barometric pressure and time. The conditions of temperature, humidity and atmospheric pressure must be considered [33,149,186,187,188]. Metabolic carts need calibration, which is made by a combustion test of ethanol, using mixes of oxygen and O2 at known concentrations [33,180]. Metabolic carts employ electrochemical [191] (polarographic [120,179]) or paramagnetic oxygen sensors which use the fact of that oxygen has higher magnetic susceptibility than some other gases [30,64,114,148,149,155,156,180]. CO2 is usually measured by infrared absorption [30,64,148,149,155,156,172,180,183,190,191]. Most used flow sensors are pneumotachographs, turbines, hot-filament anemometry or ultrasonic flow sensors [140,178]. However, most of the instruments turn progressively less exact as the inspired fraction of O2 raises higher than 50 to 60% [7]. FiO2 fluctuations can be less when oxygen comes from a bottle instead the general wall inlet [180]. Water vapor must be retired of the circuits to prevent mistakes [114]. Air leaks can cause mistakes of around 3% in the measurements.

In children with spontaneous breathing the use of these system requires the employment of plastic hoods over the head of the child, closed around the chest and shoulders [30,34,54,59,114,180], or a mouth piece with a nose close-clip or mask [46,53,54,84,68,139,140,164,149,150] with a system of no-rebreathing [149,155,156], along all the period of study. Air flow must be constant and enough to prevent the rebreathing [180,190]. So for effort tests air flow must be higher (air flows of 100 lpm have been used in adults) [191].

Plastic hoods are named canopy [34] and they are made of transparent policarbonate. There are several sizes of canopy: neonatal, pediatric (3-20 kg) and adult. Canopy has an air entrance were there must be a constant flow of 12 lpm for children less than 20 kg and 30-40 lpm for bigger ones. The inspiratory gases are monitored at this place. Canopy also has a exit where the air is extracted [30,43]. If a canopy is used, it is necessary to wait some time until the gases equilibrate inside (20, 25 and 30 minutes in studies with children of 2, 1, and 0,5 kg). The usage of calorimetry has been validated in preterm infants, with FiO2 up to 0.4 in canopy [180].

The difference between canopy and the other systems is less than 40%, but mouth pieces and mask can generate anxiety (they are bad tolerated in young children) [156]. Anxiety can modify the respiratory pattern, so it may produce artifacts.

In children with mechanical ventilation the FiO2 is analyzed in the inspiratory tubing [10,15,180] and expired volume and gases in the expiratory tubing [14]. As FiO2 mustn't be higher than 0.69, [10,15,105,179,180], this limits the use of this technique in the most critically ill children. FiO2 fluctuations seem to be less during volume-controlled ventilation than in intermittent modalities (IMV) [193]. Although measurements have been done with ventilators with continuous flow [180], these require oxygen and CO2 analyzers especially sensible and with high precision, and most of metabolic carts in the market don't fit this features [105]. Recently some authors have shown that if the child is well-sedated, the respiratory modality (pressure-regulated volume controlled ventilation (PRVC), synchronized intermittent mandatory ventilation (SIMV), or biphasic intermittent positive airway pressure/airway pressure release ventilation (BiVent)) doesn't influence significantly the results with the calorimeter E-COVX, which is marketed as the replacement of Deltatrac [188].

The association of teleexpiratory positive pressure higher than 5 or 10 cm of H20 or the elevated media pressures or high inspiratory pressures can also difficult measurements because of they can modified the gas exchange [149,178]. Besides, variations in temperature or humidity may turn into mistakes, too [149,178].

It is recommendable to connect the sensor inlet distal to humidifier because thus it serves as a mixing gas chamber, and this ensure a more constant FiO2 [14]. To prevent mixes with expiratory gases it is better putting the sensor connection far away the endotracheal tube. During the period of study neither nebulizations nor other gases apart from oxygen, nitrogen or CO2 mustn't be employed inside the circuit [9,14,15].

Open endotracheal suctioning can cause an immediate decrease in expired tidal volume resulting in inaccurate respiratory gases monitoring. But some authors have shown that the effect is low and not significant, so prolonged periods of stabilization of connected calorimeter are innecessary [194]. Nevertheless, most metabolic carts need a period of time (usually 5 minutes) after the suctioning until the restart of measurement. Open endotracheal suctioning is required regularly in mechanically ventilated patients to remove obstructive secretions and in some children can be necessary more often than the duration of the study [194].

The quantification of gases at the inspiratory tube supposes a volume loss than may produce hypoventilation in children with low tidal volumes. So it is necessary to assess that the ventilatory volumes keeps being enough after the insert of the calorimeter.

Air leaks around endotracheal tube are frequent in children, because of the use of uncuffed tubes. If there are leaks the results can be considerably altered [105,195,196] (up to 45% [195]). Some researchers, however, think that results don't seem changed if the leaks are inaudible and less than 20% [196]. An air leak is usually considered inaudible when it's not higher than 7,5% of the tidal volume [180]. Many reports haven't considered leaks [19,100,104] or they have only studied children with cuffed [9,106,112] or difference less than 5% [10] or 10% [14,15] or even 15% [9] between inspired and expired tidal volume.

Direct Calorimetry

Direct calorimetry provides a TEE assessment by the measurement of heat production in a closed ambient [7,121]. It can compute heat losses by radiation, convection and evaporation. Direct calorimetry is considered as gold standard for the measurement of TEE at laboratory [151]. The precision is about 1-2% in studies with wet and dry heat producers and 3% in a clinical trial respect to indirect calorimetry [121]. But its costs are very high [7]. A metabolic chamber costs approximately 350000 euros and it requires specialized staff.

Doubly-Labeled Water Method

This technique is considered as one of the most accurate methods to measure EE in free-living conditions [29, 160, 202,] and it gets a gold standard to compare other tools [29,41,46,51,64,65,148,150,154-157,197]. It was validated in human beings at the beginning of 80's [197,199]. It has a good correlation with indirect calorimetry as in adults [191,200] as in infants and children [191, 76,192,201]. This method provides information about average TEE during a period of 7-14 days but it can't evaluated fast variations in EE [25,150], Precision of this technique is adequate if the duration of the study is between 0.5 and 3 H_2O_2 media biological lifes [203]. Two media biological lifes suppose 3-14 days in neonates (usually 5 days [200] and children and about 14-28 days in adults [203,204]. The highest precision has been found with periods of 6 to 7 days in neonates and children and 14 days in adults [203,204].

In opposition to calorimetry, this method can provide exact reports of the EE because of physical activity [25,46,47,51,57,59,64,65], and of the pattern of the activity during a period of 7-14 days. However, it can't provide information of the pattern of physical activity during a 24-hours period [25,150]. Besides, it gives information about the rates of production of CO_2 [144,164] the water intakes and insensible losses of water [192,197].

But this method requires high technical experience [148] and a mass-spectrometer [30,65,156,1164,191,192,197-201,204,205] and the cost of oxygen-18 water [128,130,144,150,164,204], is relatively high (a dose usually last 14 days and its price is about 800-1000 euros). All this reasons have limited the sprend of this technique in current clinical practice and relegated this to clinical research. In critically ill patients is difficult to employ because the duration of studies is long and there aren't compounds of isotopes for intravenous route.

The method is based on the equilibre of the molecules of oxygen in the body water an in the expired CO2 [177,197,198,203-205]. The technique requires the oral administration of water labelled with stable, non-radiactive isotopes, 2H2O and H218O, and the measurement of their content in biological samples (for example, urine) along the time. The dosage of labelled water is made according to weight. As 2H (deuterium) is eliminated only as water, and 18O is eliminated only as water and CO2, the difference between the concentrations of 2H2O and H2 reflects the production of CO2 [177,197,198,203-205]. The production of CO2 is an index of the DER. DER are calculated using the composition of the diet by classical respirometric equations (Weir's and adaptations from this) [171,193].

The daily rate of production of CO2 can be calculated by Lifson and Mc Clintock's equation (or some modification of this, for example: rCO2 = (N/2,078)(1,01 Ko – 1,04 kh) – 0,0246 rGf (N is the total body water in mol, ko the rate of elimination of 18O, kh the rate of elimination of 2H and rGf the estimated rate of loss of isotopic fractioned water: rGf = 1,05N (1,01 ko – 1,04 kh)[200]). Elimination rates of 18O and 2H can be calculated by the 2-points method using the difference in the excess of atomic percentage (APE) of the samples of the start (s) and end (e) point and the time between the urine collection, like k = (lnAPEs – lnAPEe)/time, APE is the isotopic enrichment of the sample relative to the basal (before dose administration) sample [57,65]. This method prevents the continuous urine collection [57]. Other way is by the use of Speakman's R2 equation [202] for the calculation of the rate of production of CO2, using the dilution space of 2H [52]. Other authors calculate the production of CO2 from the dilution spaces and the rates of fractioning of deuterium and 18O, by the multipoint method and the equation: rCO2 = 0,4584 (Ko No – kh Nh), ko and kh are the rates of fractioning of 18O and deuterium and No and Nh the dilution spaces of 18O and deuterium, respectively [29].

Once the production of CO2 is know, the GET is calculated by Weir's equation [174]: GET (MJ/d) = 22,4 (1,106 V CO2 + 3,941 VO2)/23. VO2 is calculated by the division of VCO2 by the food quotient [197,200,204]. This implies the assumption of the food quotient is equivalent to the RQ (VO2/VCO2). Food quotient is calculated from diet composition, like Black et al described [193] Usually the register of the daily intake of macronutrients is done by diaries during 7 days [46,57,65,156] (sometimes in children this period is reduced) [48]. Food quotients of 0.87, 0,855, 0,855, 0,87 and 0,87 are employed for the ages of 3, 6, 9, 12 and 24 months, respectively [193]. Corrections must be made for the isotopic fractioning of the water lost by breath or skin (not by swallow) [66,156,177,197,203-205]

.

Energy Balance Studies

This method has been used as a reference method to compare other tools for the measurement of TEE [20,114,151,192]. They consist in the quantification of the energy intake and measurements of the loss or increase of weight. So they require an exhaustive compilation of all energy supports and a period of study long enough to be able to detect changes in body stores [151]. These studies have been employed in different clinical situations, as UCIP [21].

Movement Sensors

Pedometers. They are a type of movement sensors what measure the number of paces and convert them in energy expenditure by regression equations [147]. Theoretically they don't inform about other activities rather paces, but Sequeira et al [168] demonstrated that pedometers could difference between several levels of occupational activities in adults (sitting, standing). Hard work, however, is not distinguished from moderate, because first have a large proportion of static work, whom measurement is impossible with a pedometer [147].

Accelerometers. An accelerometer is a type of movement sensor capable of detect acceleration and deceleration in one or more directions [47,128,167]. When there is a movement an electric flow is generated inside the acceloremeter and this flow is proportional to the grade of acceleration of the movement [128]. Detected acceleration is converted in calories by the use of regression equations [141,143,144]. These equations usually include sex, age and antropometric features of the person [141,143,144].

An uniaxial accelerometer measures the acceleration in the vertical axis, whereas the biaxial and triaxial accelerometers are sensible to the movements in two and three dimensions, respectively [128,131,140,142]. Theoretically, a triaxial accelerometer may be more sensible to the higher range of movements of children [131]. However, differences between accelerometers have been explained by the precision of their equations rather than the sensor themselves [142].

Accelerometers were developed in 1976 by the National Institute of Health in an attempt of simplify measurement of EE during physical activity [161] But the validity and reliability of initial prototype never were documented [161]. Later models have been employed to the measurement of physical activity and TEE in free-living people [46,47,85,144,162-164,206]. They are able to keep registering data during long time [143]. These models are usually light and they hardly hinder common activities [143].

There are a lot of validation studies of these monitors [113,128,139-144,154,164,167,207] in children [132,,135,137,147,165]. These researches show a good correlation with EE measured by indirect calorimetry [113,128,130,139-144,146,147,154,165,207] or doubly-labelled water method [133,138,166]. Accelerometers correlate well with heart rate [129], questionnaires [131,132] and direct observation [131,136,137] for the measurement of physical activities. WHO keeps considering these instruments useful regard it judge that TEE measurements by doubly labelled water have almost ever substituted the estimations of TEE based in calorimetric studies or movement sensors [25].

Table 3. Commercial Models of Accelerometers employed in children.

Actigraph	Direction	Uniaxial. Vertical axis.
Model 7164, Shalimar, FL; MTI Actigraph, Fort Walton Beach, FL	Placement	Mid line[132,146,169] or hip[128,146]. Leg[128,146]. Scapular waist, anterior axilary line[86]
(formerly named MTI/CSA) [85,128,131,132,139,145,146,166].	Water resistance	No
	Intervals	Minimun 1 second. There have been used 15 seconds[85]-1 minute[132,146].
Very employed in children[85,86,131,132,145,146,166,169].	Weight	85 grams.
	Prize	819.00 the reader, 350.00 the unit.
Actiwatch	Direction	Omnidirectional[151]. Triaxial[138].
(New version Actical)[137,146,196,161,194]	Placement	Hip and leg[146]. Around the waist, monitor at basis of lumbar vertebrae[161].
Some studies in children[137,146].	Water resistance	Yes (new version).
	Intervals	1 minute is employed[137,146].
	Weight	16,5 grams.
	Prize	725.00 the reader 500.00 the software 450.00 a unit.
Wam.	Direction	Uniaxial.
(A smaller versión of Actigraph	Placement	Wrist[166], hip[129]. Lower back[46,47].
(model 7164, Computer Science	Water resistance	No
Applications, Shalimar, FL)) [46,47,129,139,166,167].	Intervals	1 second-1minute.
	Weight	43 g (5.1 x 3.8 x 1.5 cm).
Some studies in children[46,47,129,145,166].	Prize	---
Caltrac[130,134,136,142,146,165,167,206].	Direction	Vertical
(now CTI, Stay Healthy)	Placement	Lower back[167]
A lot of studies in children[33,130,134,136,146,165,206].	Water resistance	No
	Intervals	-
	Weight	78 grams.
	Prize	99.95 a unit
TriTrac-R3D[128-130,135,141,142,147,167].	Direction	Triaxial.
	Placement	Hip[141]. Scapular waist at media axilary line[128].
(Professional Products Inc.,	Water resistance	No
Madison, WI).	Intervals	1 minute is employed[130].
Some studies in children[129,130].	Weight	170 grams. 11.1 × 6.7 × 3.2 cm
	Prize	-- . 500.00 the new product RT3.
RT3.	Direction	Triaxial.
Triaxial Research Tracker	Placement	Scapular waist at medial or anterior axilary position.
(Stayhealthy Inc., Monrovia, CA)[128].	Water resistance	-
	Intervals	1 second - 1 minute.
It is the new versión of the	Weight	Around 1/3 of Tritrac
former.	Prize	500.00.
SenseWear Armband	Direction	Biaxial.
(BodyMedia Inc., Pittsburgh, PA)[139,142,143,164].	Placement	Posterior[128] or lateral[144] region of the right (left[128]) arm at media humeral position[128,143,144].
It contents an accelerometer nde	Water resistance	No (a little).
a termocoupler. It may acoplar a	Intervals	1 minute[128,144].
heart rate monitor.	Weight	85 g (85.3 mm x 53.4 mm x 19.5 mm)
Some studies with children[156].	Prize	Around 1000 euros
Actiheart[208,210].	Direction	Uniaxial (vertical).
It combine an accelerometer and	Placement	Unit is worn on two standard ECG pads on the chest[210].
a register of Heart Rate[208,210].	Water resistance	Yes.
Some studies with children[156].	Intervals	15-30-60 seconds according with information and duration of registers[210]. It can register up to 21 days.
	Weight	< 10 grams[210].
	Prize	Unit £900 Reader/Charger £350 Software £900 Shipping within Europe £12.50

Most of published researches with accelerometers don't measure REE, but they estimate it [140,143]. Fruin et al [143] studied reliability and validity of SenseWear Armband during rest and exercise, in comparison with indirect calorimetry. They didn't found any differences between measurement by SenseWear Armband and indirect calorimetry at rest, and the correlation between them was high ($r = 0.76$: `$p < 0.004$) [143]. EE estimation by SenseWear was more reproducible than EE estimation by indirect calorimetry when estimations at rest were compared ($r = 0.93$, $p < 0.001$) [143]. They attributed this higher reliability to the fact of REE estimation by Armband is based in age, sex and weigh and REE is not really measured [143].

General criticism to belt accelerometers include inability to detect arm movements [128,140,141], cyclism [128,134,140,141,153], up or downhill locomotion [128,146] and/or external work as pushing, taking up or carrying things [128,134,139,141,153],There were been found infraestimations of 8-21% of EE in up or downhill, 53-68% in bycicle [140] and higher than 35-45% in ordinary life [146] with triaxial accelerometers. Most accelerometers are not water resistant (see table 3) and so they are not able to register calories burned while swimming [86]. Accelerometers can difficult hard physical activity. Results may vary according the place where they were putted [134,139],. Another source of error can emerge from the need of convert physical activity units in EE units [134]. The employ of specific equations for each activity has been proposal to improve the exactitude of the estimations [140,142,144].

Accelerometry has been extensively used in children, most of all in obesity researches [46,85]. There is a progressively higher use of accelerometry as method of study in ill children [163,206], as there is in adults [113,162,164]. Table 2 shows a brief description of several comercial models, with references to some researches performed with them.

Heart Rate Monitoring

Heart rate has been used as an index of EE [25,55,102,147,148,150,151,153-155,157,158,208]. The relationship between heart rate and EE is usually described as lineal [148,151], as well in children [158,] in special in the heart rate range between 110-150 bpm [158], but there isn't a evidence of this relationship were strong during periods at low or very high activity [150,151,158]. Besides, the relationship between heart rate and EE is specific for each person and there are a lot of factors that can change it (anxiety, drugs, etc) [134,135]. These reasons turn controversial its useful as isolated tool. However, heart rate monitoring has been employed to measure EE in illness conditions, too [102,157].

Heart rate monitoring can also help to identify the frequency, intensity and duration of diary pattern of physical activity [148,155]. The relationship between oxygen consumption and heart rate is influenced by the proportion of active muscle mass and by the continuous or intermitent character of the exercise. Training level of the children is other factor of restriction [147,206,209] because a trained child has an higher ejection volume and thus a low heart rate for every activity and a lower response to exercise [209].

In addition to other data, heart rate monitoring has demostrated to improve estimations in some predictive models [55,102,208]. Its application at great scale is possible [150]. Heart rate monitoring is simple [150,158], cheap and well-accepted by children [150].

FLEX-HR Method. This method was developed to fight against individual features in the relationship between heart rate and EE and physical activity [135]. It consists in the calculation of a value of HR capable to discriminate between heart rate at rest and during exercise in that individual [150]. That value is named critical FLEX-HR 150. The critical FLEX-HR is obtained from regresión equations designed from the values of VO2 at rest and during several activities at laboratory [150,156]. Then minute by minute registers of heart rate are made in free living conditions [25,148,150]. In the part of day in with heart rate is beyond the FLEX-HR EE is calculated with the regression equation obtained at the high intensity activities during the calibration period [148,150,156].

Relationships between heart rate and VO2 may differ in real conditions because of factors not controlled during the calibration protocols (ambient temperature, drugs [158], posture, muscle activity intensity, fatigue, postpandrial state [134,150], day by day variations in heart rate-VO2 relationship [134]...). Children show a natural trend to hard, brief and of brusco begin physical activity patterns [134,150]. These patterns are opposite to the standardized exercise employed for the calibration of the method [150].

Several researches have found EE values obtained by FLEX-HR method as only measure comparable with media value obtained by doubly labeled water method [150,156,157] or calorimetry [155]. For this reason last FAO/WHO/UNO employed studies done by this methods [25]. Nevertheless, thought its correlation is good to group comparison, it show lack of precision at individual level [150,155,157].

Core Body Temperature

At laboratory conditions core body temperature has been employed to estimate TEE [196]. Under very controlled conditions, there is a close relationship between core body temperature and EE [159]. But in usual clinical practice there isn't good correlation with E102 [102,160]. However, when core body temperature is combined with other tools of measurement of EE, it can sum a valuable dimension, so it has been included in several regression equation [15,116]. Because of its invasive character, core body temperature is impracticable to study people at free living conditions.

Heat Flow Measurement

Heat loss measurement has been proposed as a precise/good indicator of the total of burned body calories because heat production and heat loss are a result of metabolism. Even thought its correlation with EE as independient variable is not enough to predict EE, its measurement may improve the EE estimation used in combination with other parameters [113,128,144,164]. Skin temperature has good correlation with VO2 and VCO2 during physical activity [113,128,144,164]. During last years several instruments with a heat flow sensor have been developed and commercialized, these instruments incorpored the heat flow sensor in monitors what simultaneously study several parameters (heat flow, accelorometry, skin galvanic response, skin temperature, near body temperature and, some of them, besides, heart rate). These monitors convert these data in EE by the use of regression equations.

SenseWear Pro Armband (BodyMedia Inc., Pittsburg, PA, table 3) and HealthWear Armband (Roche Diagnostics, Indianapolis) are examples of these monitors. They measure the difference in temperature between two sensors, one in touch with the skin and other in the side of monitor exposed to the ambient. From the total of thermic energy dissipated from the body they specifically register the part correspondient to convection. They are putted over the arm, between the skin and the side of the Armband exposed to the ambient [113,128,144,164] Arm location is prefer to other like front or trunk. More peripheral locations are subject to excessive movements and frequent periferic vascular changes, as wrist, hands, fingers and feet. These monitors have been employed as in healthy people as in ill individuals, but not yet in critically ill conditions.

Combined Methods

Several attempts to combine different types sensors have been made during last years in order to get an higher precision. For example, the above-mentioned SenseWear Armband and HealthWear Armband. They register approximately 21 parameterers, including heat flow, accelerometry, skin galvanic response (employed as an indicator of heat losses by evaporation [144]) , skin temperature, near body temperature [113,128,143,144]. Thus, the addition of the heart rate monitoring to a movement sensor has proven to improve the precision over each method alone [147,153,154,208], at least by a 2% [147,153,154].

In the future, the combination of REE determinations by indirect calorimetry and the employ of multisensor monitors might improve the estimations of DER in clinical practice. But whereas a simple podometer may cost only betwewn 15 and 30 euros, the most sofisticated movement sensor may cost 1000 euros.

References

[1] Lobstein T, Baur L, Uauy R. Obesity in children and young people: a crisis in public health. *Obes Rev* 2004;5(suppl 1):4-104.

[2] Baker JL,Olsen LW, Sorensen TI. Childhood body-mass index and the risk of coronary heart disease in adulthood. *N Engl J Med* 2007;357:2329-37.

[3] Weiss R, Dziura J, Burgert TS, Tamborlane WV, Taksali SE, Yeckel CW, et al. Obesity and the metabolic syndrome in children and adolescents. *N Engl JMed* 2004;350:2362-74.

[4] Berenson GS, Srinivasan SR, Bao W, Newman WP, Tracy RE, Wattigney WA. Association between multiple cardiovascular risk factors and atherosclerosis in children and young adults. The Bogalusa Heart Study. *N Engl JMed* 1998;338:1650-6.

[5] Hsu A, Heshka S, Janumala I, Song MY, Horlick M, Krasnow N, et al. Larger mass of high-metabolic-rate organs does not explain higher resting energy expenditure in children. *Am J Clin Nutr*. 2003;77:1506-11.

[6] Holliday MA. Metabolic rate and organ size during growth from infancy to maturity and during late gestation and early infancy. *Pediatrics*. 1971;47:169-79.

[7] Fung E. Estimating energy expenditure in critically ill adults and children. *AACN Clinical Issues*. 2000;11:480-97.

[8] Joosten KF, Verhoeven JJ, Hazelzet JA. Energy expenditure and substrate utilization in mechanically ventilated children. *Nutrition*. 1999;15:444-8.
[9] Briassoulis G, Venkataraman S, Thompson AE: Energy expenditure in critically ill children. *Crit Care Med*. 2000;28:1166-72.
[10] Taylor RM, Cheeseman P, Preedy V, Baker AJ, Grimble G. Can energy expenditure be predicted in critically ill children? *Pediatr Crit Care*. 2003;4:176-80.
[11] Hulst JM, van Goudoever JB, Zimmermann LJ, Hop WC, Buller HA, Tibboel D, et al. Adequate feeding and the usefulness of the respiratory quotient in critically ill children. *Nutrition*. 2005;21:192-8.
[12] Verhoeven JJ, Hazelset JA, Van Deer Voort E, Joosten KEM. Comparison of measured and predicted energy expenditure in mechanically ventilated children. *Intensive Care Med.* 1998;24:464-8.
[13] Hardy CM, Dwyer J, Snelling LK, Dallal GE, Adelson JW. Pitfalls in predicting resting energy requirements in critically ill children: A comparison of predictive methods to indirect calorimetry. *Pediatric Nutr Clin Pract*. 2002;17:182-9.
[14] Vázquez JL, Dorao P, Díez, J, Ruza F. Predicted versus measured energy expenditure by continuous, online indirect calorimetry in ventilated, critically ill children during the early postinjury period. *Pediatr Crit Care Med.* 2004;5:19-27.
[15] White MS, Shepherd RW, McEniery JA. Energy expenditure in 100 ventilated, critically ill children: Improving the accuracy of predictive equation. *Crit Care Med*. 2000;28:2307-12.
[16] Coss-Bu J, Jefferson LL, Walding DD, David Y, Smith E, Klish W. Resting energy expenditure and nitrogen balance in critically ill pediatric patients on mechanical ventilation. *Nutrition.* 1998;14:649-52.
[17] Pollack MM, Ruttimann UE, Wiley JS. Nutritional depletions in critically ill children: Associations with physiologic instability and increase quantity of care. *JPEN J Parenter Enteral Nutr.* 1985;9:309-13.
[18] Almeida-Santos L, Ruza F, Guerra JM, Alves A, Dorao P, García S, et al. Evaluación nutricional de niños con insuficiencia respiratoria aguda: Antropometría al ingreso en Cuidados Intensivos Pediátricos. *An Esp Pediatr*. 1998;49:451-5.
[19] Gebara B, Gelmini M, Sarnaik A. Oxygen consumption, energy expenditure and substrate utilization after cardiac surgery in children. *Crit Care Med*. 1992;20:1550-4.
[20] Oesterveld MJ, Van Der Kuip M, De Meer K, De Greef HJ, Gemke RJ. Energy expenditure and balance following pediatric intensive care unit admission: A longitudinal study of critically ill children. *Pediatr Crit Care*. 2006;7:147-53.
[21] Long CL, Schaffel N, Geiger JW, Schiller WR, Blakemore WS. Metabolic response to injury and illness: Estimation of energy and protein needs from indirect calorimetry and nitrogen balance. *JPEN J Parenter Enteral Nutr*. 1979;3:452-6.
[22] Raurich JM, Ibañez J, Marsé P. CO2 production and thermogenesis induced by enteral and parenteral nutrition. *Nutr Hosp*. 1996;11:108-13.
[23] Liposki J, Nelson L. Ventilatory response to high caloric loads in critically ill patients. *Crit Care Med*. 1994;22:796-802.
[24] Burke JF, Wolfe RR, Mullany CJ, Mathews DE, Bier DM. Glucose requirements following burn injury. Parameters of optimal glucose infusion and possible hepatic and respiratory abnormalities following excessive glucose intake. *Ann Surg*.1979;190:274-85.

[25] Human Energy Requirements. Report of a Joint FAO/WHO/UNU Expert Consultation. *Rome* 17-24 October 2001.
[26] Battezzati A, Viganò R. Indirect calorimetry and nutritional problems in clinical practice. *Acta Diabetol.* 2001;38:1-5.
[27] McClave S, Snider H. Use of indirect calorimetry in clinical nutrition. *Nutr Clin Pract.* 1992;7:207-21.
[28] Brandi LS, Bertolini R, Calafà M. Indirect calorimetry in critically ill patients: Clinical applications and practical advice. *Nutrition.* 1997;13:349-58.
[29] Butte NF, Wong WW, Hopkinson JM, Heinz CJ, Mehta NR, Smith EO. Energy requirements derived from total energy expenditure and energy deposition during the first two year of life. *Am J Clin Nutr.* 2000;72:1558-69.
[30] Weststrate JA, Weys PJM, Poortvliet EJ, Deurenberg P, Hautvast JGAJ. Diurnal variation in postabsorptive resting metabolic rate and diet-induced thermogenesis. *Am J Clin Nutr.* 1989;50:908-14.
[31] Boothby WM, Berkson J, Dunn HL. Studies of the energy metabolism of normal individues. *Am J Physiol.* 1936;116:468-84.
[32] Berke EM, Gardner AW, Goran MI, Poehlman ET. Resting metabolic rate and the influence of the pretesting environment. *Am J Clin Nutr.* 1992:55:626-9.
[33] Feurer ID, Mullen JL. Bedside measurements of resting energy expenditure and respiratory quotient via indirect calorimetry. *Nutr Clin Pract.* 1986;1:43-9.
[34] Cunningham JJ. Body composition as a determinant of energy expenditure: A synthetic review and a proposed general prediction equation. *Am J Clin Nutr.* 1991;54:963-9.
[35] Figueroa-Colon R, Franklin F, Goran M, Lee J, Weinsier R. Reproducibility of measurement of resting energy in prepuberal girls. *Am J Clin Nutr.* 1996;64:533-6.
[36] Harris JA, Benedict FG. A biometric study of basal metabolism in man. *Publication no. 279. Washington, DC: Carnegie Institute of Washington*; 1919. p. 1-266.
[37] Fleisch A. Le metabolisme basal standard et sa determination au moyen du "metabocalculator". *Helv Med Acta.* 1951;18:23-44.
[38] Schofield WN. Predicting basal metabolic rate, new standard and review of previous work. *Hum Nutr Clin Nutr.* 1985;39 Suppl 1:S5-41.
[39] De Lorenzo A, Tagliabue A, Andreoli A, Testolin G, Comelli M, Deurenberg P. Measured and predicted resting metabolic rate in Italian males and females, aged 18-59 y. *Eur J Clin Nutr.* 2001;55:208-14.
[40] Tershakovec AM, Kuppler KM, Zemel B, Stallings VA. Age, sex, ethnicity, body composition, and resting energy expenditure of obese African American and white children and adolescents. *Am J Clin Nutr.* 2002;75:867-71.
[41] Maffeis C, Schutz Y, Micciolo R, Zocacante L, Pinelli L. Resting metabolic rate in six-to-ten-year-old obese and nonobese children. *J Pediatr.* 1993;122:556-62.
[42] Bogardus C, Lillioja S, Ravussin E, Abbott W, Zawadzki JK, Young A, et al. Familial dependence of the resting metabolic rate. *N Engl J Med.* 1986;315:96-100.
[43] Caldwell MD, Kennedy-Caldwell C. Normal nutritional requirements. *Surg Clin North Am.* 1981;61:489-506.
[44] Talbot FB. Basal metabolism standards for children. Am J Dis Child. 1938;55:455-9.
[45] Lewis RC, Duval AM, Hiff A. Standards for the basal metabolism of children from 2-15 years of age. *J Pediatr.* 1943;23:1-5.

[46] Ekelund U, Aman J, Yngve A, Renman C, Westerterp K, Sjostrom M. Physical activity but not energy expenditure is reduced in obese adolescents: A case-control study. *Am J Clin Nutr.* 2002;76:935-41.

[47] Ekelund U, Yngve A, Brage S, Westerterp K, Sjostrom M. Body movement and physical activity energy expenditure in children and adolescents: How to adjust for differences in body size and age. *Am J Clin Nutr*. 2004;79:851-6.

[48] Schoeller DA, Jefford G. Determinants of the energy costs of light activities: Inferences for interpreting doubly labeled water data. *Int J Obes Relat Metab Disord*. 2002;26:97-101.

[49] Sun M, Gower BA, Bartolucci AA, Hunter GR, Figueroa-Colon R, Goran MI. A longitudinal study of resting energy expenditure relative to body composition during puberty in African American and white children. *Am J Clin Nutr*. 2001;73:308-15.

[50] Molnar D, Schutz Y. The effect of obesity, age, puberty and gender on resting metabolic rate in children and adolescents. *Eur J Pediatr*. 1997;156:376-81.

[51] DeLany JP, Bray GA, Harsha DW, Volaufova J. Energy expenditure in African American and white boys and girls in a 2-y follow-up of the Baton Rouge Children's Study. *Am J Clin Nutr*. 2004;79:268-73.

[52] Mifflin MD, St Jeor ST, Hill LA, Scott BJ, Daugherty SA, Koh YO. A new predictive equation for resting energy expenditure in healthy individuals. *Am J Clin Nutr*. 1990;51:241-7.

[53] Owen OE, Holup JL, D'Alessio DA, Craig ES, Polansky M, Smalley KJ, et al. A reappraisal of the caloric requirements of men. *Am J Clin Nutr*. 1987;46:875-85.

[54] Owen OE, Kavle E, Owen RS, Polansky M, Caprio S, Mozzoli MA, et al. A reappraisal of caloric requirements in healthy women. *Am J Clin Nutr*. 1986;44:1-19.

[55] Goran MI, Carpenter WH, Poehlman ET. Total energy expenditure in 4- to 6-yr-old children. *Am J Physiol*. 1993;264:E207-11.

[56] Weinsier RL, Schutz Y, Bracco D. Re-examination of the relationship of resting metabolic rate to fat-free mass and to the metabolically active components of fat-free mass in human. *Am J Clin Nutr*. 1992;55:790-4.

[57] Bandini LG, Must A, Spadano JL, Dietz WH. Relationship of body composition, parental overweight, pubertal stage, and race-ethnicity to energy expenditure among premenarcheal girls. *Am J Clin Nutr*. 2002;76:1040-7.

[58] Fontvieille AM, Harper IT, Ferraro RT, Spraul M, Ravussin E. Daily energy expenditure by five-year-old children, measured by double labeled water. *J Pediatr*. 1993;123:200-7.

[59] Wren RE, Blume H, Mazariegos M, Solomon N, Alvarez JO, Goran MI. Body composition, resting metabolic rate and energy requirements of short and normal stature, low-income Guatemalan children. *Am J Clin Nutr*. 1997;66:406-12.

[60] Molnar D, Jegees S, Erhardt E, Schutz Y. Measured and predicted resting metabolic rate in obese and nonobese adolescents. *J Pediatr*. 1995;127:571-7.

[61] Gallagher D, Belmonte D, Deurenberg P, Wang Z, Krasnow N, Pi-Sunyer FX, et al. Organ-tissue mass measurement allows modeling of REE and metabolically active tissue mass. *Am J Physiol Endocrinol Metab*. 1998;275:E249-58.

[62] Illner K, Brinkmann G, Heller M, Bosy-Westphal A, Muller MJ. Metabolically active components of fat free mass and resting energy expenditure in non obese adults. *Am J Physiol Endocrinol Metabol.* 2000;278:E306-15.

[63] Goran MI, Kaskoun M, Johnson R. Determinants of resting energy expenditure in young children. *J Pediatr*. 1994;125:362-7.

[64] Treuth MS, Butte NF, Wong WW. Effects of familial predisposition to obesity on energy expenditure in multiethnic prepubertal girls. *Am J Clin Nutr*. 2000;71:893-900.

[65] Spadano JL, Bandini LG, Must A, Dallal GE, Dietz WH. Longitudinal changes in energy expenditure in girls from late childhood through midadolescence. *Am J Clin Nutr*. 2005;81:1102-9.

[66] Roberts SS, Savage J, Coward W, Chew B, Lucas A. Energy expenditure and intakes in infants born to lean and overweight mothers. *N Engl J Med*. 1988;318:461-6.

[67] Goran MI, Carpenter WH, McGloin A, Johnson R, Hardin JM, Weisner RL. Energy expenditure in children of lean and obese parents. *Am J Physiol*. 1995;268:E917-24.

[68] Roza AM, Shizgal HM. The Harris-Benedict equation reevaluated: Resting energy requirements and the body cell mass. *Am J Clin Nutr*. 1984;40:168-82.

[69] Montgomery RD. Changes in the basal metabolic rate of the malnourished infant and their relation to body composition. *J Clin Invest*. 1962;41:1653-63.

[70] Parra A, Garza C, Garza Y, Saravia JL, Hazlewood CF, Nichols BL. Changes in growth hormone, insulin, and thiroxine values, and in energy metabolism of marasmic infants. *J Pediatr*.1973;82:133-42.

[71] Melchior JC, Raguin G, Bouvet E, Rigaud D, Matheron S, Casalino E, et al. Resting energy expenditure in human immunodeficiency virus-infected patients: Comparison between patients with and without secondary infections. *Am J Clin Nutr*. 1993;57:614-9.

[72] Müller MJ, Lautz HU, Plogmann B, Burger M, Korber J, Schmidt FW. Energy expenditure and substrate oxidation in patients with cirrhosis: The impact of cause, clinical staging and nutritional state. *Hepatology*. 1992;15:782-94.

[73] Evolution during antibiotic treatment for pulmonary exhacerbation in cystic fibrosis. *Chest*. 1993;103:1819-25.

[74] Puhakka K, Räsänen J, Elijala M, Peltola K. Metabolic effect of corrective surgery in infants and children with congenital heart disease. *Br J Anaesth*. 1993;70:149-53.

[75] Chiolero R, Flatt JP, Revelly JP, Jequier E. Effects of catecholamines on oxygen consumption and oxygen delivery in critically ill patients. *Chest*. 1991;100:1676-84.

[76] Hart DW, Wolf SE, Chinkes DL, Lal SO, Ramzy PI, Herndon DN. Beta-blockade and growth hormone after burn. *Ann Surg*. 2002;236:450-7.

[77] Noor MI, Poh BK, Zawiah H, Henry CJ. New predictive equations for estimation of basal metabolic rates of adolescents:Tropics versus temperate. *Forum Nutr*. 2003;56:250-3.

[78] Weissman C, Kemper M, Elwyn DH. The energy expenditure of the mechanically ventilated critically ill patient. An analysis. *Chest*. 1986;89:254-9.

[79] Stallings VA, Pencharz PPB. The effect of a high protein-low calorie diet on the energy expenditure of obese adolescents. *Eur J Clin Nutr*. 1992;46:897-902.

[80] Grant J, Denne SC. Effect of intermittent versus continuous enteral feeding on energy expenditure in premature infants. *J Pediatr*. 1991;118:928-32.

[81] FAO = WHO = UNU. Expert consultation on energy and protein requirements. Geneva: *WHO*, 1985.

[82] Jeevanandam M, Grote-Holman E, Chikenjii T, Askanazi J, Elwyn D, Kinney J. Effects of glucose on fuel utilization and glycerol turnover in normal and injured man. *Crit Care Med.* 1990;18:125-35.

[83] Boulanger BBR, Nayman RRD, McLean RRF, Phillips E, Rizoli SSB. What are the clinical determinants of early energy expenditure in critically injured adults? *J Trauma.* 1994;37:969-74.

[84] Maffeis C, Schutz Y, Schena F, Zaffanello M, Pinelli L. Energy expenditure during walking and running in obese and nonobese prepubertal children. *J Pediatr.* 1993;123:193-9.

[85] Trost SG, Sirard JR, Dowda M, Pfeiffer KA, Pate RR. Physical activity in overweight and nonoverweight preschool children. *Int J Obes Relat Metab Disord.* 2003;27:834-9.

[86] Riddoch CJ, Leary SD, Ness AR, Blair SN, Deere K, Mattocks C, Griffiths A, Smith GD, Tilling K. Prospective associations between objective measures of physical activity and fat mass in 12-14 year old children: the Avon Longitudinal Study of Parents and Children (ALSPAC). *BMJ* 2009;339:b4544 doi:10.1136/bmj.b4544.

[87] Fernández IS, Kurpad AV, Kilpaldi AR, Shetty PS. Resting energy expenditure and nitrogen loss after surgery in chronically undernourished patients. *World J Surg.* 1993;17:80-4.

[88] Coss-Bu JA, Jefferson LS, Walding D, David Y, Smith EO, Klish WJ. Resting energy expenditure in children in a pediatric intensive care unit: Comparison of Harris-Benedict and Talbot predictions equations with indirect calorimetry values. *Am J Clin Nutr.* 1998;67:7.

[89] Soares MJ, Shetty PS. Basal energy rates and metabolic economy in chronic undernutrition. *Eur J Clin Nutr.* 1991;45:363-7.

[90] Kurpad V, Kulkarni RN, Shetty PS. Reduced Thermoregulatory thermogenesis in undernutrition. *Eur J Clin Nutr.* 1989;43:27-31.

[91] Salas J, Dozio E, Goulet O, Marti-Henneeberg C, Moukarcel E, Ricour C. Energy expenditure and substrate utilization in the course of renutrition of malnourished children. *JPEN J Parenter Enteral Nutr.* 1991;15:288-93.

[92] Mericq V, Iñiguez G, Bazaes R, Bouwman C, Avila A, Salazar T, Carrasco F. Differences in body composition and energy expenditure in prepubertal children born term or preterm appropriate or small for gestational age. *J Pediatr Endocrinol Metab.* 2009 Nov;22(11):1041-50.

[93] García de Lorenzo A, Ortiz Leyba C. Grupo de trabajo de Metabolismo y Nutrición, Respuesta a la agresión: Valoración e implicaciones terapéuticas. 2.ª Conferencia de Consenso de la SEMIUC. *Medicina Intensiva.* 1997;21:13-28.

[94] and hypermetabolism: Effect of trauma and gender differences. *Ann Surg.* 2005;241:759-67; discussion 767-8.

[95] Barret JP, Herndon DN. Modulation of inflammatory and catabolic responses in severely burned children by early burn wound excision in the first 24 hours. *Arch Surg.* 2003;138:127-32.

[96] Powis MR, Smith K, Rennie M, Hallyday D, Piero A. Effect of major abdominal operations on energy and protein metabolism in infants and children. *J Pediatr Surg.* 1998;30:49-53.

[97] Jones MO, Pierro A, Hammond P, Lloyd DA. The metabolic response to operative stress in infants. *J Pediatr Surg.* 1993;28:1258-63.

[98] Jeevanandam M, Young DH, Ramias L, Schiller WR. Aminoaciduria of severe trauma. *Am J Clin Nutr*. 1989;49:814-22.

[99] Wang P, Li N, Li JS, Li WO. The role of endotoxin, TNF-alpha, and IL-6 in inducing the state of growth hormone insensitivity. *World J Gastroenterol*. 2002;8:531-6.

[100] Steinhorn DM, Green TP. Severity of illness correlates with alterations in energy metabolism in the pediatric intensive care unit. *Crit Care Med.* 1991;19:1503-9.

[101] Van Lanschot JJ, Feenstra BW, Vermeij CG, Bruining HA. Calculation versus measurement of total energy expenditure. *Crit Care Med*. 1986;14:981-5.

[102] Giantin V, Ceccon A, Enzi G, Sergi G, Perini P, Bussolotto M, et al. Heart rate and metabolic response to burn injury in humans. *JPEN J Parenter Enteral Nutr*. 1995;19:55-62.

[103] Ortiz Justiniano V, Diaz Troche J, Ramirez Ferrer L. Resting energy expenditure in children after major surgery. *Bol Asoc Med PR*. 1994;86:81-3.

[104] Tilden SJ, Watkins S, Tong T, Jeeevanandam M. Measured versus predicted energy expenditure in pediatric intensive care patients. *Am J Dis Child*. 1989;143:490-2.

[105] Selby AM, McCauley JC, O'Connell A, Gillis J, Gaskin KJ. Indirect calorimetry in mecanichally ventilated children: A new technique that overcomes the problem of endotracheal tube leak. *Crit Care Med*. 1995;23:365-70.

[106] Chwals WJ, Lally KP, Woolley MM, Mahour GH. Measured energy expenditure in critically ill infants and young children. *J. Surg Res*. 1988;44:467-72.

[107] Steinhorn DM. Energy expenditure in critically ill children is lower than expected. *Pediatric Res*. 1993;33:41A.

[108] Phillips R, Ott L, Young B, Walsh J. Nutritional support and measured energy expenditure of the child and adolescent with head injury. *J Neurosurg*. 1987;67:846-51.

[109] Fossyth JS, Crighton A. Low birth weight infants and total parenteral nutrition immediately after birth. Energy expenditure and respiratory quotient of ventilated and non-ventilated infants. *Arch Dis Child*. 1995;71:31-4.

[110] Brown PE, McClave SA, Hoy NW, Short AF, Sexton LK, Meyer KL. The Acute Physiology and Chronic Health Evaluation II classification system is a valid marker for physiologic stress in the critically ill patient. *Crit Care Med*. 1993;21:363-7.

[111] López-Herce J, Sánchez C, Mencía S, Santiago MJ, Carrillo A, Bellón JM. Consumo calórico en el niño crítico: relación con las características clínicas, el aporte calórico y las fórmulas teóricas de cálculo de las necesidades energéticas. *An Pediatr* (Barc). 2007;66:229-39.

[112] Vernon DD, Witte MK. Effect of neuromuscular blockade on oxygen consumption and energy expenditure in sedated, mechanically ventilated children. *Crit Care Med*. 2000;28:1569-71.

[113] Cole PJ, Lemura LM, Klinger TA, Strohecker K, McConnell TR. Measuring energy expenditure in cardiac patients using the Body Media(TM) Armband versus indirect calorimetry: A validation study. *Journal of Sports Medicine and Physical Fitness*. 2004;44:262-71.

[114] Duggan M, Milner R. The maintenance energy requirement for children: An estimate based on a study of children with infection associated underfeeding. *Am J Clin Nutr*. 1986;43:870-8.

[115] Poblete B, Romand JA, Pichard C, Konig P, Suter PM. Metabolic effect of paracetamol, metimazol or external cooling in critically ill febrile sedated patients. *Br J Anaesth.* 1997;78:123-7.

[116] Bruder N, Raynal M, Pellissier D, Courtinat C, François G. Influence of body temperature, with or without sedation, on energy expenditure in severe head-injured patients. *Crit Care Med.* 1998;26:568-72.

[117] Weissman C, Kemper M, Damask MC, Askanazi J, Hyman AI, Kinney JM. Effect of routine intensive care interactions on metabolic rate. *Chest.* 1984;86:815-8.

[118] Weissman C, Kemper M. Assessing hypermetabolism and hypometabolism in the postoperative critically ill patient. *Chest.*1992;102:1556-71.

[119] Hollyday MA, Segar WE. Maintenance need for water in parenteral fluid therapy. *Pediatrics*. 1957;19:823-7.

[120] Darrow DC. A guide to learnig fluid therapy. *Springfield III: Charles C. Thomas Co.;* 1964.

[121] Daly JM, Heymsfield SB, Head CA, Harvey LP, Nixon DW, Katzeff H, et al. Human energy requirements: Overestimation by widely used prediction equation. *Am J Clin Nutr*. 1985;42:1170-4.

[122] Thompson MA, Bucolo S, Quirk PP, Shepherd RR. Measured versus predicted resting energy expenditure in infants: A need for reappraisal. *J Pediatr*. 1995;126:21-7.

[123] Kaplan AS, Zemel BS, Neiswender KM. Resting energy expenditure in clinical pediatrics: Measured versus prediction equations. *J Pediatr*. 1995;127:200-5.

[124] Monk DN, Plank LD, Franch-Arcas G, Finn PJ, Streat SJ, Hill GL. Sequential changes in the metabolic response in critically injured patients during the first 25 days after blunt trauma. *Ann Surg*. 1996;223:395-405.

[125] Havalad S, Quaid MA, Sapiega V. Energy expenditure in children with severe head injury: Lack of agreement between mesured and estimated energy expenditure. *Nutr Clin Pract.*2006;21:175-81.

[126] Vermeij CG, Feenstra BW, Van Lanschot JJ. Day to day variability of energy expenditure in critically ill surgical patients. *Crit Care Med*. 1989;17:623-6.

[127] Adamo KB, Prince SA, Tricco AC, Connor-Gorber S, Tremblay M. A comparison of indirect versus direct measures for assessing physical activity in the pediatric population: a systematic review. *Int J Pediatr Obes*. 2009;4(1):2-27.

[128] King GA, Torres N, Potter C, Brooks TJ, Coleman KJ. Comparison of activity monitors to estimate energy cost of treadmill exercise. *Med Sci Sport Exerc*. 2004;36:1244-51.

[129] Welk GJ, Corbin CB. The validity of the Tritrac-R3D Activity Monitor for the assessment of physical activity in children. *Res Q Exerc Sport.* 1995;66:202-8.

[130] Going SB, Levin S, Harrell J, Stewart D, Kushi L, Cornell CE, et al. Physical activity assessment in American Indian schoolchildren in the Pathways study. *Am J Clin Nutr.* 1999;69 Suppl:S788-95.

[131] Janz KF, Witt J, Mahoney LT. The stability of children's physical activity as measured by accelerometry and self-report. *Med Sci Sports Exerc*. 1995;27:1326-32.

[132] Treuth MS, Sherwood NE, Butte NF, McClanahan B, Obarzanek E, Zhou A, et al. Validity and reliability of activity measures in African American girls for GEMS. *Med Sci Sports Exerc*. 2003;35:532-9.

[133] Kohl HW, Fulton JE, Caspersen CJ. Assessment of physical activity among children and adolescents: A review and synthesis. *Prev Med.* 2000;31:S54-76.

[134] Welk GJ, Corbin CB, Dale D. Measurement issues in the assessment of physical activity in children.of physical activity in children. *Res Q Exerc Sport*. 2000;71:59-73.

[135] Sirard JR, Pate RR. Physical activity assessment in children and adolescents. *Sports Med.* 2001;31:439-64.

[136] Noland M, Danner F, Dewalt K, McFadden M, Kotchen M. The measurement of physical activity in young children. *Res Q Exerc Sport*. 1990;61:146-53.

[137] Finn KJ, Specker B. Comparison of Actiwatch activity monitor and Children's Activity Rating Scale in children. *Med Sci Sports Exerc.* 2000;32:1794-7.

[138] Puhl J, Greaves K, Hoyt M, Baranowski T. Children's Activity Rating Scale (CARS): Description and calibration. *Res Q Exerc Sport*. 1990;61:26-36.

[139] Swartz AM, Strath SJ, Basset DR Jr, O'Brien WL, King GA, Ainsworth BE. Estimation of energy expenditure using CSA, Inc. accelerometer hip and wrist sites. *Med Sci Sports Exerc.* 2000;32 Suppl:S450-6.

[140] Campbell KL, Crocker PR, Mckenzie DC. Field evaluation of energy expenditure in women using Tritrac accelerometers. *Med Sci Sports Exerc*. 2002;34:1667-74.

[141] Chen KY, Sun M. Improving energy expenditure estimation by using a tri-axial accelerometer. *J Appl Physiol*. 1997;83:2112-22.

[142] Welk GJ, Blair SN, Wood K, Jones S, Thompson RW. A comparative evaluation of three accelerometry-based physical activity monitors. *Med. Sci. Sports Exerc*. 2000; 32 Suppl:S489-97.

[143] Fruin ML, Rankin JW. Validity of a multi-sensor armband in estimating rest and exercise energy expenditure. *Med Sci Sports Exerc*. 2004;36:1063-9.

[144] Jakicic JM, Marcus M, Gallagher KI, Randall C, Thomas E, Goss FL, et al. Evaluation of the SenseWear Pro Armband® to assess energy expenditure during exercise. *Med Sci Sports Exerc*. 2004;36:897-904.

[145] Trost SG, Ward DS, Moorehead SM, Watson PD, Riner W, Burke JR. Validity of the Computer Science and Applications (CSA) activity monitor in children. *Med Sci Sports Exerc.*1998;30:629-33.

[146] Puyau MR, Adolph AL, Vohra FA, Butte NF. Validation and calibration of physical activity monitors in children. *Obes Res*. 2002;10:150-7.

[147] Eston RG, Rowlands AV, Indedew DK. Validity of heart rate, pedometry and acceleromety for predicting the energy cost of children's activities. *J Appl Physiol.* 1998;84:362-71.

[148] Li R, Deurenberg P, Hautvast JGAJ. A critical evaluation of heart rate monitoring to assess energy expenditure in individuals. *Am J Clin Nutr*. 1993;58:602-7.

[149] Bassett DR Jr., Howley ET, Thompson DL, King GA, Strath SJ, McLaughlin JE, et al. Validity of inspiratory and expiratory methods of measuring gas exchange with a computerized system. *J Appl Physiol*. 2001;91:218-24.

[150] Livingstone MD, Coward WA, Prentice AM, Davies PS, Strain JJ, McKenna PG, et al. Daily energy expenditure in free-living children; Comparison of heart rate monitoring with the doubly labeled water method. *Am J Clin Nutr*. 1992;56: 343-52.

[151] Dauncey Mi, James WPT. Assessment of the heart-rate method for determining energy expenditure in man, using a whole-body calorimeter. *Br J Nutr.* 1979;42:1-13.

[152] Meijer GA, Westerterp KR, Koper H, Ten Hoor F. Assessment of energy expenditure by recording heart rate and body acceleration. *Med Sci Sports Exerc.* 1989;221:343-7.

[153] Rennie K, Roswell T, Jebb SA, Holbuen D, Wareham NJ. A combined heart rate and movement sensor; Proof of concept and preliminary testing study. *Eur J Clin Nutr.* 2000;54:409-14.

[154] Luke A, Maki KC, Barhey N, Cooper R, McGee D. Simultaneous monitoring of heart rate and motion to assess energy expenditure. *Med Sci Sports Exerc*. 1997;29:144-8.

[155] Spurr GB, Prentice AM, Murgatroyd PR, Goldberg OR, Reina JC, Christman NT. Energy expenditure from minute-by-minute heart rate recording: Comparison with indirect calorimetry. *Am J Clin Nutr*. 1988;48:552-9.

[156] Livingstone MBE, Prentice AM, Coward WA, Ceesay SM, Strain JJ, McKenna PG, et al. Simultaneous measurement of free-living energy expenditure by the doubly labeled water method and heart-rate monitoring. *Am J Clin Nutr*. 1990;52:59-65.

[157] van den Berg-Emons RJG, Saris WHM, Westerterp KR, van Baak MA. Heart rate monitoring to asses energy expenditure in children with reduced physical activity. *Med Sci Sports Exerc.* 1996;28:496-501.

[158] Rowlands AV, Eston RG, Ingledew DK. Measurement of physical activity in children with particular reference to the use of heart rate and pedometry. *Sports Med.* 1997;24:258-72.

[159] Gass EM, Gass GC. Rectal and esophageal temperatures during upper and lower body exercise. *Eur J Applied Physiol.* 1998;78:38-42.

[160] Van Marken Lichtenbelt WD, Westerterp-Plantenga MS, Van-Haydonek P. Individual variation in the relation between body temperature and energy expenditure in response to elevated ambient temperature. *Physiol Behav*. 2001;73:235-42.

[161] Colburn TC, Washburn RA, Guarini JJ, Simmons NW. An ambulatory activity monitor with solid state memory. *Instrument Society of America, Transactions; BM* 76322. p. 117-22.

[162] Pitta F, Troosters T, Spruit MA, Probst VS, Decramer M, Gosselink R. Characteristics of physical activities in daily life in chronic obstructive pulmonary disease. *Am J Respir Crit Care Med.* 2005;(171):972-7.

[163] Perini M, Fiocchi A, Bollini F, Lanfranchi S, Zarcone D. Relationship between active and total energy expenditure in Sydenham's Chorea: *A clinical case. Abstract presented at Grado (Gorizia, Italia, May 18-21st, 2005) to the Italian Association of Clinical Neurologists (Associazione Nazionale Neurologi Ospedalieri).*

[164] Mignault D, St.-Onge M, Karelis AD, Allison DB, Rabasa-Lhoret R. Evaluation of the Portable HealthWear Armband. *Diabetes Care*. 2005;28:225-7.

[165] Maliszewski A, Freedson P, Ebbeling C. Validity of the Caltrac accelerometer in estimating energy expenditure, and activity in children and adults. *Pediatr Exerc Sci.* 1991;3:41-151.

[166] Ekelund U, Sjöström M, Yngve A, et al. Physical activity assessed by activity monitor and doubly labeled water in children. *Med Sci Sports Exerc*. 2001;33:275-81.

[167] Westerterp KR. Physical activity assessment with accelerometers. *Int J Obes.* 1999;23 Suppl 3:S45-9.

[168] Sequeira, MM, Rickenbach M, Wietlisbach V, Tullen B, Schutz Y. Physical activity assessment using a pedometer and its comparison with a questionnaire in a large population survey. *Am J Epidemiol*. 1995;142:989-99.

[169] Kriemler S, Zahner L, Schindler C, Meyer U, Hartmann T, Hebestreit H, Brunner-La Rocca HP, van Mechelen W, Puder JJ. Effect of school based physical activity

programme(KISS)on fitness and adiposity in primary schoolchildren: cluster randomised controlled trial. *BMJ* 2010;340:c785. doi:10.1136/bmj.c785
[170] Ferrannini E. The theoretical bases of indirect calorimetry:A review. *Metabolism.* 1988;37:287-301.
[171] Weir JB. New methods for calculating metabolic rate with special reference to protein metabolism. *J Physiol.* 1949;109:1-9.
[172] Porter C, Cohen NH. Indirect calorimetry in critically ill patients: Role of the clinician, dietitian in interpreting results. *J Am Diet Assoc.* 1996;96:49-57.
[173] Saffle J, Larson C, Sullivan J. A randomized trial of indirect calorimetry-based feedings in thermal injury. *J Trauma.* 1990;30:776-83.
[174] Livesey G, Elia M. Estimation of energy expenditure, net carbohydrate utilization, and net fat oxidation and synthesis by indirect calorimetry: Evaluation of errors with special reference to the detailed composition of fuels. *Am J Clin Nutr.* 1988;47:608-28.
[175] Torun B, Cabrera Santiago MI, Viteri FE. Protein requirements of preschool children: Obligatory nitrogen losses and nitrogen balance measurements using cow's milk. *Arch Latinoam Nutr.* 1981;31:571-85.
[176] Jensen CL, Butte NF, Wong WW, Moon JK. Determining energy expenditure in preterm infants: Comparison of 2H218O method and indirect calorimetry. *Am J Physiol.* 1992;32:R685-92.
[177] Schoeller DA, Taylor PB, Shay K. Analytic requirements for the doubly labeled water method. *Obes Res.* 1995;3 suppl:15-20.
[178] Weissman Ch, Kemper M. Metabolic measurement in the critical ill. *Crit Care Clin.* 1995;11:169-97.
[179] Smyrnios NA, Curley FJ, Shaker KG. Accuracy of 30-minute indirect calorimetry studies in predicting 24-hour energy expenditure in mechanically ventilated, critically ill patients. *JPEN J Parenter Enteral Nutr.* 1997;21:168-74.
[180] Thuren P, Phillips RR, DeMarie M, Hoffenberg A, Bronstein M, Spedale S, et al. Technical and methologic considerations for performance of indirect calorimetry in ventilated and nonventilated preterm infants. *Crit Care Med.* 1997;25:171-80.
[181] Mickell JJ. Urea nitrogen in critically ill children. *Pediatrics.* 1982;70:949-55.
[182] Williams RR, Fuenning CR. Circulatory indirect calorimetry in the critically ill. *JPEN J Parenter Enteral Nutr.* 1991;15:509-12,comment in JPEN. 1992;16:497; JPEN. 1992;16:595; JPEN.1993;17:398.
[183] Benotti P, Bistrian B. Metabolic and nutritional aspects of weaning from mechanical ventilation. *Crit Care Med.* 1989;17:181-5.
[184] Oh TE, Bhatt SS, Lin ES, Hutchinson RRC, Low JM. Plasma catecholamines and oxygen consumption during weaning from mechanical ventilation. *Intensive Care Med.* 1991;17:199-203.
[185] Frankenfield D, Smith JS, Cooney R. Accelerated nitrogen loss after traumatic injury is not attenuated by achievement of energy balance. *JPEN J Parenter Enteral Nutr.* 1997;21:324-9.
[186] Feenstra BW, Van Lanschot JJ, Vermeij CG. Artifacts in the assessment of metabolic gas exchange. *Intensive Care Med.* 1986;12:312-5.
[187] White MS, Shepherd RW, McEniery JA. Energy expenditure measurements in ventilated critically ill children: Within- and between-day variability. *JPEN J Parenter Enteral Nutr.* 1999;23:300-4.

[188] Briassoulis G, Michaeloudi E, Fitrolaki DM, Spanaki AM, Briassouli E. Influence of different ventilator modes on VO2 and VCO2 measurements using a compact metabolic monitor. *Nutrition* 2009;25:1106–1114.
[189] Frankenfield D, Sarson GY, Blosser SA, Conney RN, Smith JS. Validation of an abbreviated indirect calorimetry protocol in critically ill patients. *J Am Coll Nutr.* 1996;15:397-402.
[190] Damak M, Askanazi J, Wiessman Ch, Elwyn D, Kinney J. Artifacts in measurement of resting energy expenditure. *Crit Care Med.* 1983;11:750-2.
[191] Schoeller DA, Webb P. Five-day comparison of the double labeled water method with respiratory gas exchange. *Am J Clin Nutr.* 1984;40:153-8.
[192] Roberts SB, Coward WA, Noshia V, Schlingenseipen K, Lucas A. Comparison of the double labeled water method with indirect calorimetry and a nutrient balance study for simultaneous determination of energy expenditure, water and metabolizable energy intake in premature infants. *Am J Clin Nutr.*1986;44:315-22.
[193] Black AE, Prentice AM, Coward WA. Use of food quotients to predict respiratory quotients for the doubly-labelled water method of measuring energy expenditure. *Hum Nutr Clin Nutr.* 1986;40C:381-91.
[194] Briassoulis G, Briassoulis P, Michaeloudi E, Fitrolaki DM, Spanaki AM, Briassouli E. The effects of endotracheal suctioning on the accuracy of oxygen consumption and carbon dioxide production measurements and pulmonary mechanics calculated by a compact metabolic monitor. *Anesth Analg.* 2009 Sep;109(3):873-9.
[195] Chwals WJ, Lally KP, Woolley MM. Indirect calorimetry in mechanically ventilated infants and children: Measurement accuracy with absence of audible airleak. *Crit Care Med.* 1992;20:768-70.
[196] Dietrich K, Romero M, Conrad S. Effects of gas leak around endotracheal tubes on indirect calorimetry measurement. *JPEN J Parenter Enteral Nutr.* 1990;14:408-13.
[197] Schoeller DA, Van Santen E. Measurement of energy expenditure in humans by doubly labeled water. *J Appl Physiol.* 1982;53:955-9.
[198] Schoeller DA. Recent advances from application of doubly labeled water to measurement of human energy expenditure. *J Nutr.* 1999;129:1765-8.
[199] Klein PD, James WPT, Wong WW, Irving CS, Murgatroyd PR, Cabrera M, et al. Calorimetric validation of the doubly labeled water method for determination of energy expenditure in man. *Hum Nutr Clin Nutr.* 1984;38:95-106.
[200] Schoeller DA, Ravussin E, Schutz Y, Acheson KJ, Baertschi P, Jequier E. Energy expenditure by doubly labeled water: Validation in humans and proposed calculation. *Am J Physiol.*1986;250:R823-30.
[201] Jones P, Winthrop A, Schoeller D, Swyer PP, Smith JJ, Filler RR, et al. Validation of doubly labeled water for assessing energy expenditure in infants. *Pediatr Res.* 1987;21:242-6.
[202] Speakman JR, Nair SK, Goran MI. Revised equations for calculating CO, production from doubly labeled water in humans. *Am J Physiol.* 1993:264:E912-7.195.
[203] Lifson N, McClintock R. Theory of use of the turnover rates of body water for measuring energy and material balance. *J Theor Biol.* 1966;12:46-74.
[204] Schoeller DA. Energy expenditure from doubly labeled water: Some fundamental considerations in humans. *Am J Clin Nutr.* 1983;38:999-1005.

[205] Schoeller DA, Leitch CA, Brown C. Doubly labeled water method: In vivo oxygen and hydrogen isotope fractionation. *Am J Physiol.* 1986;251:R1137-43.

[206] Takken T, Van der Net J, Cuis WW, Helders PJ. Physical activity and health related physical fitness in children with juvenile idiopathic arthritis. *Ann Rheum Dis.* 2004;63:112.

[207] Patel SA, Slivka WA, Sciurba FC. Validation of a Wearable Body Monitoring Device in COPD. *Poster presented at 2004 ATS annual meeting (Orlando).*

[208] Zakeri IF, Adolph AL, Puyau MR, Vohra FA, Butte NF. Multivariate adaptive regression splines models for the prediction of energy expenditure in children and adolescents. *J Appl Physiol.* 2010 Jan;108(1):128-36.

[209] Riddoch CJ, Boreman CAG. The health-related physical activity of children. *Res Q Exerc Sport.* 1995;65:84-8.

[210] CamNtech Actiheart.htm. *Downloaded from Camntech.com* on 2010, April, 20th.

In: Advances in Medicine and Biology Research, Vol. 14 ISBN: 978-1-61761-930-4
Editor: Leon V. Berhardt

Chapter II

Pompe Disease: Symptoms, Diagnosis and Treatment

Jayesh J. Sheth *

Institute of Human Genetics, Foundation for Research in Genetics and Endocrinology (FRIGE), FRIGE House, Satellite, Ahmedabad-380 015. India

Abstract

Pompe disease is a rare neuromuscular glycogen storage disorder with a variable frequency of 1:138,000 in Caucasians, 1:40,000 in USA and higher in other ethnic groups. It is one of the 41 lysosomal storage disorders occurring in children and adults due to deficiency of lysosomal hydrolase acid-α-Glucosidase (acid maltase). This results in the accumulation of glycogen in numerous tissues with marked deposition in cardiac, skeletal and hepatic tissues of infants. The clinical presentation varies among infants and adults. Infants commonly present with muscle weakness, hypotonia and floppy baby with delayed motor milestones; whereas in adult onset, proximal muscle weakness with gait abnormalities is more distinct. Respiratory insufficiency with obvious cardiomyopathy is seen in infantile group but is less common in adults. The early presenting time of the disease in our centre is found to be 4.5 months to 3.5 years with an average frequency of 3.1% among five hundred and sixty six children with neuroregression and developmental delay. In older children and adults, diagnostic delay averages to 7 years. Initial diagnostic tests for cardiomyopathy include X-ray of the chest, EEG and cardiac echo for specific Pompe pattern and elevated serum Creatine Kinase (CPK). Muscle biopsy is often performed for glycogen accumulation. However, normal values in adults do not rule out this possibility. Hence, the confirmatory diagnosis is made by estimating acid-Glucosidase from skin fibroblasts or muscle tissue. Though collection of both tissues are invasive and time consuming estimating enzyme activity from leucocytes, using natural acid-Glucosidase inhibitor (acarbose) is found to be more appropriate. Infantile Pompe disease has virtually undetectable activity in fibroblasts and 30-40% in leucocytes.

* Corresponding author: Institute of Human Genetics, Foundation for Research in Genetics and Endocrinology (FRIGE), FRIGE House, Satellite, Ahmedabad-380 015. India, Tele: +91-79-26921414/65128444, Fax: +91-79-26921415, Email: jshethad1@gmail.com

However, a newer method using ratio of acid-Glucosidase activity with and without acarbose of less than 0.2 is found to be as sensitive and specific as the fibroblast study. Mutation analysis of GAA gene is the recommended method for confirming diagnosis as it additionally helps in carrier identification and prenatal diagnosis in the subsequent pregnancy. The availability of precise diagnosis and awareness has made it possible to offer enzyme replacement therapy at an early age. The naturally occurring form of human GAA that acts like a native enzyme is found to be effective in degrading glycogen by catalysing the hydrolysis of lysosomal glycogen and compensates for the GAA deficiency seen in Pompe disease. Reversal of cardiomyopathy and measurable motor gains have been observed in 70-100% of the study subjected reported to date. With the availability of specific diagnosis and therapeutic measures, timely diagnosis and treatment as well as appropriate genetic counseling for the future, reproductive decisions can be provided.

KeyWords: Lysosomal Storage disorders, Pompe disease, Glycogen storage disorders, myopathy.

Introduction

Pompe disease is a rare, progressive lysosomal storage disorder caused due to the deficiency of hydrolase acid-glucosidase [1]. It was first observed by the Dutch Pathologist J.C. Pompe [2] in 1932 in an 8-month-old girl, who died of cardiac hypertrophy due to the presence of glycogen vacuoles in cardiac muscles and other tissues. Because of the storage material, the disease is also known as glycogen storage disease type II or acid maltase deficiency. However, a common nomenclature is needed to recognize the disease and hence, it has been proposed to name the disease as Pompe disease both for infantile or neonatal form as well as the adult form [3].

The deficiency of the enzyme is caused by mutations in the gene that encodes for alpha-glucosidase (GAA) located on the chromosome 17q25.3. It cleaves alpha 1,4 and 1,6 linkages in glycogen to release glucose [4]. Mutation in GAA gene leads to the accumulation of glycogen in different tissues of the body with most serious effect on the cardiac, skeletal and smooth muscles. It is an extremely heterogeneous disease that can present itself in early infancy, childhood or adulthood; and the rate of disease progression and extent of organ involvement depends on the amount of functioning enzyme present.

The disease is inherited in an autosomal recessive manner and often presents as a fatal muscular disease with variable frequency depending on the ethnic group and the geographic area studied. The reported incidences are 1:138,000 in white population [5], 1:31,000 in Afro-Americans [6] and 1:50,000 in Chinese population [7]. However, the combined incidence of Pompe disease is estimated to be approximately 1:40,000 [5] with higher incidences in adults [8]. The reported incidences of Pompe disease in adults is likely to be much higher and is more common than amyotrophic lateral sclerosis, myasthenia gravis and inflammatory myopathies, which we suspect and a high index of suspicion is needed for an early diagnosis and treatment [4].

Clinical Presentation

The clinical presentation encompasses a range of phenotypes that include variable degree of myopathy with different ages of onset. In a series of patients with Pompe disease, the age range was 5 months to 3.5 years [9-11]. Marked hypotonia, respiratory distress and myopathy are the commonest features in infantile Pompe disease, as shown in Figure-1. The symptoms in the affected child may appear in the first few months of life to late sixties and based on the age of onset, different nomenclatures have been used. It is classified as infantile which progresses rapidly and late onset or non-classic form of the disease in childhood or juvenile and adult form of the disease which presents late progress slowly [10,11].

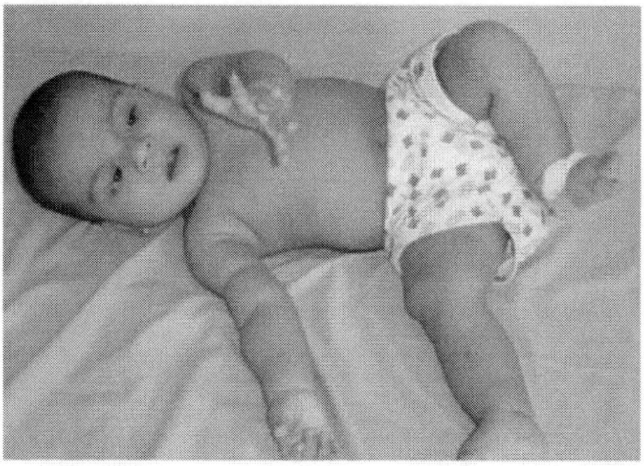

Source: Author's centre

Figure 1. 11 Month-old male child with delayed milestone, repeated respiratory tract infections, breathlessness and GAA activity of 10% of the normal in leucocytes.

The infantile form is most severe as described by Pompe, with prominence of hypotonia and cardiomegaly with likely death prior to one year of age [12]. In childhood or late infantile form, a predominance of skeletal muscle involvement mostly related to mobility and limb-girdle weakness is observed [11,13]. In late onset or adult form of the disease, it is present with initial muscle weakness as early as after 1 yr of age to 20s or 30s to as late as 71 yrs of age [11,14,15].

Infantile onset of Disease

The disease usually appears between the age of 8-12 weeks after birth [12, 16]. In 55% cases in the Dutch series and 44% of cases reported in literature, the disease was manifested with a floppy baby phenotype with difficulty in feeding and failure to thrive by generalized hypotonia. The early age of onset in the author's centre was 2.5 months to 11 months with hypotonia, infiltrative cardiomyopathy and respiratory distress in the affected infant. Severe cardiomyopathy, arrthymias and cardiac failure were observed in 15% and 23% cases while respiratory complications were observed in 40% and 27% of the cases of Dutch and literature series respectively [12] Hepatomegaly may be variable. Affected infants usually do not achieve major milestones such as sitting, rolling over and standing. Table 1 describes the differential diagnosis of Infantile Pompe disease and overlapping common clinical signs and symptoms.

Table 1. Differential diagnosis of infantile onset Pompe Disease.

Differential Diagnosis	Shared signs and symptoms
Peroxisomal disorders, Acute Werdnig-Hoffman disease (SMA-1), Hyperthyroidism, Congenital muscular dystrophy, Danon Disease, Lipid storage disease, Hereditary polyneuropathies, Congenital or transient neonatal myasthenia gravis.	Hypotonia and muscle weakness
Spinal Muscular Atrophy 1	Absent reflexes and progressive proximal myopathy
Peroxisomal Disorders, Mitochondrial-respiratory chain disorders	Hepatomegaly
Myocarditis, Idiopathic hypertrophic cardiomyopathy, Endocardial fibroelastosis, GSD type IIIa and IV, Danon Disease	Cardiomyopathy

Source: Katzin L. W. et al, Kishnani P. et al. [4,38]

The cardiac hypertrophy usually begins in utero and becomes significant in the first few months of life. Echocardiography shows markedly enlarged heart with increased thickness of ventricular walls and interventicular septum. With progressive cardiomyopathy, thickening of left ventricular occurs over time with eventual obstruction to left ventricular outflow with hypertrophy developing over time [17]. Infants may present with symptoms of bronchiolitis with underlying neuromuscular disorder [18]. Electrocardiogram of these children includes a shortened PR interval, an increased QT dispersion (QTd) and large left ventricular (LV) voltages [19]. True Wolf-Parkinson White syndrome has also been reported in some patients [20]. Mechanical ventilation is often needed by 6 months of age with a median age of death of 8.7 month due to respiratory or cardiac failure [21]. In a multicentric study of 168 children with infantile Pompe disease, the ventricular free survival was 16.9% at 12 months, 8.5 % at 18 months and 4.9% at 24 months with overall survival rate of 25.7%, 14.3% and 9.0 % respectively [16].

Patients with classical infantile form have virtually no residual activity of alpha-Glucosidase due to deleterious mutations in the GAA gene. However, infants surviving beyond one year of age tend not to manifest signs until after 6 months of age and have residual activity of a-Glucosidase in skin fibroblasts [16]. Although cardiomyopathy together with muscular symptoms are common, a report of five patients by Hers in 1963 [22] had shown absence of cardiomegaly. Similarly, 5 of 20 patients [23] were reported to have profound muscular weakness in the absence of cardiomegaly. In a study from Spain, Pascual [24] observed an absence of cardiomyopathy in early onset of Pompe disease with slow progression, and termed it as atypical infantile form of the disease.

In a survey of 20 Dutch patients and 133 multicentric cases in literature, median age of first symptom for infantile Pompe disease ranged from 1.6-2.0 months with median age at death of 6-8.7 months [12]. Only two patients were reported to have an age of death above 1.5 yrs [25,26]. Cardiomyopathy, respiratory distress and muscle weakness were the most common signs observed in 88%, 78% and 63% of infants studied by two different groups [11,6]. Feeding difficulty was also observed in 55% of Dutch series against 44% cases in literature [12,13,15,16]. The other less frequently reported symptoms like hepatomegaly, splenomegaly and macroglosia were observed in 90% and 29%, 15% and 6% and 45% and 29% in Dutch series and literature survey respectively [12,16]. Though limited information is available about ocular findings in Pompe disease, strabismus and ptosis have been recently observed [27].

Childhood or Juvenile or Adult form of Disease (Late onset or Non Classic form of the Disease)

The Childhood or late onset form of the disease overlaps clinically with atypical infantile and classical infantile forms [3,18,24]. The disease may occur in infancy or late childhood with signs of exercise intolerance, poor performance in school sports and limb girdle weakness with rigid spine syndrome [10]. In the series of 38 patients, slowly progressive proximal myopathy with trunk weakness and myalgia followed by asymmetric hypertrophy of the thigh were observed [10]. A recent literature review of 225 patients of non-classic Pompe disease [11] demonstrated loss of muscle function in almost 95% patients, while 44% of patients suffered from respiratory difficulties, 28% from artificial ventilation and 3% from cerebral aneurysms. There is also a correlation between the age of onset and a particular subset of symptoms. An enlarged tongue, hypotonia and scoliosis were found in patients with onset of symptoms before the age of one. Delayed motor milestones, feeding problems and hepatomegaly were reported in patients with onset of first symptoms less than six years of age by Winkel group [11] and the author has made similar observations. More than 50% of patients have indicated mobility problem in childhood and severe loss of pulmonary function can occur in patients with minor mobility problem, indicating the importance of pulmonary function monitoring independent of severity of myopathy [11,13,15,29,30]. Finally, one third of patients require artificial ventilation and death usually ensues by 6 yrs of age for patients presenting before one year (0.9 - 24 years); and by 25 years for patients presenting between 1 and 18 years of age [11]. Disease severity such as patients requiring a wheelchair or a ventilator is mainly determined by the disease duration and not by age. However, in the series of Hagemans nearly 25% of children under the age of 15 had a severe course of the disease requiring respiratory, ambulatory and nutritional support [29].

Secondary musculoskeletal impairments (contractures, deformities, lordosis, kyphosis, scoliosis, local pseudo hypertrophy) were more commonly observed in age of onset of one year and greater with absence of hypertrophic cardiomyopathy [15,11]. Adult patients usually show muscular weakness in their 20s or 30s [10,29] up to as late as 71 years of age [11]. Almost 50% of these patients report respiratory insufficiency or exertional dyspnea as the presenting sign with exertional myalgia and fatigue; and rarely with cardiac arrhythmias [10,15,31]. The muscular weakness is more seen as limb girdle myopathy with pelvic muscles being weaker than the shoulder muscles. The femoral muscles are less affected than the pelvic muscles. There is also a compensatory waddling gait and lumbar hyperlordosis due to pelvic girdle weakness. Proximal arm and distal leg weakness fitting a scapuloperoneal distribution are also seen in the adult form of the disease with less affected upper extremities than the lower extremities. However, trunk extensors and abdominal muscles are frequently affected [15]. Non-specific symptoms like pain and fatigue are commonly seen in adult onset Pompe. In an extensive study by Hagerman and his colleagues, 76% of patients complained of fatigue and 46% reported pain more often in the legs. Decrease in functional activities with increased respiratory support has also been observed in patients who are followed over two years time [30,31]. The mean age of death was 44.9 years (25-66 years) in patients who were diagnosed after 18 years of age and the most common cause of death was respiratory insufficiency [11].

Other rare complications in late onset Pompe are vascular involvement of intracranial blood vessels. This happens due to accumulation of glycogen in vascular smooth muscles resulting in aneurysm and rupture of basilar artery, internal carotid artery and medical cerebral arteries [1,3].

Table 2. Differential diagnosis of late onset Pompe Disease.

Differential Diagnosis	Shared signs and symptoms
Limb-girdle muscular dystrophy, Becker muscular dystrophy, Mitochondrial myopathy, Scapnloperoneal syndrome	Progressive muscle weakness in pelvis, legs and shoulders, behind knees proximal muscle weakness.
GSD type IIIa, IV, V and VII	Hypotonia, hepatomegaly, muscle weakness and increased creatine kinase (CK).
Amyotrophic lateral sclerosis SMA type 2 and 3 keneddy disease	Anterior born cell disorders
Vitamin B-12 deficiency, Hereditary neuropathies, Diabetes, Chronic inflammatory demyelinating poly neuropathy	Peripheral neuropathy
Myasthenia Gravis, congenital myasthenia	Generalized muscle weakness and neuromuscular junction disorders
Danon Disease	Hypertrophic cardiomyopathy, Skeletal muscle myopathy, Vacuolar glycogen storage
Rigid spine syndrome	Spinal rigidity and lower back pain

Source: Katzin L. W. et al., Kishnani P. et al. [4,38]

In a series of 225 patients by Winkel and colleagues, the median age in late onset Pompe was 24 years, start of wheelchair use was by 16 years and ventilation at 34 years with better prognosis [11]. However, due to heterogeneous phenotypes the mean delay in medical consultation has been found to be 10 years and the medical community has not reduced the time lag in diagnosis in recent years [10]. The onset of disease may occur at any age in childhood or adulthood predominantly with slow progressive proximal myopathy with or without respiratory failure [32]. Therefore, high degree of clinical suspicion and correct diagnosis of adult or non-classical form of Pompe disease requires additional attention (Table 2).

In both forms of Pompe disease, muscular weakness and respiratory distress are the most common presenting signs with prominence of cardiomyopathy in infantile form of the disease (Table 3), while the rest are variable.

Table 3. Clinical Phenotypes of Infantile and Late onset Pompe disease.

Phenotype	Number of Cases in Percentage	
	Infantile	Late onset
Muscular weakness/ Hypotonia	52-96% (46.67%)[#] 52-96% (46.67%)[#]	95% 11%
Cardiomyopathy/ Cardiomegaly and Left Ventricular Hypertrophy	88-100% (60.0%)[#]	5%
Respiratory Distress	41-78% (20.0%)[#]	44%
Hepatomegaly	29-90% (46.67%)[#]	4%
Enlarged tongue	29-62%	9%
Absent deep tendon reflexes	33-35%	16%
Feeding difficulties	57%	6%
Failure to thrive	53% (40.0%)[#]	-

Source: Hirschhorn and Reuser, Winkel LP et al,, Van den Hout et al [1,11,12], #:Author's centre observation

Diagnosis

Diagnostic procedures for Pompe disease are more extensive because of non-specific presenting phenotypes. They include simple blood chemistry, electromyography, muscle biopsy, heart and lung evaluation, enzyme analysis and molecular diagnosis by mutation detection in GAA gene. Figure-2 briefly describes the approach to systemic diagnosis of Pompe disease.

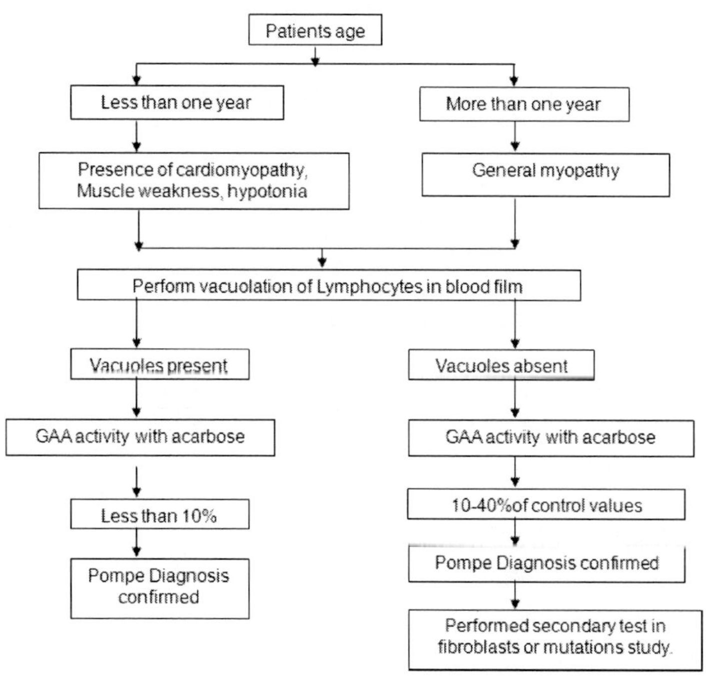

Figure 2. Algorithm of Pompe Diagnosis.

Blood Chemistry Study

More than 90% of all patients with Pompe disease have an elevated level of serum creatine kinase (CPK) [11,21,33]. In 10% of the juvenile and adults with Pompe disease, the level of CK can be as high as 15 times that of the normal [34]. Other enzymes like aspartate aminotransferase (ASAT), alanine aminotransferase (ALAT) and lactate dehydrogenase (LDH) are elevated with disease progression in infants. However, normal levels do not exclude the disease [4,11]. In a series of 225 cases with non-classic Pompe disease, normal levels of CPK with elevated aspartate aminotransferase, alanine aminotransferase were observed in two patients with disease onset after 18 years of age without hepatomegaly. In his series, CK was elevated in 91%, LDH in 96%, ALAT in 94% and ASAT in 95% of the non classic Pompe disease [11].

Electromyography and Electrocardiography

Electromyography reveals myopathic pattern in all types except in non-classic form with late onset disease where study may be reported as normal. An increased insertional and spontaneous myotonic or pseudomyotonic discharge activity, complex repetitive discharges, fibrillation potentials and positive sharp waves are observed more commonly in infants [15]. In atypical Pompe disease, high interventicular septal diastolic dimension, a high left ventricular mass index and asymmetric septal hypertrophy with mitral valve regurgitation with 23.4% decrease of the ejection fraction has been reported in a five year old Korean boy [35].

Muscle Biopsy

The glycogen accumulation in muscle tissue can be visualized as vacuoles with Periodic Acid Schiff (PAS) staining in infantile disease. Often autophagic vacuoles with highly condensed granular inclusions are also seen with mitochondrial abnormalities in adult onset Pompe disease [36].

However, in a large series of non-classical Pompe, about 20% of cases have shown normal glycogen in muscle morphology and vacuoles are not seen in muscle biopsy [11]. Non specific changes such as internalized nuclei, muscle fiber size variation and splitting or an increase in connective tissue are seen in adult form of the disease due to displacement or compression of normal cellular organelles and damage to muscle fibers [34]. Neurogenic atrophy with scattered narrow angular fibers is also seen in rare cases of Pompe explaining glycogen accumulation in anterior horn cells [4].

Measurement of Enzyme Activity (Alpha-1,4-Glucosidase)

Acid alpha-Glucosidase (GAA) is present in all tissues and cells. Hence, measurement of its activity is the hallmark of diagnosis. The enzyme assay can be performed in leucocytes and fibroblasts obtained from skin biopsy or muscle tissue. Of all these, muscle tissue and skin fibroblasts reliably demonstrates absence of GAA activity in infantile from [21] with high residual activity in non classical form of the disease [37]. Although both methods take several weeks for diagnosis, muscle biopsy requires anesthesia, making it a less preferred tissue for the study [38].

Leucocytes obtained from whole blood are widely used for diagnosis of Lysosomal enzyme deficiencies and are the choice for GAA activity using artificial substrate 4-MUG. The consensus group recommendation is to perform GAA activity in a blood sample in a patient older than one year with general myopathy without clear symptoms of Pompe disease even if muscle biopsy is taken for histological examination [38]. However, leukocyte assay is error prone giving false negative results due to presence of additional alpha-Glucosidase known as maltase-glucoamylase (MGA, E.C. 3.2.1.20) in neutrophils. Therefore, competitive inhibition by acarbose was demonstrated in GAA assays in dried blood spots (DBS) [39].This is a rapid, non invasive and reliable method and has been used as an initial screening method. A recent consensus method for prompt and reliable diagnosis of Pompe disease recommends measuring GAA activity in dried blood spot (DBS) in the presence of acabose [40]. If the dried blood spot demonstrates significantly reduced activity of GAA, measuring GAA activity either from fibroblasts or muscle tissue provides a second confirmatory test.

The use of acarbose inhibition for GAA activity in blood lymphocytes has been found to be sensitive and specific with 100% positive predictive values [41]. In this series, 13 of 14 patients with infantile Pompe have demonstrated clear deficiency with one patient showing intermediate activity of GAA. In the author's laboratory, use of acarbose inhibition method and ratio of GAA with and without acarbose is found to be highly specific in the diagnosis of Pompe disease. In our series of 31 children in the age range of 2.5 months to 3.5 years, ratio of GAA with and without acarbose less than 0.2 was found to be confirmative for the infantile Pompe disease, while carrier status in the parents of affected child showed an overlapping ratio with control group (Figure-3). In a recent study of Pompe disease screening, GAA activity was measured in DBS at pH 3.8 and at pH 7.0; and in lymphocytes at pH 4.0 and 7.0 with and without acarbose. Seven infantile and 29 late onset Pompe patients were reliably identified by both methods [42]. This suggests that fluorometric method for the assessment of GAA activity in DBS and leucocytes provides reliable diagnosis without increasing false positive rate. Additionally, DBS can be used for neonatal screening of Pompe disease with leukocyte study as a confirmative method for GAA activity [42].

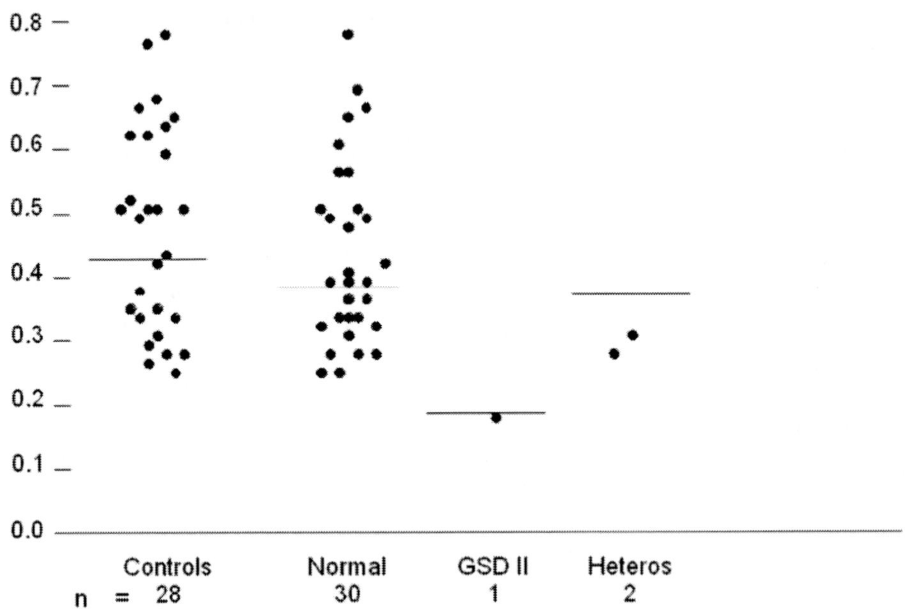

Figure 3. Ratio *α*-glucosidase in leucocytes +/- acarbose.

Mutation Study

Though the identification of disease causing mutation in GAA gene is not essential as per the international consensus meeting for Pompe disease [40], it provides a better alternative for family screening and genetic counseling where diagnosis based on enzyme study is not clear. This is more important when an index case diagnosis is not confirmed and screening of a parent is needed for carrier testing. As with other Lysosomal storage disorders, carrier testing using enzyme activity is less reliable for Pompe disease and in such cases genotype of the family is extremely useful for genotype/phenotype correlation and prenatal diagnosis.

Genetic testing using targeted mutation analysis can be used where disease-causing mutations are known for the population. However genetic testing is not 100% sensitive and large deletions, duplications, multiple exon insertions and sequence alterations may not be picked up on targeted analysis. More than 200 mutations are known for GAA gene and are catalogued by the Erasmus University Medical centre, Netherlands [43]. Recent study in Italian patients has shown that despite the common genotype, patients in their study presented with wide variability in residual GAA activity, age of onset of clinical signs and rate of disease progression. This suggested that other genetic/environmental factors may be responsible for the phenotype variability [44]. In 40 of 225 published cases of late onset Pompe, the most common mutation in GAA gene was IVS1 (-13T-G) in 78% while it was 42% in Caucasians. The age range of patients having this mutation was 2-53 years [44, 45]. In infantile Pompe, of the 20 cases reported 34% cases of Dutch and 7% of US origin have shown deletion in one base pair (Del525T). While exon 18 deletion is seen in 25% of Dutch and Canadian infantile Pompe disease patients, it accounts for only 5% of the cases in the US [46,47]. Two of 20 infantile Pompe cases from Netherlands have shown frame shift mutation

(2741AG- CAGG), two with point mutation (1933G-C and 2303C-G) [48], while in three cases of infantile Pompe disease with very low activity of GAA, mutation could not be identified [12]. In many African-American and African, Chinese, Turkish and European cases of Infantile Pompe R854X, 2741AG-CAGG and G925A mutations are seen more frequently [38].

Prenatal Diagnosis

Pompe disease was one of the first Lysosomal disorders for which prenatal diagnosis was attempted. Highly reliable study for prenatal diagnosis has been demonstrated from direct Chorionic Villus sampling (CVS) and amniotic fluid cells, where enzyme estimation, and mutation study of GAA gene can both be studied [49-54].

In the first report using direct CV and cultured CV cells from Japan, the authors have found intermediate activity of the GAA. However, further confirmation of normal fetus was made by measuring enzyme activity from cultured CV cells [49]. A larger study in 111 CVS from UK had demonstrated that membrane bound vacuoles could successfully diagnose 31 positive cases of storage disorders including GSD type II [52]. A recent study from Spain has shown that ultra structural findings of CVS correlate well with the biochemical and genetic results, confirming the role of CVS ultra structure in prenatal diagnosis [51]. The author's laboratory has also made successful use of direct CV cells in the prenatal diagnosis of one case of Pompe disease; and a follow up study confirmed the normal status of the child. Using maltose as a substrate, the study from USA could diagnose two affected fetus and the diagnosis was subsequently confirmed [55]. Another study has shown glycogen deposition in second trimester fetus with detection of homozygous D645E mutation in GAA gene, further confirming the role of CVS ultra structure as a reliable method for prenatal diagnosis [54].

Therapeutic Approaches

Like other Lysosomal storage disorders, Pompe disease is a multi-system disorder requiring multidisciplinary approach in its management. It requires biochemical and clinical geneticists, cardiologists, pulmonologists, neuromuscular specialists, neurologists, physiotherapists and respiratory specialists with a clear understanding of the disease, its manifestations and challenges, which mainly include psychological and emotional impact on the family members. The primary supportive approach such as physiotherapy and respiratory support may help to improve the quality of life till the decision of enzyme replacement therapy (ERT) is made [56].

Until recently, there has been no specific treatment for Pompe disease. However, with the first trial of ERT in January 1999 followed by many others, ERT is found to be safe and efficient using recombinant human GAA (rhGAA) [57-64]. Several open ended clinical trials in infantile and late onset Pompe disease have shown significant effect on survival and improvement in the quality of life by decreasing cardiomegaly and improving cardiac and skeletal muscle functions.

Effect of ERT in Infantile Pompe Disease

The first clinical trial of ERT was carried out in Rotterdam [57-59] using rhGAA obtained from transgenic rabbit milk and later on using CHO derived rhGAA (Myozyme). Van den Hout and colleagues have treated four infantile Pompe cases with ERT in the age range of 2.5 to 8 months. Initial intravenous dosage of 15-20 mg/kg weekly with an increment of 40 mg/kg weekly resulted in the normalization of GAA activity with decreased muscle glycogen in only one child of 3 months although all four children survived till four years of age whereas life expectancy in untreated patient is less than one year [57-59]. Reduction in hypertrophic cardiomyopathy and improvement in motor function was observed in the youngest child. Nonetheless, one of the children became ventilator dependant at the age of 2 years and died at 4 years with intractable fever, unstable blood pressure and coma [59]. Similar observations were made in two infants of 3.1 and 5.9 months in Germany [61,62]. Both infants received rhGAA from the milk of transgenic rabbits for 48 weeks in a dosage of 40 mg/kg weekly intravenously. There was an overall improvement in cardiac function, left ventricular mass and skeletal muscle function. Over the study course of 10 months, ERT was well tolerated. There was an increase in GAA activity in muscles with improvement in motor function as measured by Alberta infant motor scale, but glycogen storage reduction was seen only in one infant.

Larger trial was carried out using 20-40 mg of rhGAA biweekly in 18 infants with Pompe disease, where improvement in cardiomyopathy as indicated by reduction in the left ventricular mass was observed. All children had shown an improvement in the Pompe PEDI score - a tool specifically designed for functional skill of Pompe disease [65,66]. Thirteen children in the study had shown an improvement in motor function as determined by Alberta Infant Motor Scale (AIMS). In overall survival study, 15 of 18 children reached to 18 months versus 1 of 61 historical control patients. Although there was no clinical difference in response to the treatment of 20 or 40 mg/kg, infusion related reactions like rash, fever, urticaria and decreased oxygen saturation were common with higher dose. This can be managed by simple approaches such as slowing or interrupting the infusion [66].

In 2006, Myozyme - phase II multicentric trial in eight infants from 2.7-14.6 months for 52 weeks was published by Kishnani and colleagues [66]. The dosage schedule was 10-20 mg/kg of rhGAA weekly or bi-weekly for up to 153 weeks in surviving infants. Similar to all previous studies, the significant effect was on cardiac muscles with an improvement in left ventricular mass index. The storage and clearance of glycogen from muscle varied widely and those with low glycogen levels, mild ultra structural muscle damage, high proportion of type I muscle fibers and young age at baseline studies showed good histological response [67]. Six of eight infants were alive after 52 weeks of therapy and five were free of ventilation support. An improvement in motor function was seen in five of these and three were able to walk independently. Four of the six infants died during extension period bringing total deaths to six. Two surviving infants have shown significant reduction in skeletal muscle glycogen and current status of these two children is not known [66].

In a recently released report of 21 infants (3-43 months old with median age of 13 months), Myozyme was given for a period of 168 weeks. At the end of the study, 71% of children were alive, 44% (7/16) remained ventilator free; and the risk of death and invasive ventilation was reduced to 79% and 58% respectively. Left ventricular mass improved in all patients and 62% (13/12) achieved a new motor milestone. At the end of the study 86% of patients (18/21) gained functional independence, while 95% (19/21) patients developed IgG antibodies to rhGAA [68]. A potential long-term complication of fractures has also been reported in four cases receiving ERT and the physician involved in the care should be familiar with this complication [69].

Effect of ERT in Late onset Pompe Disease

Studies about therapeutic response in late onset or Juvenile form of Pompe disease are limited. A three-year study carried out in three patients of 11, 16 and 32 years with Myozyme (CHO derived enzyme) in a weekly dose of 10-20 mg/kg [70] has shown stabilization of pulmonary function after 72 weeks of treatment. The youngest patient of 11 years who was in a wheelchair before treatment was able to ride a bike and play sports; while in two other patients there was an improvement in the quality of life and less disability. Though GAA activity remained low and glycogen was only slightly decreased in all after 12-24 weeks of therapy, other biochemical parameters such as CPK, LDH, transaminases were decreased in parallel with clinical accomplishments. There was also an improvement in distant muscles group than the proximal muscles. The extension period of five years study in these patients has shown improvement in the quality of life and stabilization of muscles and pulmonary functions [71].

In a review of 18 juvenile patients [72], most of them demonstrated improvement in their quality of life after ERT. At baseline, 17 out of 18 patients required the use of ventilator support, and after ERT 10 out of 17 had an improvement in respiratory function. Thirteen of 18 patients had shown improvement in motor functions, while the remaining 5 were stabilized after ERT. An earlier intervention in the disease treatment had shown better improvement in clinical parameters. An observational study was carried out in three patients with Pompe disease (3years 8 month, 2 years 8 months and 19 years 9 months) using Myozyme, with variable dosage of 10-40 mg/kg bi-weekly. The least affected patient of 3 years had an improvement in muscle function, while second patient of 2 years showed initial muscle function improvement with plateau effect at 114 week even after increasing the dosage of the drug. The eldest patient had severely compromised muscle function and died suddenly after 20 weeks of ERT [73]. Thus, prognosis of late onset Pompe is very much dependant on the time of diagnosis and treatment initiation, as has been observed in the Japanese study [74]. They have reported pulmonary veno-occlusive disease (PVOD) as a cause of pulmonary hypertension in Pompe disease in a 28 year old man whose condition at the start of ERT was worst and resulted in bad prognosis. Larger trial of ERT was carried out in 44 patients with late onset Pompe disease (21 to 69 years) by Strothotte and colleagues [75]. After one year of treatment with 20mg/kg aglucosidase alfa, a significant improvement of the six minute walk test was seen (from 341$\pm$ 149m at baseline to 393 $\pm$ 157m) with a significant mean decrease of 10.5% in CPK.

In a recent randomized placebo-controlled trial study of 90 patients with aglucosidase alfa [76] who were 8 years of age or older, an improved walking distance and stabilization of pulmonary function over an 18 month period was observed. All studied groups received intravenous aglucosidase alfa 20 mg/kg of body weight with two primary end points of walk test of 6 minutes and percentage of forced vital capacity (FVC). Infusion associated reactions like urticaria, hyperhidrosis, chest discomfort, vomiting and increased blood pressure occurred in 5-8% of the patients.

The overall results using rhGAA clearly shows that there is a dramatic effect of ERT in infantile Pompe disease with modest response in late onset Pompe disease. Several open label clinical trials have shown that ERT is an effective treatment in infantile Pompe disease with improved quality of life in late onset form of the disease.

A recent study at Duke University has shown an improvement in bilateral ptosis with higher doses of ERT in children with Pompe disease [77]. However, the serious ethical questions about the cost-benefit ratio of ERT and chronicity of the disease need consideration [78].

Safety and Limitations of ERT

In the study of Kishnani and her colleague [66,79], majority of patients developed immunoglobulin G (IgG) antibodies to alpha Glucosidase at different times of treatment. However, all adverse events due to ERT were mild to moderate and no ERT related death has been reported till date. Infusion associated reactions were more common in patients receiving higher dosage (40mg/kg) than in patients receiving 20mg/kg dose of the ERT. A subset of patients with no residual activity of GAA protein (cross reactive immunological material negative-CRIM negative) was found to have higher titer of anti-human acid alpha-glucosidase (hGAA) antibodies on ERT. This has a direct correlation with poor or short-lived response to ERT in Pompe disease [60, 66]. While those with CRIM positive patient, the response of ERT was better and it is likely that in addition to the inactivation of the enzyme, specific antibodies may be interfering with the targeting enzyme [59]. Thus CRIM negative status of the patient is a major challenge and the full implications of the anti-hGAA antibodies remain to be investigated, besides limitations of the therapy.

The major improvements observed so far in preclinical and clinical studies are the effects on cardiac pathology and overall survival. However, the effect on skeletal muscles is less seen than expected due to less M6P receptors in skeletal muscles than heart cells. Limited glycogen clearance may be true for other tissues such as motor neurons; and persistence of glycogen storage in motor neurons may account for the neurological symptoms such as distal foot drop in infants who survive longer after ERT. Other limitation is the inefficient delivery of rhGAA to lysosomes and the resistance of muscle fibers to ERT [80, 81].

Future Perspectives and Experimental Therapies

Gene therapy for Pompe disease is in preclinical stage and has been explored by several *in vitro* studies using Adenovirus (Ad) and Adeno viral vectors (AVV), expressing human GAA [82,83]. The rationale of gene therapy is to transfer the copy of unaffected acid alpha-glucosidase cDNA to the patient's somatic cells. The site specific response in skeletal muscles rather than other distant muscle tissues has been observed with Ad and AVV [84]. Similarly, various ways of administration of Ad vector encoding human GAA into mice have been attempted – these include intramuscular, intracardiac and intravenous, and none have resulted in phenotypic cross correction in distant noninjected muscles [85,86]. However, the correction of glycogenosis type II was demonstrated in human and murine GAA deficient cells by muscle specific lentiviral vector recently [87]. Similarly, a single intramuscular injection of recombinant AAV-1 vectors expressing glycogenin (shGYG) into new born GSD II mice showed significant glycogen reduction, which seems to be the new perspective for the treatment of Pompe disease by modulating glycogen synthesis using RNA interference [88]. Unlike skeletal muscles, the liver was found to be the efficient site for vector transduction. Intravenous administration of Ad vector encoding hGAA into the GAA-KO mice demonstrated significant glycogen reduction in the heart, diaphragm and quads [89]. However, the long-term efficacy was hampered by the demonstration of anti-hGAA antibodies within days of vector injection. Both Ad and AAV based gene therapy have advantages and disadvantages and in the near future, gene therapy does not seem to be possible. Few studies have explored the potential of lentiviral mediated expression of GAA in hematopoietic stem cells (HSC) in mouse model of Pompe disease with stable chimerism of 35% HSC cells, with over expression of GAA and major clearance of glycogen in heart, diaphragm, spleen and liver cells without any adverse events. This study is a novel ex vivo HSC gene therapy for Pompe disease and is worthwhile to pursue as an alternative to ERT [90,91].

Other therapeutic approaches such as enzyme enhancement therapy (EET) seem to be promising for certain mutations in the GAA gene [92]. A significant increase in GAA activity was observed in fibroblasts obtained from patients carrying L552P and G549R mutations with Pompe disease using two imino sugars deoxynojirimycin (DNJ) and N-(n-butyl)deoxynojirimycin(NB-DNJ), but not in those carrying other mutations (A445P,L355P,R375L). Another inhibitor of GAA - D-Glucose, was also found to increase the GAA activity in fibroblast cell lines from late onset Pompe patients. However, the genetic defects were not identified in these cases [93,94].

Another approach of using rhGAA conjugated with oligosaccharide ligand bearing mannose 6 phosphate residues (M6P) has significantly increased affinity for cation independent mannose 6-phosphate receptor (CI-MPR). This has resulted in an increased delivery of the therapeutic enzyme, a second generation of the rh-GAA (neo-rhGAA); and delivery of this preparation resulted in greater clearance of glycogen from all affected muscles compared with the currently available drug [95]. Similarly, intraperitoneal injection of hyluronidase prior to ERT increased GAA activity in heart, diaphragm, kidney and quadriceps in GAA-KO mice [96].

Last but not the least, nutritional and high protein, low-carbohydrate diet along with regular exercise showed improved prognosis in patients with late onset Pompe disease with slower rate of muscle deterioration [97,98].

Conclusion

Pompe disease is a continuous clinical spectrum of progressive muscle weakness in infants and adults. In infants, it presents with cardiomyopathy, respiratory complications and marked hypotonia due to myopathy. In late onset or adult form, it presents with predominance of slow progressive proximal muscle weakness with or without respiratory failure. The development of dried blood spot alpha-glucosidase activity assay has made it possible to offer new born screening for this disease. The importance of making diagnosis in adults is more critical due to availability of ERT and various methods of enzyme assay of GAA using acarbose as an inhibitor of the alpha-glucosidase in neutrophils, ratio of GAA with and without acarbose and mutation detection in GAA gene, which can make accurate diagnosis of the disease. ERT in infants, juvenile and adults shows promising results without many adverse events. The best results of ERT are obtained when treatment is started early in the course of the disease, before irreversible damage to muscles and cardiac tissue have occurred due to glycogen accumulation. However, the cost of ERT and the quality of life after therapy are the major challenges; and an improved technique of enzyme delivery, gene therapy using various vectors and modified stem cell therapy are the future prospects. Currently, the time lag in the diagnosis of Pompe disease must be shortened so that patients are benefited by ERT, which can be achieved by detailed knowledge of the natural course of the disease and by accurate diagnosis.

Acknowledgments

Sincere thanks to Prof Fensom, Mr..Sashi Vaza, Mrs. Marie Jackson of Guy Hospital for providing an opportunity to learn the techniques; the Indian council of Medical Research for grant support to carry out the work; Mr. Mehul Mistry and Mr. Nrupesh Oza for technical assistance.

References

[1] Hirschhorn R., Reuser A. J. J. Glycogen storage disease type II; acid glucosidase deficiency. In: Scriver C.R., Beaudet A.L.,Sly W., Valle D. (eds). *The metabolic and molecular bases of inherited disease.* 8th ed. McGraw–Hill: New York, 2001.

[2] Pompe, J. C. (1932). Over idiopathische hypertrophie van het hart. Ned. *Tijdschr. Geneeskd,* 76, 304-312.

[3] Di Rocco M., Buzzi D., Taro M. (2007). Glycogen storage disease type II: clinical overview. *Acta Myologica.* 26: 42-44.

[4] Katzin L. W., Amato A. A. (2008). Pompe Disease: A Review of the Current Diagnosis and Treatment Recommendations in the Era of Enzyme Replacement Therapy. *J Cli Neuromuscul Disord,* 9(4), 421-431.

[5] Ausems M.G.E.M. Verbiest J., Hermans M.M.P., Kroos M.A., Beemer F.A., Wokke J.H., Sandkuijl L.A., Reuser A.J., van der Ploeg A.T. (1999). Frequency of glycogen storage disease type II in The Netherlands; implications for diagnosis and genetics counseling, *Eur J Hum Genet*, 7, 713-716.

[6] Hirschhorn R., Huie M.C., Caggana M., Butler J., Pass K.A., Hirschhorn *J.N.* (2004). Increased frequency of Pompe disease (infantile onset glycogen storage disease type II) in Afro-Americans. *Pediatr. Res*. 55, 272A.

[7] Lin C.Y., Hwang B., Hsiao K.J., Jin Y.R. (1987) Pompe's disease in Chinese and prenatal diagnosis by determination of alpha-glucosidase activity. *J Inherit Metab Dis.* 10(1), 11-17.

[8] Ausems M.G., ten Berg K., Kroos M.A., van Diggelen O.P., Wevers R.A., Poorthuis B.J., Niezen-Koning K.E., van der Ploeg A.T., Beemer F.A., Reuser A.J., Sandkuijl L.A., Wokke J.H. (1999). Glycogen storage disease type II: birth prevalence agrees with predicted genotype frequency, *Community Genet.* 2(2-3), 91-96.

[9] Sheth J., Patel P., Sheth F., Shah R (2004). Lysosomal Disorders: Our experience, *Ind Pediatr.* 41(3), 260-266.

[10] Müller-Felber W., Horvath R., Gempel K., Podskarbi T., Shin Y., Pongratz D., Walter M.C., Baethmann M., Schlotter-Weigel B., Lochmüller H., Schoser B. (2007). Late onset Pompe disease: clinical and neurophysiological spectrum of 38 patients including long-term follow-up in 18 patients, *Neuromuscul Disord.* 17(9-10), 698-706.

[11] Winkel L.P.E., Hagemans M.L.C., van Doorn P.A., Loonen M.C.B., Hop W.J.C., Reuser A.J.J., vander Ploeg A.T. (2005).The natural course of non-classic Pompe's disease: a review of 225 published cases, *J Neurol* 252, 875-884.

[12] van den Hout H.M., Hop W., van Diggelen O.P., Smeitink J.A., Smit G.P., Poll-The B.T., Bakker H.D., Loonen M.C., de Klerk J.B., Reuser A.J., van der Ploeg A.T. (2003). The natural course of infantile Pompe's disease: 20 original cases compared with 133 cases from the literature. *Pediatr,* 112(2), 332-340.

[13] Hagemans M.L., Winkel L.P., Hop W.C., Reuser A.J., Van Doorn P.A., Van der Ploeg A.T. (2005). Disease severity in children and adults with Pompe disease related to age and disease duration. *Neurology*, 64(12), 2139-2141.

[14] Müller-Felber W., Horvath R., Gempel K., Podskarbi T., Shin Y., Pongratz D., Walter M.C., Baethmann M., Schlotter-Weigel B., Lochmüller H., Schoser B. (2007). Late onset Pompe disease: clinical and neurophysiological spectrum of 38 patients including long-term follow-up in 18 patients. *Neuromuscul Disord*, 17(9-10), 698-706.

[15] Laforêt P., Nicolino M., Eymard B., Puech J.P., Caillaud C., Poenaru C., Fardeau M. (2000). Juvenile and adult onset acid maltase deficiency in France: genotype-phenotype correlation. *Neurology*, 55, 1122-1128.

[16] Kishnani P.S., Hwu W.L., Madel H., Nicolino M., Yong F., Corzo D. (2006). A retrospective multinational, multicentric study on the natural history of infantile-onset Pompe disease, *J Pediatr*, 148(5), 671-676.

[17] Seifert B.L., Snyder M.S., Klein A.A., O'Loughlin J.E., Magid M.S., Engle M.A. (1992). Development of obstruction to ventricular outflow and impairment of inflow in glycogen storage disease of the heart: serial echocardiographic studies from birth to death at 6 months, *Am Heart J*, 123(1), 239-242.

[18] Dixon C.A., Anderson J.B., Ruddy R.M., Cripe L.H. (2010). Infantile onset Pompe disease: A diagnosis not to miss, *Pediatr Emerg Care*, 26, 293-295.

[19] Ansong A.K., Li J.S., Nozik-Grayck E., Kravitz R.M., Idriss S.F., Kanter R.J., Rice H., Chen Y.T., Kishnani P.S. (2006). Electrocardiographic response to enzyme replacement therapy, *Genet Med*, 8(5), 297-301.

[20] Bulkley B.H., Hutchins G.M. (1978). Pompe's disease presenting as hypertrophic myocardiomyopathy with Wolf-Parkinson-White syndrome, *Am Heart J*, 96, 246-52.

[21] Kishnani P.S., Hwu W.L., Mandel H., Nicolino M., Yong F., Corzo D., Infantile-Onset Pompe Disease Natural History Study Group. (2006). A retrospective, multinational,

multicenter study on the natural history of infantile-onset Pompe disease. *J Pediatr*, 148(5), 671-676.

[22] Hers H.G. (1963). Alpha-Glucosidase deficiency in generalized glycogenstorage disease (Pompe's disease), *Biochem J,* 86(1),11-16.

[23] Engel A.G., Gomez M.R., Seybold M.E., Lambert E.H. (1973). The spectrum and diagnosis of acid maltase deficiency. *Neurology*, 23:95-106.

[24] Pascual S.I. (2009). Phenotype variations in early onset Pompe disease: diagnosis and treatment results with myozyme, *Adv Exp Med Biol*, 652, 39-46.

[25] Garancis J.C. (1968). Type II glycogenosis. Biochemical and electron microscopic study. *Am J Med,* 44(2), 289-300.

[26] Yamamoto T., Eguchi A., Okudaira M., Suzuki E., Yokoyama T., Tanabe J. (1960) Glycogen storage disease of the heart. First case in Japan, *Am J Cardiol*, 5, 556-559.

[27] Yanovitch T.L., Banugaria S.G., Proia A.D., Kishnani P.S. (2010). Clinical and histologic ocular findings in Pompe disease, *J Pediatr Ophthalmol Strabismus*, 47(1), 34-40.

[28] Pellegrini N., Laforet P., Orlikowaski D., Pellegrini M., Caillaud C, Eymard B, Raphael J.C., Lofaso F. (2005). Respiratory insufficiency and limb muscle weakness in adults with Pompe's disease, *Eur Respir J,* 26, 1024-1031.

[29] Hagemans M.L., Winkel L.P, Van Doorn P.A., Hop W.J., Loonen M.C., Reuser A.J., Van der Ploeg A.T. (2005). Clinical manifestation and natural course of late-onset Pompe's disease in 54 Dutch patients, *Brain*, 128(Pt 3), 671-677.

[30] Hagemans M.L.C., Hop W.J.C., van Doorn P.A., Reuser A.J., Van der Ploeg A.T., (2006). Course of disability and respiratory function in untreated late onset Pompe disease, *Neurology,* 66, 581-583.

[31] Hagemans M.L., van Schie S.P., Janssens A.C., van Doorn P.A., Reuser A.J., van der Ploeg A.T. (2007). Fatigue: an important feature of late-onset Pompe disease. *J Neurol*, 254(7), 941-945.

[32] Van der Beek N.A.M.E., Hageman M.L.C., van der Ploeg A.T., Reuser A.J.J., van Doorn P.A. (2006). Pompe disease (Glycogen storage disease type II): clinical features and enzyme replacement therapy, *Acta Neurol Belg*, 106, 82-86.

[33] Ausems M.G., Lochman P., Van Diggelen O.P., Ploos vanAmstel H.K., Reuser A.J., Wokke J.H. (1999). *A diagnostic protocol for adult onset glycogen storage disease type II,* 52, 851-853.

[34] Barohn R.J., Mcvey A.l., DiMauro S. (1983). Adult acid maltase deficiency, *Muscle Nerve*, 6, 672-676.

[35] You Honn J., Baik L.E., Chang S.S., Dong H.L. (2007). Clinical improvement in a case of atypical infantile onset Pompe disease with ERT, *Korean J Pediatr*, 50, 213-217.

[36] Schoser B.G.H., Muller-Hockert L., Horvath R., Gempel K., Pongratz D., Lochmuller H.(2007). Adult onset glycogen storage disease type II: clinic-pathological phenotype revisited, *Neuropathol Appl Neurobiol*, 33, 544-559.

[37] Kallwass H., Carr C., Gerrein J., Titlow M., Pomponio R., Bali D., Dai J., Kishnani P.S., Skrinar A., Corzo D., Keutzer J. (2007). Rapid diagnosis of late onset Pompe disease by fluorometric assay of alpha-glucosidase activities in dried blood spots, *Mol Genet Metab*, 90, 449-452.

[38] Kishnani P.S., Steiner R.D., Bali D., Berger K., Byrne B.J., Case L.E., Crowley J.F., Downs S., Howell R.R., Kravitz R.M., Mackey J., Marsden D., Martins A.M.,

Millington D.S., Nicolino M., O'Grady G., Patterson M.C., Rapoport D.M., Slonim A., Spencer C.T., Tifft C.J., Watson M.S. (2006). Pompe disease diagnosis and management guideline, *Genet Med*, 8(5), 267-88.

[39] Li Y., Scottt C.R., Chamoles N.A., Ghavami A., Pinto B.M., Turecek F., Gelb M.H. (2004).Direct multiplex assay of Lysosomal enzymes in dried blood spots for new born screening, *Clin Chem*, 50, 1785-1796.

[40] Winchester B., Bali D., Bodamer O.A., Caillaud C., Christensen E., Cooper A., Cupler E., Deschauer M., Fumic K., Jackson M., Kishnani P.S., Lacerda L., Ledvinova J., Lugowska A., Lukacs Z., Maire I., Mandel H., Mengel E., Muller-Felber W., Piraud M., Reuser A., Rupar T., Sinigerska I., Szlago M., Verheijen F., van Diggelen O.P., Wuyts B., Zakharova E., Keutzer J. (2008). Methods for a prompt and reliable laboratory diagnosis of Pompe disease: Report from an international consensus meeting, *Mol Genet Metab*, 93, 275-281.

[41] Jack R.M., Gordon C., Scott C.R., Kishnani P.S., Bali D. (2006). The use of acarbose inhibition in the measurement of acid alpha-glucosidase activity in blood lymphocytes for the diagnosis of Pompe disease, *Genet Med*, 8, 307-312.

[42] Lukacs Z., Cobos P.N., Mengel E., Hartung R., Beck M., Deschauer M., Keil A., Santer R. (2010). Diagnostic efficacy of the fluorometric determination of enzyme activity for Pompe disease from dried blood specimens compared with lymphocytes-possibility for newborn screening, *J Inherit Metab Dis*, 33, 43-50.

[43] *http://www2.eur.nl/fgg/ch1/pompe* assessed on 26th May 2010

[44] Pittis M.G., Filocamo M. (2007). Molecular genetics of late onset glycogen storage disease II in Italy, *Acta Myologica*, 26, 67-71.

[45] Raben N., Plotz P., Byrne B.J. (2002). Acid alpha glucosidase deficiency (glycogenosis type II, Pompe disease), *Curr Mol Med*, 2, 145-66.

[46] Hirschhorn R., Huie M.L. (1999). Frequency of mutations for glycogen storage disease type II in different populations: The del525T and delta exon 18 mutations are not generally common in white populations, *J Med Genet*, 36, 85-86.

[47] Van der kraan M., Kroos M.A., Joosse M., Bijvoet A.G., Verbeet M P.,Kleijer W J. (1994). Deletion of exon 18 is a frequent mutation in glycogen storage disease type II, *Biochem Biophys Res Commun*, 203, 1535-1541.

[48] Hermans M.M., van Leenen D., Kroos M.A., Reuser A.J. (1997). Mutation detection in glycogen storage disease type II by RT-PCR and automated sequencing. *Biochem Biophy Res Commun,* 241, 414-418.

[49] Ezaki M., Sugiyama K., Wada Y., Suzumori K. (1987). The first trimester prenatal diagnosis of Pompe's disease at risk, *Jpn J Hum Genet*, 32, 267-269.

[50] Fensom A.H., Benson P.F., Blunt S., Brown S. P., Coltart T.N. (1976). Amniotic cell 4-methylumbelliferyl-alpha-glucosidase activity for prenatal diagnosis of Pompe's disease. *J Med Genet*, 13, 148-149.

[51] San Millian B., Teijeira S. Dominguez C., Vieitez I., Navarro C. (2010). Chorionic villi ultrastructure in the prenatal diagnosis of glycogenosis type II. *J Inherit Metab Dis* Feb. 16 [epub ahead of print]

[52] Flower D.J., Anderson G., Vellodi A., Malone M., Sebire N.J. (2007). Electron microscopy of chorionic villus samples for prenatal diagnosis of Lysosomal storage disorders, *Ultrastruc Pathol*, 31(1), 15-21.

[53] Phupong V., Shuangshoti S., Suttiruangwong P., maneesri S., Nuayboonma P., Shotelersuk V. (2005). Prenatal diagnosis of Pompe disease by electron microscopy, *Arch Gynercol Obstet*, 271, 259-261.

[54] Chen C.P., Lin S.P., Tzen C.Y., Tsai F.J., Hwu W.L. Wang W. (2004). Detection of a homozygous D645E mutation of the acid alpha-glucosidase gene and glucogen deposition in tissues in a second-trimester fetus with infantile glycogen storage disease type II, *Prenat diagn,* 24(3), 231-232.

[55] Park H.K., Kay H.H., Rosell A.M., Lanman. J., Chen Y T. (1991). Prenatal Diagnosis of Pompe's disease (type II glycogenosis) in chorionic villus biopsy using maltose as a substrate, *Prenatal Diag*, 12,169-173.

[56] van Der Ploeg A.T. (2005). Monitoring of pulmonary function in Pompe disease: a muscle disease with new therapeutic perspectives, *Eup Respir J*, 26, 984-985.

[57] Van den Hout J.M., Reuser A.J., Vulto A.G., Loonen M.C., Cromme-Dijkhuis A., Van der Ploeg A.T. (2000). Recombinant human alpha-glucosidase from rabbit milk in Pompe patients, *Lance*t, 356, 397-398.

[58] Van den Hout J.M., Reuser A.J., de Klerk J.B., Arts W.F., Smeitink J.A., Van der Ploeg A.T. (2001). Enzyme therapy for Pompe disease with recombinant human alpha-glucosidase from rabbit milk, *J Inherit Metab Dis*, 24, 266-274.

[59] Van den Hout J.M., Kamphoven J.H., Winkel L.P., Arts W F., De Klerk J.B., Loonen M.C., Vulto A.G., Dijkhuis A.C., Kuperus N.W., Hop W., Hirtum H.V., Van Diggelen O.P., Boer M., Kroos M., Van Doorn P., Van der Voort E., Sibbles B., Van Corven E., Brakenhoff J.P., Van Hove J., Smeitink J.A.M., De Jong G., Reuser A.J., Van der Ploeg A.T. (2004). Long tern intravenous treatment of Pompe disease with recombinant human alpha-glucosidase with milk, *Pediatr,* 113, e448-e457.

[60] Amalfitano A., Bengur A.R., Morse R.P., Majure J.M., Case L.E., Veerling D.L., Mackey J. (2001). Recombinant human acid alpha-glucosidase enzyme therapy for infantile glycogen storage disease type II: results of a phase I/II clinical trial, *Genet Med* 3, 132-138.

[61] Klinge L., Straub V., Neudorf U, Voit T. (2005). Enzyme replacement therapy in infantile Pompe disease: results of a ten-month follow-up study. *Neuropediatr*, 36, 6-11.

[62] Klinge L., Straub V., Neudorf U., Schaper J., Bosbach T., Gorlinger K., Wallot M., Richards S., Voit T. (2005). Safety and efficacy of recombinant acid alpha-glucosidase (rhGAA) in patients with classical infantile Pompe disease: results of a phase II clinical trial, *Neuromuscul Disord*, 15, 24-31.

[63] Kishnani P.S., Nicolino M., Voit T., Rogers R C., Tsai A C., Waterson J., Herman G E., Amalfitano A., Thurberg B.L., Richards S., Davison M., Corzo D., Chen Y.T. (2006). Chinese hamster ovary cell-derived recombinant human acid alpha-gluocosidase in infantile-onset Pompe disease, *J Pediatr*, 149, 89-97.

[64] But W.M., Lee S.H., Chan A.O., Lau G.T.C. (2009). Enzyme replacement therapy for infantile Pompe disease during the critical period and identification of a novel mutation, *Hong Kong Med J*, 15(6), 474-477.

[65] Haley S.M., Fragala M.A., Aseltine R., Ni P. (2003). Development of a disease-specific disability instrument for Pompe disease, *Pediatr Rehabil,* 6, 77-84

[66] Kishnani P.S., Corzo D., Nicolino M., Byrne H., Mandel H., Hwu W L., Leslie N. (2007). Recombinant human acid ∝-gluocosidase. Major clinical benefits in infantile-onset Pompe disease, *Neurology,* 68, 99-109.
[67] Thurberg B.L., Lynch M.C., Vaccaro C., Afonso K., Chun Hui T., Bossen A., Kishnani P.S., O'Callaghan M. (2006). Characterization of pre- and post-treatment pathology after enzyme therapy for Pompe disease, *Lab Invest,* 86, 1208-1220.
[68] Nicolino M., Byrne B., Wraith J.E., Leslie N., Mandel H., Freyer D.R., Arnold G.L., Pivnick E.K., Ottinger C.J., Robbinson P.H., Loo J.C., Smitka M., Jardine P., Tato L., Chabrol B., Mccanbless S., Kimura S., Mehta L., Bali D., Skrinar A., Morgen C., Rungachari L., Corzo D., Kishnani P.S. (2009). Clinical outcomes after long term treatment with agluocidase alfa in infants and children with advanced Pompe disease, *Genet Med,* 11, 210-219.
[69] Case L.E., Hanna R., Frush D.P., Krishnamurthy V., DeArmey S., Mackey J., Boney A., Morgen C., Corzo D., Bouchard S., Weber T.J., Chen Y.T., Kishnani P.S. (2007). Fractures in children with Pompe disease: a potential long-term complication, *Pediatr Radiol*, 37, 437-445.
[70] Winkel L.P.F., Van den Hout J.M.P., Kamphoven J.H.J., Disseldrop J.M.A., Remmerswaal M., Arts W.F.M., Loonen M.C.B., Vulto A.G., Van Doorn P.A.V., De Jong G., Hop W., Smit G.P.A., Shapira S.K., Boer M.A., van Diggelen O.P., Reuser A.J.J., Van der Ploeg A.T. (2004). Enzyme replacement therapy in late-onset Pompe's disease: a three-year follow-up, *Ann Neurol*, 55, 495-502.
[71] Capelle C.V. (2008). Pompe disease in children early to mid stage. Clinical Theraputics. 30:S11-S12
[72] van der Ploeg A., Marsden D.L. (2007). Response to enzyme replacement therapy in 18 juvenile and adult patients with severe Pompe disease, *AAN abstracts* SC01.003.
[73] Rossi M., Parenti G., Della C.R., Romano A., Mansi G., Agovino T., Rosapepe F., Vosa C. (2007). Long-term enzyme replacement therapy for Pompe disease with recombinant human alpha-glucosidase derived Chinese hamster ovary cells, *J Child Neurol,* 22, 565-573.
[74] Kobayashi H., Shimada Y., Ikegami M., Kawai T., Sakurai K., Urashima T., Ijima M., Fujiwara M., Kaneshiro E., Ohashi T., Eto Y., Keiko I., Osawa M., Kyosen S.O., Ida H. (2010). Prognostic factors for the late on-set Pompe disease with enzyme replacement therapy: From our experience of four cases including an autopsy case, *Mol Genet Metabol*, 100, 14-19.
[75] Strothotte S., Strigl-Pill N., Grunert B., Kornblum C., Eger K., Wessig C., Deschauer M., Breunig F., Glocker F X., Vielhaber S., Brejova A., Hılz M., Reiners K., Muller-Felber W., Mengel E., Spranger M., Schoser B. (2010). Enzyme replacement therapy with alglucosidsae alpha in 44 patients with late-onset glycogen storage disease type II: 12- month results of an observational critical trial. *J Neurol*. 257: 91-7.
[76] van der Ploeg A.T., Clemens P.R., Corzo D., Escolar D.M. Florence J., Groeneveld G.J., Hersen S., Kishnani P.A., Laforet P., Lake S.L., Lange D.J., Leshner R.T., Mayhew J.E., Morgen C., Nozaki K., Park D.J., Pestronk A., Rosenbloom B., Skrinar A., van Capelle C.I., van der Beek N.A., Wasserstein M., Zivkovic S.A. (2010). A randomized study of Alglucosidase Alfa in late-onset Pompe disease, *NEJM*, 362, 1396-1406.

[77] Yanovitch T.L., Casey R., Banugaria S.G., Kishnani P.S. (2010). Improvements of bilateral ptosis on higher dose enzyme replacement therapy in Pompe disease. *J Neuroophthalmol* April 16 (Epubmed ahead of print)

[78] Wagner K. (2007) Enzyme replacement for infantile Pompe disease. The first step towards a cure, *Neurology,* 68, 88-89.

[79] Kishnani P.S., Corzo D., Leslie ND., Gruskin D., van der Ploeg A., Clancy J.P., Parini R., Morin G., Beck M., Bauer M.S., Jokic M., Tsai C.E., Tsai B.W.H., Morgan C., O'Meara T., Richards S., Tsao E.C., Mandel H. (2009). Pediatric research, Early treatment with alglucosidase alpha prolongs long-term survival of infants with Pompe disease. *Pediatr Res*, 66: 329-335.

[80] Beck M. (2009). Alglucosidase alfa: Long term use in the treatment of patients with Pompe disease, *Ther Clin Risk Mang*, 5, 767-772.

[81] Raben N., Fukuda T., Gilbert A.L., de Jong D., Thurberg B.L., Mattaliano R.J., Meikle P., Hopwood J.J., Nagashima K., Nagaraju K., Plotz P.H. (2005). Replacing alpha-gluucosidase in Pompe disease: recombinant and transgenic enzymes are equipotent, but neither completely clears glycogen from type II muscle fibers, *Mol Ther*, 11, 48-56.

[82] Zaretsky J.Z., Candotti F., Boerkoel C., Adams E.M., Yewdell J.W., Blaese R.M., Plotz P.H. (1997). Retroviral transfer of acid alpha-glucosidase cDNA to enzyme-deficient myoblasts results in phenotypic spread of the genotypic correlation by both secretion and fusion, *Hum Gene Ther*, 8, 1555-1563.

[83] Pauly D.F., Fraites T.J., Toma C., Bayes H.S., Huie M.L., Hirschhorn R., Plotz P.H., Raben N., Kessler P.D., Byrne B.J. (2001) Intercellular transfer of the virally derived precursor from of acid alpha-glucosidase corrects the enzyme deficiency in inherited cardioskeletal myopathy Pompe disease, *Hum Genet Ther,* 12, 527-538.

[84] Ding E., Hu H., Hodges B.L., Migone F., Serra D., Xu F., Chen Y.T., Amalfitano A. (2002). Efficacy of gene therapy for a prototypical Lysosomal storage disease (GSD-II) is critically dependent on vector dose, transgene promotor, and the tissues targeted for vector transduction, *Mol Ther,* 5, 436-446.

[85] Sun B., Zhang H., Franco L.M., Brown T., Bird A., Schneider A., Koeberl. (2005). Correction of glycogen storage type II by an adeno-associated virus vector containing a muscle-specific promotor, *Mol Ther,* 11, 889-898.

[86] Raben N., Lu N., Nagaraju K., Rivera Y., Lee A., Yan B., Byrne B., Meikle P., Umapathysivam K., Hopwood J., Plotz P. (2001). Conditional tissue-specific expression of the acid alpha-glucosidase (GAA) gene in the GAA knockout mice: implimentation for therapy, *Hum Mol Genet*, 10, 2039-2047.

[87] Richard E., Douillard-Guilloux G., Batista L., Calliard C. (2008). Correction of glycogenosis type 2 by muscle-specific lentil viral vector, *In Vitro Cell Dev Biol Anim,* 44(10), 397-406.

[88] Douillard-Guilloux G., Raben N., Takikita S., Batista L., Caillaud C., Richard E. (2008). Modulation of glucogen synthesis by RNA interference: towards a new therapeutic approach for glycogenosis type II, *Hum Mol Genet*, 17, 3876-3886.

[89] Xu F., Ding E., Migone F., Serra D., Schneider A., Chen Y.T., Amalfitano A. (2005). Glycogen storage in multiple muscle of old GSD-II mice can be rapidly cleared after a single intravenous injection with a modified adenoviral vector expressing hGAA, *J Gene Med,* 7, 171-178.

[90] Guilloux G.D., Richard E., Batista L., Caillaud C. (2009). Partial phenotypic correction and immune tolerance induction to enzyme replacement therapy after hematopoietic stem cell gene transfer of alpha-glucosidase in Pompe disease. *The J Gene Medicine.* 11:279-287.

[91] van Til N.P., Stok M., Aerts Kaya F.S., de Waard M.C., Farahbakhshian E., Visser T.P., Kroos M.A., Jacobs E.H., Willart M.A., van der Wegen P., Schotle B.J., Lambrecht B.N., van der Ploeg A.T., Reuser A.J., Verstegen M.M., Wagemaker G. (2010). Lentiviral gene therapy of murine hematopoietic stem cells aleliorates the Pome disease phenotype, *Blood,* April 12 (epub ahead of print)

[92] Parenti G., Zuppaldi A., Gabriela Pittis M., Rosaria Tuzzi M., Annunziata I., Meroni G., Porto C., Donaudy F., Rossi B., Rossi M., Filocamo M., Donati A,. Bembi B., Ballabio A., Andria G. (2007). Pharmacological enhancement of mutated alpha-glucosidase activity in fibroblasts from patients with Pompe disease, *Mol Ther*, 15, 508-514.

[93] Kakavanos R., Hopwood J.J., Lang D., Meikle P.J., Brooks D.A. (2006). Stabilizing normal and mis-sense variant alpha-glucosidase, *FEBS Lett*, 580, 4365-4370.

[94] Porto C., Cardone M., Fontana F., Rossi B., Tuzzi M.R., Tarallo A., Barone M.V., Andria G., Parenti G. (2009). The pharmacological chaperone N-butyldeoxynojirimycin enhances enzyme replacement therapy in Pompe disease fibroblasts, *Mol Ther*, 17, 964-971.

[95] Zhu Y., Jiang J.L., Gumlaw N.K., Zhang J., Bercury S.D., Ziegler R.J., Lee K., Kudo M., Canfield W.M., Edmunds T., Jiang C., Mattaliano R.J., Cheng S.H. (2009). Glycoengineered assayed alpha-glucosidase with improved efficacy correcting the metabolic aberrations and motor function deficits in a mouse model of Pompe disease, *Mol Ther,* 17, 954-963.

[96] Matalon R., Surendran S., Campbell G.A., Michals-Matalon K., Tyring S.K., Grady J., Cheng S., Kaye E. (2006). Hyaluronidase increases the bio distribution of assayed alpha-1,4 Glucosidase in the muscle of Pompe disease mice; an approach to enhance the efficacy of enzyme replacement therapy, *Biochem Biophys Res Commun,* 350, 783-787.

[97] Mark R. (2008). The role of physiotherapy and exercise in the management of patients with Pompe disease, *Clin Ther*, 30, 520-523.

[98] Slonim A.E., Bulone L., Goldberg T., Minikes J., Slonim E., Galanko J., Martiniuk F. (2007). Modification of the natural history of adult-onset acid maltase deficiency by nutrition and exercise therapy, *Muscle Nerve*, 35, 70-77.

In: Advances in Medicine and Biology Research, Vol. 14 ISBN: 978-1-61761-930-4
Editor: Leon V. Berhardt

Chapter III

Review Insulin Resistance: Intra-Uterine Growth Retardation, Life Style, Genetic Susceptibility, Prevention and Treatment

Vincent van Ginneken* and Robert E.Poelmann
Department of Anatomy and Embryology, Leiden University Medical Center (LUMC)
P.O.Box 9600, 2300 RC, Leiden, The Netherlands

Abstract

Overweight (BMI > 25) and obesity (BMI > 30), in our modern world of food abundance has become epidemic and is associated with a whole range of health problems like hypertension, diabetes-2, and an excess of cardiovascular and renal diseases. We will describe intra-uterine growth retardation due to maternal undernutrition, life-style and genetic susceptibility as causes for insulin resistance and β-cell dysfunction. The 'fetal' or 'early' origins of adult disease hypothesis provided evidence for the association between the perturbation of the early nutritional environment and the major risk factors (hypertension, insulin resistanc , and obesity) for cardiovascular disease, diabetes, and the metabolic syndrome in adult life. In this review we give some epidemiological examples from the "Dutch Hunger Winter" a war induced famine between August 1944 and April 1946 in Amsterdam with less than 1,000 calories per day from government food rations. Epidemiological studies demonstrate that exposure to famine during gestation resulted in increases in impaired glucose tolerance, obesity, coronary heart diseases, atherogenic lipid profile, hypertension, microalbuminuria, schizophrenia, antisocial personality and affective disorders. Maternal undernutrition during gestation has important effects on health in later life, but it is hypothesized that the timing of the nutritional insult determines which organ system is affected. Maternal undernutrition can express itself in the second or third generation. Obesity due to diet and a sedentary lifestyle may result in

*Corresponding E-mail: Vincent.vanGinneken@Wur.nl

excess adipose tissue which actively participates in the integration of whole-body energy and fuel metabolism by the secretion of many hormones like leptin, adiponectin and resistin. Life style features like a) disturbance in the autonomous nervous system (biological clock), b) gender, c) socioeconomic status, d) race and e) stress all may be contributing factors in the origin of metabolic syndrome and insulin resistance. Genetic factors influence the different components of the Metabolic Syndrome. Type 2 diabetes is a polygenic disorder. The "thrifthy gene theory" states that it can be inherited from our ancestors living in an environment with unstable food supplies and famine. For them it could increase their probability of survival if they could maximize their storage of food surpluses. With the term "thrifthy genes" one can think about underlying hormonal pathways and metabolic routes like growth hormone and insulin. Inflammation, a second characteristic of insulin resistance, can be caused by Interleukin-6 (IL-6) and Tumor necrosis factor-α (TNF-α). Peroxisome proliferator-activated receptor-gamma (PPAR-gamma) is a transcription factor involved in regulating genes in adipogenesis and, by implication, insulin action. Diet, exercise and preventing stress are the first-line strategies in the management of type 2 diabetes. Exciting results for medical glycaemic control are found in (thiazolidinediones (TZD eg. Pioglitazone). They are agonist for the PPARγ and also have an effect on the recently discovered hormone resistin which probably can link obesity to diabetes-2.

Keywords: insulin resistance, intra-uterine undernutrition, life style, genetics, PPAR, thiazo-idinediones, Pioglitazone.

Metabolic Syndrome

Overweight (BMI > 25) and obesity (BMI > 30), especially visceral fat accumulation, are associated with a whole range of health problems like hypertension, diabetes-2, and an excess of cardiovascular and renal diseases. The risks of obesity are known for more than 80 years, but since in our modern world of food abundance it has become epidemic, the economic impact of the obesity-related risks is becoming larger and larger (Egan and Julius, 2004).

Due to the excessive costs brought by obesity-induced health problems, there is great interest in the mechanisms by which obesity causes these problems. Hepatic steatosis which is often related to obesitas has become a research area of interest due to its close association with obesity and metabolic syndrome. In the USA nonalcoholic fatty liver disease (NAFLD) affects approximately 30 million Americans (Browning & Horton 2004), 20 million Americans have diabetes-2 and the number of diabetics is increasing by 5% per year (Feldman et al. 2001, Boulton et al. 2005). More in general, it is estimated that about 47 million US residents have the metabolic syndrome (Ford et al. 2002). In the United States, among adults, the estimated annual cost attributable to obesity-related diseases is about 100 billion dollar. Among children and adolescents annual hospital costs related to obesity were 127 million dollar during 1997-1999 (www.aaperd.org/thepulse/archives/Pulse-Vol3no4.pdf)). Because the USA has 281,4 million citizens and the European community (EU-12) 298,1 million citizens and there is nearly no difference in estimated prevalence of Metabolic Syndrome between the continents the annual societal costs in the EU may have similar figures. A lot of research has been extended on this subject. About 10 years ago, the concept of 'syndrome X,' also called 'insulin resistance syndrome', was introduced by Reaven (1993). In this syndrome the cluster of changes associated with resistance to insulin-mediated glucose uptake plays an important role in the etiology and clinical course of patients with non-insulin-dependent diabetes, high blood pressure, and coronary heart disease (Reaven, 1993). In 1998, a unifying definition was proposed by the WHO organisation and the insulin resistance syndrome was renamed the 'metabolic syndrome' instead of the 'insulin resistance syndrome' because it had not been established that insulin resistance is the cause of all components present in the syndrome (Heilbronn *et al.*, 2004). The metabolic syndrome is a term that denotes a constellation of cardiovascular risk factors related to insulin resistance and obesity with a centripetal fat pattern. There are different definitions of the metabolic syndrome provided by the World Health Organisation (in 1999) and by the National Cholesterol Education Program (NCEP; in 2001) but any of them predict a significant risk of developing diabetes as well as an excess of coronary heart disease and total cardiovascular disease (Egan and Julius, 2004).

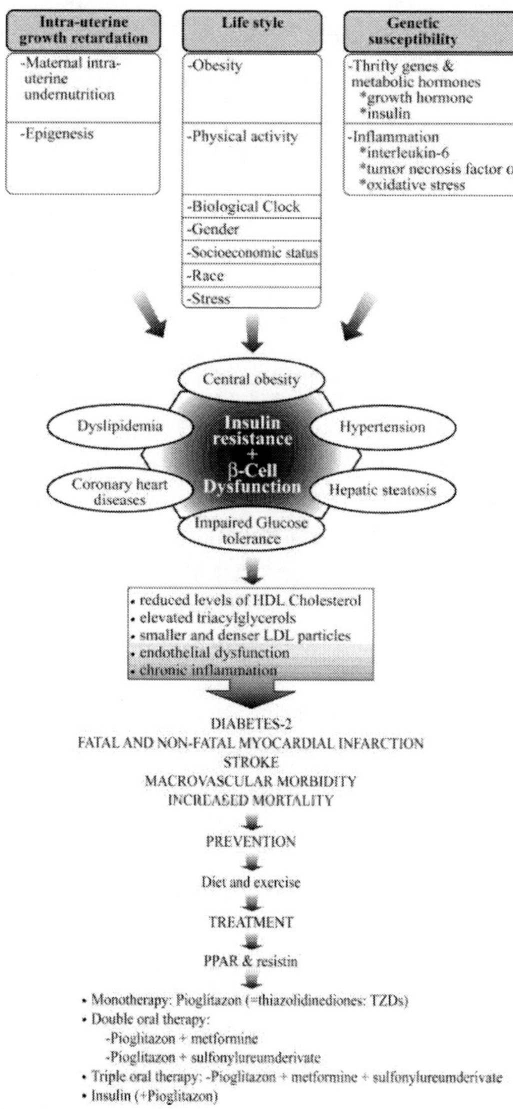

Figure 1. The complicated factors contributing to Diabetes 2, its prevention and treatment.

1. Intra-Uterine: Growth Retardation, Epigenesis

Reduced Fetal Growth and Low Birth Weight

The 'fetal' or 'early' origins of adult disease hypothesis was originally put forward by David Barker and colleagues and stated that environmental factors, particularly nutrition, act in early life to program the risks for adverse health outcomes in adult life. This hypothesis has been supported by a worldwide series of epidemiological studies that have provided evidence for the association between the perturbation of the early nutritional environment and the major risk factors (hypertension, insulin resistance , and obesity) for cardiovascular disease, diabetes, and the metabolic syndrome in adult life (McMillen & Robinson 2005).

A study of Valdez turned out that reduced fetal growth and a low birth weight have been associated with an increased risk of cardiovascular disease, type 2 diabetes and features of the metabolic syndrome developing later on (Valdez et al. 1994). In another study it turned out that children with a low birth weight showed significant more hyper-insulinemia and insulin resistance compared to controls with normal birth weight (Jaquet et al. 2000). A probable explanation could be the ‘thrifty phenotype’ theory (Hales et al. 1991). The theory suggests that low birth weight, due to intra-uterine undernutrition, might have the consequence of developmental adaptations in certain tissues (for example endocrine pancreas, adipose tissue and muscle fiber) predisposing individuals to cardiovascular and metabolic disturbances in adult life.

Not only low birth weight can induce insulin resistance, but also the other way round, in case of genetically determined insulin resistance. This means that genetically determined insulin resistance could result in low fetal growth and insulin resistance later on. This principle is known as the ‘fetal insulin hypothesis’ (Hattersley & Tooke 1999).

Several studies have been performed on the effects of maternal intrauterine undernutrition on offspring birth weights in a cohort of women born between August 1944 and April 1946 in Amsterdam, The Netherlands. This period included "the Dutch Hunger Winter", a war-induced famine. Undernutrition was defined separately for each trimester of pregnancy as an average supply of less than 1,000 calories per day from government food rations (Lumey and Stein 1997).

Maternal birth weight itself was decreased after third trimester intrauterine exposure but not after first trimester exposure. The authors suggest that there may be long-term biological effects, even into the next generation, of maternal intrauterine undernutrition which do not correspond to the effects on the mothers' own birth weights (Lumley and Stein 1997). This was later confirmed in studies on impaired insulin secretion by comparing the glucose tolerance in people who had been prenatally exposed to famine compared with people unexposed to famine (de Rooij et al. 2006). Kyle and Pichard (2006) stated in their review that exposure to famine during gestation resulted in increases in impaired glucose tolerance, obesity, coronary heart diseases, atherogenic lipid profile, hypertension, microalbuminuria, schizophrenia, antisocial personality and affective disorders. However in other studies no statistically significant differences were observed between participants exposed and unexposed to famine in utero in the mean profile of cortisol response to psychological stress protocol (de Rooij et al. 2006). Also greater prevalence of metabolic syndrome (de Rooij et al. 2007), the HPA axis activity at adult age (de Rooij et al. 2006) and increased arterial vessel wall stiffness (Painter et al. 2007) were unaffected. Differences between studies possibly can be ascribed to the period of maternal undernutrition during gestation. Painter et al. (2005) found more coronary heart disease, raised lipids, and altered clotting and more obesity after exposure to famine in early gestation compared to those not exposed to famine. Exposure in mid gestation was associated with obstructive airways disease and microalbuminuria. While decreased glucose tolerance was found in people exposed to famine in late gestation. This study shows that maternal undernutrition during gestation has important effects on health in later life, but that the timing of the nutritional insult determines which organ system is affected (Painter et al. 2005).

That the timing of undernutrition during the gestation period may be important for *homo sapiens* may be explained by a study which found differences in brain development between rodents and humans. In the normal adult rodent and primate, *arcuate nucleus* (ARH) neurons function as conduits for transmitting metabolic hormonal signals into the hypothalamic circuitry that modulates feeding and energy expenditure. Grove et al. (2005) have shown that ARH projections do not fully develop until the 3rd postnatal week in the rodent. This is in stark contrast to the nonhuman primate (NHP) in which ARH projections develop during the 3rd trimester of pregnancy. This species difference suggest that maternal diet and health are likely key factors for the development of ARH projections in the primate, whereas the postnatal environment (i.e., diet) would be more important in the rodent (Grove et al. 2005).

In contrast to the 'Dutch famine' studies were the effect of starvation was studied on the offspring, other authors investigated the effect of maternal adiposity on fetal development. Yanjnik (2004) found that the body composition of Indian babies was influenced by maternal adiposity before pregnancy and by aspects of maternal nutritional intake and circulating nutrient concentrations during pregnancy. Increased nutrition in the face of a genetic predisposition increases maternal insulin resistance in late pregnancy and promotes fetal adiposity even in absence of marked hyperglycaemia (Yanyik 2004). In the study of Kaati *et al.* (2002), investigating historical data in families of an isolated parish village in North-Sweden the researchers found that seasonal overfeeding has transgenerational effects resulting in cardio-vascular disease and diabetes in the descendants of the second or third generation.

There are no strong paternal determinants of adiposity at birth (Yanyik 2004). In contrast, in the study of Pembrey et al. (2006) it was demonstrated that these aspects can also become transgenerational via the male line depending if food abundance occurs during the prepubertal period.

So undernutrition and overfeeding can result in an increased risk of cardiovascular disease, type 2 diabetes and features of the metabolic syndrome in the second or third generation. This implies a new insight in the interplay between inheritance and environment in health and development.

2. Life Style: Obesity, Sexe Differences, Race, Physical Activity and Psychosocial Stress

Life style (weight gain and sedentary lifestyle): Physical activity has an impact on many of the components of the metabolic syndrome. An increase of physical activity results usually in weight loss and reduction of body fat. Exercise training has been associated with favorable changes in serum triglyceride and HDL cholesterol concentrations and with improvement in insulin sensitivity (Knowler et al. 2002). The opposite is that low cardio respiratory fitness (little physical activity) is associated with an increased clustering of the metabolic abnormalities associated with the metabolic syndrome. A moderate grade of physical activity reduces the likelihood of the development of the metabolic syndrome (Laaksonen et al. 2002). A very important aspect is that of food intake because a surplus of fatty acids, and namely unsaturated fatty acids, can cause dyslipidemia (ref).

Obesity due to diet and a sedentary lifestyle may result in excess adipose tissue. Adipose tissue is not just 'an organ' designed for passive storage and release of triacylglycerols (TGs). The abundant abdominal fat in most patients with fatty liver disease produces proinflammatory cytokines (Rudin & Barzilai, 2005). In addition, adipose tissue also actively participates in the integration of whole-body energy and fuel metabolism by the secretion of many hormones like leptin, adiponectin and resistin (den Boer et al. 2004).

Disturbance in the autonomous nervous system: The origin of the metabolic syndrome is probably in a disturbance in the balance between physical activity and food intake. The biological clock is a part of the autonomous nervous system and is located in the suprachiasmatic nucleus of the hypothalamus (SCN) (Kreier et al. 2007).

It has two functions a) a balance between sympathetic to parasympathetic activity in the body, b) regulation of the diurnal rhythm.

a) Ad The biological clock regulates the rhythmic change from sympathetic to parasympathetic activity in the body and adapts the metabolism to this rhythm. In our life style we constantly change between active and inactive periods, instead of one long active (day) and one long inactive period (night). The physical activity in general is reduced, so there is little fluctuation between activity and inactivity. This together results in conflicting information sent to the biological clock The biological clock adapts metabolism to the sympathetic/parasympathetic rhythm. Because this rhythm is disturbed, the biological clock will also be disturbed and this results in an increased parasympathetic activity on the visceral adipose tissue and an increased sympathetic activity on the heart and skeletal muscles. The increased parasympathetic activity on the visceral adipose tissue results in an increased level of insulin secretion and an increase of fat. The increased sympathetic activity on the heart and skeletal muscles results in hypertension, and an induced glucose uptake in skeletal muscles. In normal situation the sympathetic and parasympathetic activity alternate, but in patients the sympathetic and parasympathetic activity works at the same time on different locations in the body (ref).

b) Furthermore in a study Kreier et al. (2007) provide the hypothesis that the 'biological clock', which normally prepares us each morning for the coming activity period, is altered due to a modern life style of low activity during the day and late-night food intake. So research indicated in aged and young patients with type 2 diabetes indications for a malfunctioning of the biological clock. While elderly patients have an impaired function of the SCN due to the degeneration of neurons, the researchers propose that in younger subjects the clock loses its 'feeling' for internal and external rhythms caused by the modern lifestyle. Sleeping late and less coupled with constant metabolic excess alter both internal and external environmental stimuli to the brain. In response to these alterations, the rhythm of the biological clock is disrupted which may lead to the metabolic syndrome and type 2 diabetes (Kreier et al. 2007).

Explanation for differences between sexes: For a long time, cardiovascular disease (CVD), one of the symptoms of metabolic syndrome, has been seen as a 'male' disease due to men's higher absolute risk compared with women, but the relative risk in women of CVD morbidity and mortality is actually higher. In Europe CVD kills a higher percentage of women (55%) than men (43%) (Petersen et al. 2005).

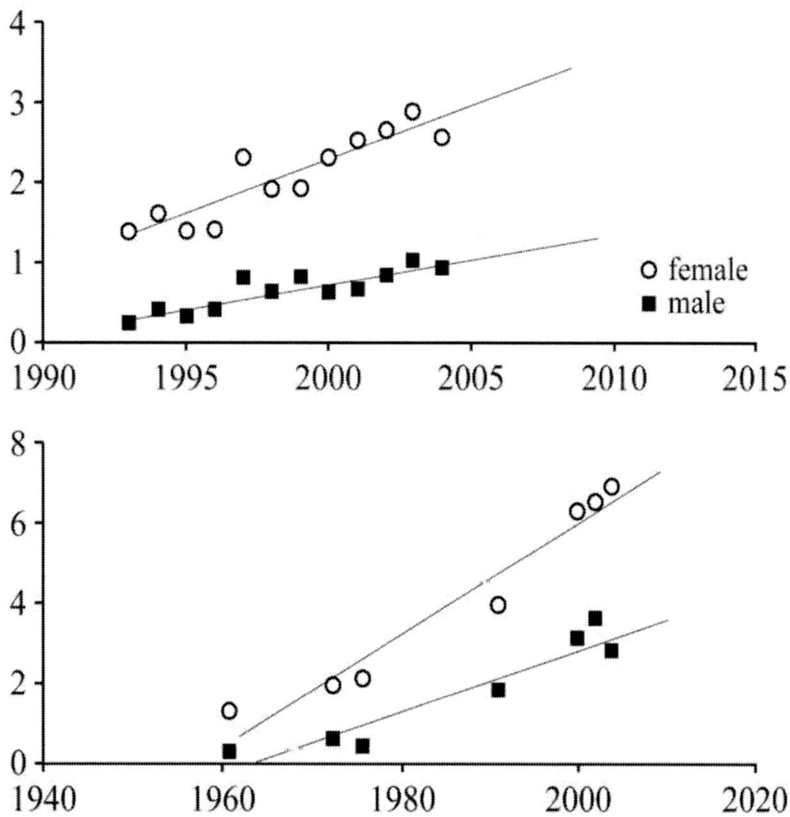

(Source IOTF 2007 forecast from the Health Survey for England data)

Figure 2. The rise in morbid obesity in the UK (top) and the USA (bottom). % morbidity obesity UK (top) and the USA (bottom). Top figure: increasing prevalence of morbid obesity in England (BMI => 40 kg/m2), Bottom figure: faster rising morbid obesity in the USA.

Compared with men, CVD risk in women is increased to a greater extent by some traditional factors (e.g., diabetes, hypertension, hypercholesterolemia, obesity), and socioeconomic and psychosocial factors also seem to have a higher impact on CVD in women. Women with diabetes have 2.6 times the risk of dying from coronary heart disease than women without diabetes compared with a 1.8-fold risk among men with diabetes (Wengner 2003). Obesity is a huge problem, it has increased almost 3-fold over the past decades. Nevertheless there are no significant differences between girls and boys in the prevalence of overweight and obesity (Möller-Leimkühler 2007). However body composition and fat distribution are different between boys and girls across ages, and these differences may influence the relation between fatness and lipids (Möller-Leimkühler 2007). The prevalence of smoking is similar in girls and boys (Möller-Leimkühler 2007) but girls report less physical activity than boys, both before and during adolescence (Higgens, 2003). For biochemical compounds at menopause LDL is increasing. This effect is thought to be partly the result of advancing age and declining levels of estrogen (Möller-Leimkühler 2007). There is some evidence to suggest that high levels of triglycerides are a significant independent risk factor for CVD in both sexes, but more so in women than men (Castelli et al. 1992). There is a clear socioeconomic gradient in risk of CVD, with those in the lowest socioeconomic strata have the highest risk of cardiovascular events (Winkleby & Cubbin 2003).

Other factors between the sexes are women's coronary vessels tend to be smaller than those of men which makes them more difficult to revascularise percutaneously as well as surgically (Jacobs 2003).

Socioeconomic status*:* is associated with obesity and components of the metabolic syndrome (even after multivariate analysis taking account of smoking, alcohol (ab)use and physical excersice (Brunner et al. 1997). Also a poor standard of education is negatively correlated with an elevated risk of the metabolic syndrome (in middle aged women) (Wamala et al. 1999). The fact is that mortality due to CVD is higher in people with low socioeconomic status. It is possible that the metabolic syndrome got its place in the causative pathway.

Overweight among Racial/Ethnic Groups, Differences between Races

Boys: In figure 3 the top panel gives the Prevalence of overweight (Sex and aged specific BMI) for boys from three racial/ethnic groups. Overweight has increased for all children and adolescents over time. NHANES data indicate disparities among racial/ethic groups. The most recent NHANES data (2003-2004) showed that for boys, aged 12-19 years the prevalence rate of overweight was slightly higher among adolescent non-Hispanic white boys (19.1%) than among non-Hispanic black boys (18.5%) and Mexican American boys (18.3%) (Hedley *et al.* 2004, Ogden *et al.* 2006).

Data from NHANES (1988-1994) through NHANES (2003-2004) showed that adolescent non-Hispanic white and black boys experienced larger increases in the prevalence of overweight (7.5% and 7.8% respectively) compared to the increase among Mexican American boys (4.2%) (Hedley *et al.* 2004, Ogden *et al.* 2006).

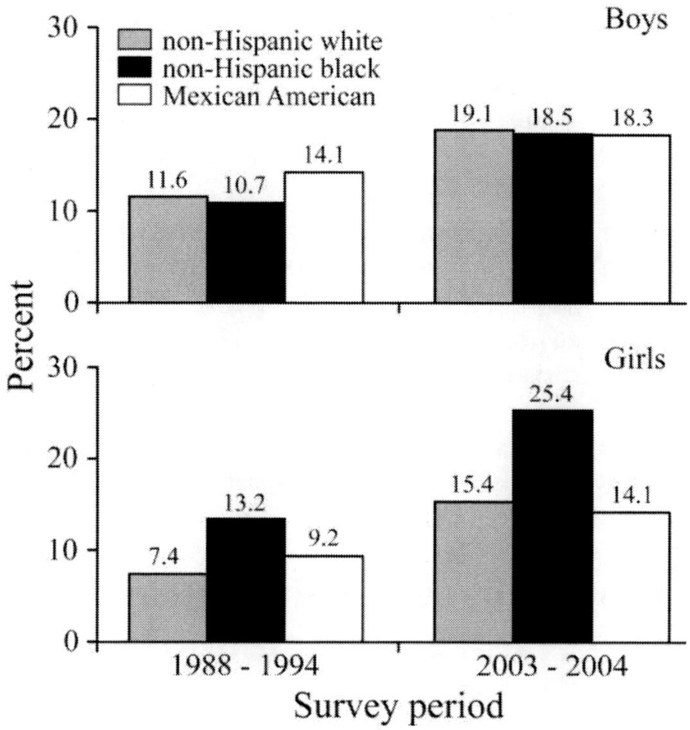

Figure 3. Differences between ethnical groups for obesity in the USA for the period 1988-1994 versus 2003-2004 split up to gender.

Girls: In figure 3 the bottom panel give the Prevalence of overweight (Sex and aged specific BMI) for girls from three racial/ethnic groups. Overweight has increased for all children and adolescents over time. NHANES data indicate disparities among racial/ethic groups. The most recent NHANES data (2003-2004) showed that for girls, aged 12-19 years the Non-Hispanic black girls had the highest prevalence of overweight (25.4%) compared to that of non-Hispanic white (15.4%) and Mexican American (14.1%) girls. (Hedley *et al.* 2004, Ogden *et al.* 2006). Data from NHANES (1988-1994) through NHANES (2003-2004) showed that non-Hispanic black adolescent girls experienced the largest increases in the prevalence of overweight (12.2%) compared to non-Hispanic white adolescent (8.0%) and Mexican American adolescent (4.9%) girls (Hedley *et al.* 2004, Ogden *et al.* 2006).

Stress activates the sympathoadrenal system and the hypothalamic-pituitary-adrenocortical (HPA) axis. Increased sympathetic activity might contribute to several aspects of the metabolic syndrome, such as glucose uptake in skeletal muscle, lipolysis and development of hypertension (Hjemdahl 2002). A complex relationship is proved between the HPA axis function, obesity and metabolic parameters suggesting an etiological role for these factors in the development of the metabolic syndrome (Bjorntorp & Rosmond 2000). People suffering of the metabolic syndrome had an increased urinary excretion of cortisol metabolites and normetanephrine and lower heart rate variability (Brunner et al. 2002). This means that chronic stress results in elevated glucocorticoid levels (cortisol), and this is a strong indication that psychosocial factors, like stress, can contribute or even cause the metabolic syndrome (Boulogne & Vantyghem 2004).

Glucocorticoids are interesting hormones to study in relation to metabolic syndrome because 'chronic stress', as a consequence of our 'modern' life style results in elevated glucocorticoid levels. These may be a contributing or even causative factor for metabolic syndrome (Boulogne & Vantyghem 2004). Evidence for this hypothesis comes from two different studies. Cortisol replacement by diet indicated it was a necessary condition for the development of diet-induced obesity, alterations in insulin sensitivity and TG lipoprotein metabolism (Mantha et al. 1999). In another study with transgenic mice with overexpression of 11-β hydroxysteroid dehydrogenase (11-β HSD) in adipose tissue (resulting in increased glucocorticoid production), Metabolic Syndrome (with concomitant features: visceral obesity, hyperlipidaemia and insulin restistance) could be induced (Masuzaki et al. 2001). Recently an interaction between corticosterone and insulin has been demonstrated. Glucocorticosteroids provoke a dose related increase in total caloric intake in combination with increasing insulin concentrations resulting in obesity (La Fleur 2004). Still the interactions between hepatic steatosis, elevated corticosteroids and insulin levels (pancreas) are not understood.

3. Genetic Susceptibility: "Thrifty Genes and Inflammation"

Genetic factors influence the different components of the Metabolic Syndrome, for example insulin resistance. Type 2 diabetes is a polygenic disorder (Gloyn et al. 2006). Insulin resistance clusters within families, 45% of first degree relatives of patients with type two diabetes are insulin resistant compared to 20% of individuals without a family history of diabetes (Groop et al. 1996). The estimation of heritability in obesity has varied (20-90%) depending on whether the results are based on twin, adoption or family studies (Maes et al. 1996). Another aspect is the combination Diabetes Melitus and waist to hip ratio: first degree relatives of patient with type 2 diabetes have an increased waist to hip ratio compared to subjects without family history type 2 diabetes (Groop et al. 1996). Heritability also influences hypertension (Levy et al. 2000) triacylglycerols and HDL cholesterol (Snieder et al. 1999) and micro-albuminuria (Forsblom et al. 1999).

Obesity, type 2 diabetes and the Metabolic Syndrome seem to develop in societies with a fast transition from rural to urban lifestyle. In 1962 Neel (Neel 1962) proposed the 'thrifty gene' theory. He suggested that our ancestors living in an environment with unstable food supplies and famine would increase their probability of survival if they could maximize their storage of food surpluses (Neel 1962). So genetic selection would favor energy storing genes in an environment like in a hunter-gatherer society. If this 'storing energy' genotype which was advantageous in the past, is expressed in the western lifestyle with abundant food supply, it becomes disadvantageous, by causing obesity and diabetes. In this theory thrifty genes could predispose to the metabolic syndrome. It is not clear which genes exactly belong to this genotype, because not all genes are already detected and/or analyzed. But the genes that regulate fat storage, body weight, fat distribution, lipolysis, FFA metabolism, fuel oxidation, insulin resistance and skeletal muscle glucose metabolism are definitely belonging to this genotype (ref).

Probably one of the most important genes that play an important role in heredity obesity is the gene for production of ghrelin. The equilibrium between insulin, leptin and ghrelin are responsible for normal appetite and normal body weight. Research on a group of 70 very obese British children showed that 10 percent has a defect in this gene. The underlying mechanism is still unknown (Saad et al. 2002). In another research it was shown that the concentration of the hormone PYY (peptide YY3-36), which reduces the release of ghrelin, is lower in obese people than in non-obese people (Baterham et al. 2003).

With the term "thrifthy genes" one can think about underlying hormonal pathways and metabolic routes. We will discuss growth hormone and insulin. Also adipose tissue should be considered as endocrine organ because adipocytes are known to secrete hormones. The list of proteins and factors is growing rapidly and would fall beyond the scope of this review. For an excellent review see (Heilbronn et al. 2004). In this review we will discuss Interleukine-6 (IL-6) and Tumor necrosis factor-α (TNF-α) as inflammatory agents. Furthermore new medicines "the Thiazoladinediones" of which the most important is “Pioglitazone Hydrochloride” have an impact on resistin, a recently discovered hormone.

Metabolic hormones: Insulin and growth hormone are the major hormones involved with glucose metabolism. The association of chronic liver disease (CLD) with impaired glucose metabolism has been known since 1906 (Naunyn 1906). Acquired GH resistance can be observed from the very early stages of CLD (Picardi et al. 2006) GH resistance even worsens in parallel with the progression of liver disease (Picardi et al. 2003). Knowledge about the effect of alterations of the GH-IGF-1 axis on the blood glucose control is a prerequisite in understanding the role of an impaired liver on GH and glucose regulation

Also, impaired insulin signaling due to a certain gene (Foxo1) has emerged as a possible unifying mechanism for various common abnormalities of type diabetes-2. It can be hypothesized that there are two critical contributors to diabetes-2: a) increased glucose production, b) decreased insulin production. Previously it was thought that these processes were independent abnormalities caused by another problem in diabetes, the resistance of new fat cells to insulin. But it seems that these processes are regulated through a gene Foxo1 in the liver (Nakai et al. 2002; Langhans 2003). In a recent study it was demonstrated that when the production of the gene Foxo 1 was lowered the level of glucose in diabetic mice declined to normal levels via two mechanisms: First, the liver reduced glucose production and secondary cells in the pancreas regained their ability to make insulin, the hormone which is important in removal of glucose from the bloodstream (Nakai et al. 2002).

Interleukin-6 (IL-6): Both obesity and insulin resistance are considered chronic inflammatory states. (Arkan *et al.*, 2005) Thus in both conditions, just like in reaction to an infection or an injury, cytokines (IL-6 and TNF-α) are released from the site of tissue injury (Tilg & Hotamisligil 2006). These cytokines promote an acute-phase response. This response is characterised by the production of a range of proteins, primarily from hepatocytes (liver cells), but also from other cells such as monocytes (precursors to macrophages), fibroblasts (connective tissue cells) and adipocytes (fat cells). (Arkan *et al.*, 2005). Production of cytokines is one of the earliest events in many types of liver injury, triggering the production of other cytokines that together recruit inflammatory cells and initiate a healing process in the liver that includes fibrogenesis (Tilg & Hotamisligil 2006).

Tumor Necrosis Factor α (TNF-α): The cytokine which is accepted widely as a cytokine involved in the inflammation reaction due to fat accumulation in the liver is tumour necrosis fator α. (TNF-α). TNF-α is increased in obese subjects. It impairs insulin signalling by serine phosphorylation of insulin receptor substrates. (Bays *et al.*, 2004). From different researches the conclusion has been made that TNF-α is an important mediator of insulin resistance through its ability to influence the tyrosine kinase activity of the insulin receptor. TNF-α is being over expressed in humans with obesity and correlates with the body mass index. TNF-α is able to impair insulin release by inducing insulin resistance in pancreatic β cells and by stimulating the release of mediators that are toxic for these cells. Thus, TNF-α exerts local pancreatic effects, and the induction of insulin resistance may be more related to the local tissue TNF-α concentration than to the levels in plasma. (Bays *et al.*, 2004)

For example, Miyazaki *et al.* (2003) found the following correlations with TNF-α, insulin resistance, obesity and T2DM. At first, an increase in circulating TNF-α concentration is associated with peripheral insulin resistance and increased plasma glucose and insulin levels prior to the onset of type 2 diabetes. Secondly, serum TNF-α levels are elevated in type 2 diabetic patients and are unrelated to the degree of obesity (BMI). And finally, the increased serum TNF-α concentration in type 2 diabetic individuals is not correlated with the severity of insulin resistance. This lack of correlation may be explained by the presence of hyperglycemia, hyperinsulinemia or other factors which have a stronger impact on peripheral insulin sensitivity in type 2 diabetes mellitus.

In addition, TNF-α polymorphisms have been reported to influence the susceptibility to different hepatic diseases, including alcoholic steatohepatitis, and increasing evidence suggests that TNF-α is involved in the pathogenesis and progression of liver diseases of different etiology. The rise of TNF- α may be related to liver fibrosis, and might promote liver fibrosis (Browning, 2003).

Oxidative stress: The pathological changes that develop in e.g. hepatocytes are often ascribed to oxidative stress due to fatty-acid oxidation (Fernandez-Checa & Kaplowitz 2005). It has been proved that the transition from steatosis to steatohepatitis is established by the role of reactive oxygen species (ROS) (Sanyal et al. 2001). This oxidative stress may result in a mitochondrial dysfunction (Pessayre, 2001), but increased oxidation of long chain fatty acids is reportedly a major source of ROS in NASH; free fatty acids are also toxic (Day, 2002; Sanyal et al. 2001). The distinction between steatosis and steatohepatitis, and the assessment of the severity of the disease rely on liver histology alone (Ratziu et al, 2005).

Key Players in Insulin Resistance: PPAR-Gamma

In the pathogenesis of insulin resistance, some key players, essential molecules in the process can be introduced like Peroxisome proliferator-activated receptor-gamma (PPAR-gamma). This is a molecule of recent investigation leading to a new generation of medicines the Thiazoladinediones of which the most important is Pioglitazone Hydrochloride. Peroxisome proliferator-activated receptor-gamma (PPAR-gamma) is a transcription factor involved in regulating genes involved in adipogenesis and, by implication, insulin action. (LeRoith and Zick, 2001)

In recent years great efforts have been directed toward understanding the molecular mechanisms that govern adipogenesis and, therefore, body fat mass. PPAR-gamma, the third member of a subset of nuclear receptors, which also includes PPAR-alpha and PPAR-delta, is now recognised to be central to this process. (ref. in Gurnell *et al.*, 2003) Human PPAR-gamma plays a role in the induction of adipose tissue differentiation. Loss of function mutation in PPAR-gamma results in lipodystrophy and administration of PPAR-gamma antagonists promotes adipogenesis. (Gurnell *et al.*, 2003)

PPARs posses a ligand-binding domain and a DNA-binding domain. PPARs bind to cognate DNA elements called PPAR response elements (PPREs) in the 5'-flanking region of target genes. PPARs, like other nuclear hormone receptors, form protein-protein interactions with a variety of nuclear proteins known as coactivators and corepressors, which mediate contact between the PPAR-RXR heterodimer, chromatin, and the basal transcriptional machinery and which promote activation and repression of gene expression, respectively. Although is has been possible to generate several high affinity synthetic PPAR-gamma ligands, e.g. thialzolidinedione (TZD), true endogenous ligands *in vivo* are still to be found (Rosen and Spiegelman, 2001).

Also, PPAR-gamma enhances insulin sensitivity in both muscle and liver. (LeRoith and Zick, 2001) In a review Gurnell *et al.* (2003) estimate likely that the dramatic diminution in peripheral limb and gluteal fat found in individuals with PPAR-gamma mutations contributes to their insulin-resistant phenotype. They suggest that even the residual adipose tissue depots in these individuals are dysfunctional, perhaps resulting in exposure of skeletal muscle and liver to unregulated fatty acid fluxes and thereby impairing insulin action in these sites. Thiazoladinediones (TZD), a synthetic PPAR-gamma ligand is used clinically as an insulin sentisizer in patients with type 2 diabetes (Rosen and Spiegelman, 2001).

At last, PPAR-gamma activation also reduces IL-6 and TNF-alpha levels, cytokines that are involved in the pathogenesis of insulin resitance. There is evidence that TZDs might repress a fat cell-produced factor called resistin, which induces insulin resistance (Rosen and Spiegelman, 2001).

Resistin

Resistin is a recently discovered hormone secreted by white and brown adipose tissue, monocytes, bone marrow and other tissues (Tejero et al. 2005).

Resistin received its name from the original observation (by Steppan in 2001) that it induced insulin resistance in mice, and could potentially link obesitas to diabetes (Steppan et al. 2001). Resistin belongs to the family of resistin-like molecules (RELMs), also known as "found in inflammatory zone" (FIZZ). The FIZZ/RELM family consists of 4 members, each of which has a conserved 11-cysteine pattern at the C terminus. (X11-C-X8-C-X-C-X3-C-X10-C-X-C-X-C-X9-CC-X3-6-END.) Each of these FIZZ proteins has unique tissue distribution, and both resistin (FIZZ-3) and FIZZ-1 are expressed in adipose tissue. FIZZ was initially discovered in mice, in which it is predominantly expressed by adipocytes. In contrast, macrophages, rather than adipocytes, appear to be the most important source of resistin in human subjects (Rea et al. 2004).

Resistin thought to be a possible link between adiposity and insulin resistance by having an possible metabolic effect on insulin sensitivity in murine models (Steppan et al. 2001). In mice, elevation of resistin levels leads to insulin resistance and glucose intolerance and symptoms of type II diabetes (Tejero et al. 2005). But levels of resistin have been reported to be either increased, unchanged, or decreased in murine and human obesity and type II diabetes (Fantuzzi 2005).

While some studies have reported that resistin expression is almost undetectable in human adipose tissue other studies have demonstrated that resistin expression is increased in abdominal compared to thigh fat depots in obese humans (Tejero et al. 2005).

However, recent data indicate that stimulation of macrophages *in vitro* with endotoxin or proinflammatory cytokines leads to a marked increase in resistin production (Lehrke et al. 2004).

And administration of endotoxin to human volunteers is associated with dramatically increased circulating resistin levels (Lehrke et al. 2004). Thus in human subjects resistin seems to act as a critical mediator of the insulin resistance associated with sepsis and possibly other inflammatory conditions. Only a few reports have investigated the effects of resistin in the modulation of inflammatory responses, showing that resistin upregulates expression of MCP-1, as well as vascular cell adhesion molecule 1 and ICAM-1, in endothelial cells.As mentioned above, the adhesion molecule-upregulating effects of resistin are antagonized by adiponectin (Fantuzzi 2005).

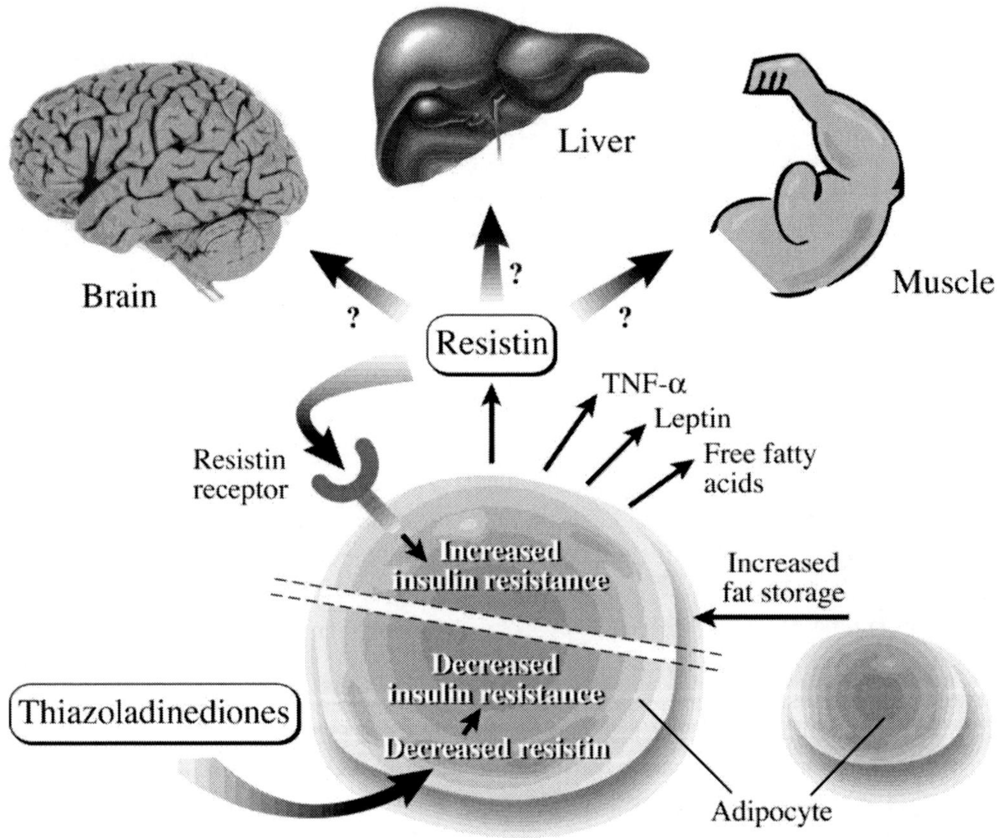

Figure 4. The target tissues of resistin.

The human resistin gene is located on chromosome 19p13.3. Several polymorphisms of this gene have been identified in humans, and associated with traits related to obesity and insulin sensitivity (Cao and Hegele 2001).

The complex nature of obesity and its related disorders indicates that the single candidate gene approach may not be the most appropriate for the study of these conditions. Thus, the genome scan approach has been applied to investigate the genetic component in obesity in a variety of populations (Clement et al. 2002).

Resistin gene expression is induced during adipocyte differentiation, and the resistin polypeptide is specifically expressed and secreted by adipocytes. Resistin circulates in mouse serum, and its level is increased markedly in both genetic and diet-induced obesity. Immuno neutralization improves blood glucose and insulin action in this model of type II diabetes. By contrast, administration of resistin impairs glucose tolerance and insulin action in normal mice. In mouse, a single mRNA of roughly 750 residues is robustly expressed in white adipose tissue but not in several other mouse tissues. Resistin expression is greater in white adipose tissue than in brown adipose tissue, where resistin mRNA is barely detectable. Resistin mRNA levels varied as a function of white adipose depot and gender, with the highest level of expression in female gonadal fat (Fantuzzi 2005).

4. Prevention and Treatment

The key factor for understanding the metabolic syndrome is insulin resistance and β-cell dysfunction (Campell 2004) and particularly the relative failure of insulin to exert its multiple biological effects on carbohydrate and lipid metabolism. Hulbert et al. (2005) propose two mechanisms to explain the initiation of the metabolic syndrome: a) decreased membrane polysaturation may act to decrease the activity of the major energy-consuming processes of the cell such as reducing the flux of ions and protons and subsequently the energy needed to maintain ionic homeostasis, b) membrane fatty acid composition influences insulin action *via* alteration of membrane proteins specifically associated with the action of insulin. Such mechanism would include modulation both of receptor affinities and translocation to the membrane of nutrient transporters (Hulbert et al 2005).

Taking away the underlying mechanism, can (partly) prevent the metabolic syndrome.

Diet, exercise and preventing stress are the first-line strategies in the management of type 2 diabetes; however, few patients maintain glycaemic control using this regime alone.

A good start could possibly be treatment with statins or ace inhibitors (to lower cholesterol and blood pressure (Faggiotto & Paoletti 1999). Other correlated features of the metabolic syndrome will then decrease, for example hepatic steatosis and insulin resistance. Of course there is a range of medicines tot treat metabolic syndrome, from ones with blood lipid decreasing effect until ones that improve glycemic control.

Currently available drug treatments, used alone or in combination, include oral glucose-lowering agents:

a) Sulphonylureas and glinides: which increase insulin secretion. They are the first-line oral antihyperglycaemic medications (OAMs) for treating hyperglycaemia in patients with Type 2 diabetes who have not responded to appropriate dietary and exercise modifications. Sulphonylureas improve glycaemic control with a high rate of success as primary agents (60-70%). As a class, sulphonylureas have no consistent effects on serum lipids in people with Type 2 diabetes (Tan et al. 2004). Examples of sulpho-nylureas are Glimepiride (Deroase et al. 2004)

b) α-glucosidase inhibitors: which retard digestion of complex carbohydrates (Tan et al. 2004).
c) metformin: which reduces endogenous hepatic glucose production (Einhorn et al. 2000).
d) Insulin has multiple effects. In addition to its primary effects on glucose homeostasis, insulin also promotes a number of other cellular events including regulation of ion and amino acid transport, lipid metabolism, glycogen synthesis, gene transcription and mRNA turnover, protein synthesis and degradation, and DNA synthesis. The major target tissues of insulin are skeletal muscle, adipose tissue, and liver. The anabolic hormone insulin plays a key role in the storage of ingested fuel and in normal cellular growth and differentiation so a tight control of this hormone is important for the homeostasis of the organism (Cheatham and Kahn, 1995) and a suppletion in case of a shortage is an important drug treatment.

A new group of pharmacological oral agents used for the treatment of type 2 diabetes mellitus is Pioglitazone (thiazolidinedione (TZD)). They are agonist for the peroxisome proliferator activated receptor-γ (PPARγ), a nuclear receptor involved in the transcription of genes that modulate carbohydrate and lipid metabolism and insulin regulation (Tan et al. 2004 Campell 2004). For the understanding of PPAR-gamma in insulin resistance the reader should see the paragraph genetic factors (see above). TZDs appear to address the dual defects of insulin resistance and β-cell dysfunction. The glucose-lowering effect of pioglitazone in patients with type 2 diabetes is related to its ability to reduce insulin resistance in tissues in which the majority of glucose uptake occurs (skeletal muscle, adipose tissue and liver) (Campell 2004). By reducing insulin resistance, pioglitazone lowers fasting and postprandial blood glucose concentrations, circulating free fatty acids (FFAs) and insulin levels and also hepatic glucose production may decline (Campell 2004). Additionally, data suggest that TZDs may preserve pancreatic β-cell function by a mechanism that is independent of the correction of glucose toxicity (Campell 2004). Furthermore Pioglitazone reduces the composite of all-cause mortality, non-fatal myocardial infarction, and stroke in patients with type 2 diabetes who have a high risk of macrovascular events (Dormandy et al. 2005). Pioglitazone can be given as monotherapy, double therapy (with metformin or a Sulphonylureum), or as triple therapy (Pioglitazone + metformin + Sulphonylureum).

Finally, Pioglitazone can be given in combination with insulin in patients. The combination of Pioglitazone in combination with metformin significantly improved glycated hemoglobin (HbA_{1c}) and fasting plasma glucose (FPG), with positive effects on serum lipid levels and no evidence of drug-induced hepatoxicity. These effects were maintained for > 1.5 years (Einhorn et al. 2000). The combination of Pioglitazone in combination with Glimepride (from the latest second generation sulfonylurea (SU)), resulted in a significant improvement in lipid and lipoproteins variables in patients who did not respond previously adequately on diet intervention or a sulfonylurea or metformin as medicine solely (Derosa et al. 2004). The combination of Pioglitazone with Sulfonylurea therapy improved glycated hemoglobin (HbA_{1c}) and fasting plasma glucose levels with beneficial effects on serum triglyceride and

HDL-cholesterol levels (Kipnes et al. 2001). Finally, in a group of patients with type 2 DM whose disease was inadequately controlled with insulin monotherapy, the combination of Pioglitazone with insulin further improved their glycemic control (Mattoo et al. 2005).

In conclusion: we demonstrated that the intra-uterine environment, life-style and genetic susceptibility all three can contribute to diabetes-2. Diet, exercise and preventing stress are the first-line strategies in the management of type 2 diabetes; however, few patients maintain glycaemic control using this regime alone. Therefore new medicines are developed for which the Thiazoladinediones (TZD), the most important "Pioglitazone hydrochloride", a synthetic PPAR-gamma ligand is used clinically as an insulin sentisizer in patients with type 2 diabetes and gives exciting results.

References

Arkan, M.C.; Hevener, A.L.; Greten, F.R.; Maeda, S.; Li, Z.W.; Long, J.M.; Wynshaw-Boris, A.; Poli, G.; Olefsky, J.; and Karin, M. (2005) IKK-beta links inflammation to obesity-induced insulin resistance. *Nat Med*, 11, No. 2, 191-198

Batterham, R.L.; Cohen, M.A.; Ellis, S.M.; Le Roux, C.W.; Withers, D.J.; Frost, G.S.; Ghatei, M.A.; Bloom, S.R. (2003). Inhibition of food intake in obese subjects by peptide YY3-36. N Engl J Med. 349:941-948

Bays, H.; Mandarino, L.; De Fronzo, R.A. (2004). Role of adipocyte, Free fatty Acids, and etopic fat in the pathogenesis of Type 2 Diabetes Mellitus: peroxisomal proliferator-activated receptor agonist provide a rational therapeutic approach., *J Clin Endocrinol Metab.,* 89: 463-478.

den Boer, M.; Voshol, P.J.; Kuipers, F.; Havekes, L.M.; Romijn, J.A. (2004). Hepatic steatosis: a mediator of the metabolic syndrome. Lessons from animal models. *Arterioscler.Thromb.Vasc.Biol.* 24: 644-649.

Bjorntorp, P. and Rosmond, R. (2000). The metabolic syndrome: a neuroendocrine disorder? *BJN*, 83(Suppl.1):S49-S57

Boulogne, A. and Vantyghem, M.C. (2004). The physiopathology of insulin resistance. *Presse Med.* 33:666-672

Boulton, A.J.; Vinik, A.L.; Arezzo, J.C.; Bril, V.; Feldman, E.L.; Freeman, R. et al. (2005). Diabetic neuropathies: a statement by the American Diabetes Association. *Diabetes Care* 28: 956-962.

Browning, J.D. and Horton, J.D. (2004). Molecular mediators of hepatic steatosis and liver injury, *The Journal of Clinical Investigation*, 114:147-152

Brunner, E.J.; Marmot, M.G.; Nanchahal, K.; Shipley, M.J.; Stansfeld, S.A.; Juneja, M.; Alberti, K.G.M.M. (1997). Social inequality in coronary risk: central obesity and the metabolic syndrome: evidence from the W II study. *Diabetologia*, 40:1341-1349

Brunner, E.J.; Hemigway, H.; Walker, B.R.; Page, M.; Clarke, P.; Juneja, M.; Shipley, M.J.; Kumari, M.; Andrew, R.; Seckl, J.R.; Papadopoulos, A.; Checkley, S.; Rumley, A.; Lowe, G.D.O.; Stansfeld, S.A.; Marmot, M.G. (2002). Adrenocortical, autonomic, and inflammatory causes of the metabolic syndrome: nested case-control study. *Circulation* 106: 2659-2665.

Campbell, I.W. (2004). Long-term glycaemic control with pioglitazone in patients with type 2 diabetes. *Int.J.Clin.Pract.* 58: 192-200.

Cao, H. and Hegele, R.A. (2001). Single nucleotide polymorphisms of the resistin (RSTN) gene. J Hum Genet. 46:553-555

Castelli, W.P.; Anderson, K.; Willson, P.W. et al. (1992). Lipid and risk of coronary heart disease: The Framingham Study. *Ann.Epidemiol.* 2: 23-28.

Cheatham, B. and Kahn, C.R. (1995). Insulin Action and the Insulin Signaling Network. *Endocr. Rev.* 16: 117-142.

Clement, K. Boutin, P. and Froguel, P. (2002). Genetics of obesity. *Am J Pharmacogenomics*, 2:177-187

Day, C.P. (2002). Pathogenesis of steatohepatitis. *Clin.Gastroeneterol.* 16: 663-678.

Derosa, G.; Cicero, A.F.; Gaddi, A.; Ragonesi, P.D.; Fogari, E.; Bertone, G.; Ciccarelli, L.; Piccinni, M.N. (2004). Metabolic effects of pioglitazone and rosiglitazone in patients with diabetes and metabolic syndrome treated with glimpepiride: a twelve-month, multicenter, double blind, randomized, controlled, parallel-group trial. *Clin.Therm.* 26: 744-754.

Dormandy, J.A.; Charbonnel, B.; Eckland, D.J.; Erdmann, E.; Massi-Benedetti, M.; Moules, I.K.; Skene, A.M.; Tan, M.H.; Lefèbvre, P.J.; Murray, G.D.; Standl, E.; Wilcox, R.G.; Wilhelmsen, L.; Betteridge, J.; Birkeland, K.; Golay, A.; Heine, R.J.; Korányi, L.; Laakso, M.; Mokán, M.; Norkus, A.; Pirags, V.; Podar, T.; Scheen, A.; Scherbaum, W.; Schernthaner, G.; Schmitz, O.; Skrha, J.; Smith, U.; Taton, J.; PROactive investigators (2005). Secondary prevention of macrovascular events in patients with type 2 diabetes in the PROactive Study (PROspective pioglitAzone Clinical Trial in macroVascular Events): a randomized controlled trial. *Lancet* 366: 1279-1289.

Egan, B.M., and Julius, S. (2004). Role of sympathic overactivity in the pathophysiology of the metabolic syndrome. *Dialogues in cardiovascular Medicine*, 9: 143-157

Einhor, D.; Rendell, M.; Rosenzweig, J.; Egan, J.W.; Mathisen, A.L.; Schneider, R.L.; for the Pioglitazone 027 Study group (2000). Pioglitazone Hydrochloride in combination with Metformin in the treatment of Type 2 Diabetes Mellitus: A randomized, placebo-controlled study. *Clinical Therapeutics* 22:1395-1409.

Faggiotto A. and Paoletti R. (1999). Statins and blockers of the angiotensin system: vascular protection beyond their primary mode of action. *Hypertension* 34 :987-996

Fantuzzi G. (2005). Adipose tissue, adipokines, and inflammation. *Clin Immunol,* 15:911-919

Feldman, E.L.; Stevens, M.J.; Russel, J.W. and Greene, D.A. (2001). Diabetic neuropathy. In: Principles and Practice of Endocrinology and Metabolism (Becker KL ed), Lippinicott Williams & Wilkins, Philadelphia, 1391-1399.

Fernandez-Checa, J.C. and Kaplowitz, N. (2005). Hepatic mitochondrial glutathione: transport and role in disease and toxicity. *Appl.Pharmacol.* 204: 263-273.

la Fleur, S.E.; Akana, S.F.; Manalo, S.L. and Dallman, M.F. (2004). Interaction between corticosterone and insulin in obesity: regulation of Lard intake and fat stores. *Endocrinology* 145:2174-2185.

Ford, E.S.; Giles, W.H. and Dietz, W.H. (2002). Prevalence of the metabolic syndrome among US adults. Findings from the third national health and nutrition examination survey. *JAMA* 287:356-359.

Forsblom, C.; Kanninen, T.; Lehtovirta, T.; Kaprio, J.C.S.; Groop, J. (1999). Albumin excretion is an inherited trait in families with NIDDM?. *Diabetologia* 42:1359-1366

Gloyn, A.L.; Mackay, D.J.G.; Weedon, M.N.; McCarthy, M.I.; Walker, M.; Hitam, G.; Knight, B.A.; Owen, K.R.; Hattersley, A.T. and Frayling, T.M. (2006). Assessment of the role of common variation in the Transient Neonatal Diabetes Mellitus (TNDM) region in Type 2 Diabetes. A Comparative Genomic and tagging single nucleotide polymorphism approach. *Diabetes* 55: 2272-2276.

Groop, L.; Forsblom, C.; Lehtovirta, M.; Tuomi, T.; Karanko, S.; Nissén, B.; Ehrnström, B.O.; Forsén, B.; Isomaa, B.; Snickars, B. and Taskinen, M.R. (1996). Metabolic consequences of a family history of NIDDM (The Botnia Study): evidence for sex-specific parental effects. *Diabetes* 45:1585-1593

Grove, K.L.; Grayson, B.E.; Glavas, M.M.; Xiao, X.Q. and Smith, M.S. (2005). Development of metabolic systems. *Physiol.Behav*. 86: 646-660.

Gurnell, M.; Savage, D.B.; Krishna, V.; Chatterjee, K.; and O'Rahilly, XX (2003). The metabolic syndrome: peroxisome proliferator-activated receptor-gamma and its therapeutic modulation. *J Clin Endocrinol Metab*. 88: 2412-

Hales, C.N.; Barker, D.J.; Clark, P.M.; Cox, L.J. and Osmond, C. and Winter, P.D. (1991). Fetal and infant growth and impaired glucose tolerance at age of 64. *BMJ*, 303:1019-1022

Hattersley, A.T. and Tooke, J.E. (1999). The fetal insulin hypothesis: an alternative explanation of the association of low birth weight with diabetes and vascular disease. *Lancet,* 353:1789-1792

Hedley, A. A.; Ogden, C.L.; Johnson, C.L.; Carrol, M.D.; Curtin, L.R. and Flegal, K.M. (2004). Prevalance of overweight and obesity among US children, adolescents, and adults, 1999-2002. *JAMA* 291: 2847-2850.

Heilbronn, L.; Smith, S.R. and Ravussin, E. (2004) Failure of fat cell proliferation, mitochondrial function and fat oxidation results in ectopic fat storage, insulin resistance and type II diabetes mellitus. *Int. J Obes. Relat. Metab. Disord*. 28 : S12-S21

Hjemdahl, P. (2002). Stress and the metabolic syndrome, an interesting but enigmatic association. *Circulation,* 106:2634-2636

Hulbert, A.J.; Turner, N.; Storlien, L.H. and Else, P.L. (2005). Dietary fats and membrane function: implications for metabolism and disease. *Biol. Rev*. 80: 155-169.

Jacobs, A.K. (2003). Coronary revascularization in women. Sex revisited. *Circulation* 107: 375-377.

Jaquet, D.; Gaboriau, A.; Czernichow, P.; Levy-Marchal, C. (2000). Insulin resistance early in adulthood in subjects with intrauterine growth retardation. *J. Clin. Endocrinol. Metab.* 85:1401-1406

Kaati, G.; Bygren, L.O. and Edvinsson, S. (2002). Cardiovascular and diabetes mortality determined by nutrition during parents' and grandparents' slow growth period. *European Journal of Human Genetics* 10:682-688.

Kipness, M.S.; Krosnick, A.; Rendell, M.S.; Egan, J.W.; Mathisen, A.L. and Schneider, R.L. (2001). Pioglitazone Hydrochloride in combination with sulfonylurea therapy improves glycemic control in patients with Type 2 Diabetes Mellitus: a Randomized, Placebo-controlled study. *Am.J.Med.* 111: 10-17.

Knowler, W.C.; Barrett-Connor, E.; Fowler, S.E.; Hamman, R.F.; Lachin, J.M.; Walker, E.A. and Nathan, D.M. (2002). Diabetes Prevention Program Research Group: Reduction in the incidence of type 2 diabetes with lifestyle intervention or metformin. *N Engl J Med.* 346:393-403

Kreier, F.; Kalsbeek, A.; Sauerwein, H.P.; Fliers, E.; Romijn, J.A. and Buijs, R.M. (2007). "Diabetes of the elderly" and type 2 diabetes in younger patients: Possible role of the biological clock. *Experimental Gerontology* 42: 22-27.

Kyle, U.G. and Pichard, C. (2006). The Dutch Famine of 1944-1945: a patho-physiological model of long-term consequences of wasting disease.*Curr.Opin. Clin.Nutr.Metab.Care* 9: 388-394.

Laaksonen, D.E.; Lakka, H.M.; Salonen, J.T.; Niskanen, J.K.; Raunamaa, R. and Lakka, T.A. (2002). Low levels of leisure-time physical activity and cardiorespiratory fitness predict development of the metabolic syndrome. *Diabetes Care*, 25:1612-1618

Langhans, W. (2003). Role of the liver in the control of glucose-lipid utilization and body weight. *Current Opinion in Clinical Nutrition & Metabolic Care* 6: 449-455.

Lehrke, M.; Reilly, M.P.; Millington, S.C.; Iqbal, N.; Rader, D.J. and Lazar, M.A. (2004). An Inflammatory Cascade Leading to Hyperresistinemia in Humans. *PLoS Med.*, 2:e45

Levy, D.; DeStefano, A.L.; Larson, M.G.; O'Donnell, C.J.; Lifton, R.P.; Gavras, H. Cupples, L.A. and Myers, R.H. (2000). Evidence for a gene influencing blood pressure on chromosome 17. Genome scan linkage results for longitudinal blood pressure phenotypes in subjects from the Framingham Heart Study. *Hypertension*, 36:477-483

Lumey, L.H. and Stein, A.D. (1997). Offspring birth weights after Maternal Intrauterine Undernutrition: a comparison within sibships. *American Journal of Epidemiology* 146: 810-819.

Maes, H.H.; Neale, M.C.; Eaves, L.J. (1996). Genetic and environmental factors in relative body weight and human adiposity. *Behaviour Genetics,* 27:325-351

Mantha, L.; Palacios, E. and Deshaies, Y. (1999). Modulation of triglyceride metabolism by glucocorticoids in diet-induced obesity. *Am J Physiol.* 277(2Pt2):R455-R464

Masuzaki, H.; Paterson, J.; Shinyama, H.; Morton, N.M.; Mullins, J.J.; Seckl, J.R.; and Flier, J.S. (2001). A transgenic model of visceral obesity and the Metabolic Syndrome.*Science* 294:2166-2170.

Mattoo, V.; Eckland, D.; Widel, M.; Duran, S.; Fajardo, C.; Strand, J.; Knight, D.; Grossman, L. Oakley, D.; Tan, M. and H6EMC-GLAT study group (2005). Metabolic efects of pioglitazone in combination with insulin in patients with type 2 diabetes mellitus whose disease is not adequqtely controlled with insulin therapy: results of a six-month, randomized, double-blind, prospective, multicenter, parallel-group study. *Clin. Ther.* 27: 554-567.

McMillen, C.; Robinson, J.S. (2005). Developmental origins of the Metabolic Syndrome: Prediction, Plasticity, and Programming. *Physiol. Rev.* 85: 571-633.

Möller-Leimkühler, A.M. (2007). Gender differences in cardiovascular disease and comorbid depression. *Dialogues Clin. Neurosci*. 9: 71-83.

Nakai, J.; Biggs, W.H.; Kitamura, T. ; Cavenee, W.K. ; Wright, C.V.E. ; Arden, K.C. ; Acilli, D.(2002). Regulation of insulin action and pancreatic beat-cell function by mutated alleles of the gene encoding forkhead transcription factor Foxol. *Nat.Genet*. 32:245-253.

Naunyn, B. (1906). *Der Diabetes Mellitus*. Nothnangels Handbuch. A. Holder: Wien.

Neel, J, (1962). Diabetes mellitus: a "thrifty" genotype rendered detrimental by "progress". *Am. J. Hum. Genet*. 14: 353-362.

Ogden, C.L.; Carroll, M.D.; Curtin, L.R.; McDowell, M.A.; Tabak, C.J. and Flegal, K.M. (2006). Prevalance of overweight and obesity in the United States, 1999-2004, *JAMA* 295: 1549-1555.

Painter, R.C.; Roseboom, T.J. and Bleker, O.P. (2005). Review. Prenatal exposure to the Dutch famine and disease in later life: An overview. *Reproductive Toxicology* 20: 345-352.

Painter, R.C.; De Rooij, S.R.; Bossuyt, P.M.M.; de Groot, E.; Stok, W.J.; Osmond, C.; Barker, D.J.; Bleker, O.P. and Roseboom, T.J. (2006). Maternal nutrition during gestation and carotid arterial compliance in the adult offspring: the Dutch famine birth cohort. *Journal of Hypertension* 25: 533-540.

Pembrey, M.E.; Bygren, L.O.; Kaati, G.; Edvinsson, S.; Northstone, K.; Sjöström, M. Golding, J. and the ALSPAC Study Team (2006). Sex-specific, male-line transgenerational responses in humans. *European Journal of Human Genetics* 14:159-66.

Pessayre, D.; Berson, A.; Fromenty, B.; Mansouri, A. (2001) Mitochondria in steatohepatitis. *Semin.Liver.Dis*. 21:57-69.

Petersen, S.; Petro, V.; Rayner, M.; Leal, J.; Luengo-Fernandez, R. and Gray, A. (2005). European cardiovascular disease statistics: 2005 edition. *London British Heart Foundation.*

Picardi, A.; D'Avola, D.; Gentilucci, U.V.; Galati, G.; Fiori, E.; Spataro, S.; Afeltra, A. (2006). Diabetes in chronic liver disease: from old concepts to new evidence. *Diabetes Metab.Res.Rev.* 22: 274-283.

Ratziu, V.; Charlotte, F.; Heurtier, A. ; Gombert, S. ; Giral, P.;Bruckert, E.; Grimaldi, A.; Capron, F.; Poynard, T. (2005). Sampling variability of liver biopsy in nonalcoholic fatty liver disease. *Gastroenterology* 128: 1898-1906.

Rea ,R.; and Donnelly, R. (2004). Resistin: an adipocyte-derived hormone. Has it a role in diabetes and obesity? *Diabetes Obes Metab,* 6:163-170

Reaven,G.M. (1993). Role of insulin resistance in human disease (syndrome X): an expanded definition. *Annu. Rev. Med.* 44 : 121-131.

de Rooij, S.R.; Painter, R.C.; Philips, D.I.W.; Osmond, C.; Michels, R.P.J.; Bossuyt, P.M.M.; Bleker, O.P. and Roseboom, T.J. (2006). Hypothalamic-pituitary-adrenal axis activity in adults who were prenatally exposed to the Dutch famine. *European Journal of Endocrinology* 155: 153-160.

de Rooij, S.R.; Painter, R.C.; Philips, D.I.W.; Osmond, C.; Tanck, M.W.T..; Bossuyt, O.P. and Roseboom, T.J. (2006). Cortisol responses to psychological stress in adults after prenatal exposed to the Dutch famine. *Psychoneuroendocrinology* 31: 1257-1265.

de Rooij, S.R.; Painter, R.C.; Philips, D.I.W.; Osmond, C.; Michels, Godsland, I.F.; R.P.J.; Bossuyt, P.M.M.; Bleker, O.P. and Roseboom, T.J. (2006). Impaired insulin secretion after prenatal exposure to the Dutch famine. *Diabetes Care* 29: 1897-1901.

de Rooij, S.R.; Painter, R.C.; Holleman, F.; D.I.W.; Bossuyt, P.M.M. and Roseboom, T.J. (2006). The metabolic syndrome in adults prenatally exposed to the Dutch famine. *Am.J.Clin.Nutr*. 86: 1219-1224.

Roseboom, T.J.; van der Meulen, J.H.P.; Osmond, C.; Barker, D.J.P.; Ravelli, A.C.J. and Bleker, O.P. (2000). Plasma lipid profiles in adults after prenatal exposure to the Dutch famine. *Am.J.Clin.Nutr*. 72: 1101-1106.

Rosen ED, and Spiegelman BM. (2001). PPAR-gamma: a nuclear regulator of metabolism, differentiation, and cell growth. *J Biol Chem*. 276: 37731-.37734.

Rudin, E. and Barzilai, N. (2005). Inflammatory peptides derived from adipose tissue. *Immun. Ageing* 2:1 doi:10.1186/1742-4933-2-1.

Saad, M.F.; Bernaba, B.; Hwu, C.M.; Jinagouda, S.; Fahmi, S.; Kogosov, E. and Boyadjian, R. (2002). Insulin regulates plasma ghrelin concentration. *J. Clin. Endocrinol. Metab.* 87:3997-4000

Sanyal, A.J.; Campbell-Sargent, C.; Mirshahi, F.; Rizzo, W.B.; Contos, M.J.; Sterling, R.K.; Luketic, V.A.; Shiffman, M.L.; Clore, J.N. (2001). Nonalcoholic steatohepatitis: association of insulin resistance and mitochondrial abnormalities. *Gastroenterol.* 120: 1183-1192.

Steppan, C.M.; Bailey, S.T.; Bhat, S.; Brown, E.J.; Banerjee, R.R.; Wright, C.M.; Patel, H.R.; Ahima, R.S. and Lazar, M.A. (2001). The hormone resistin links obesity to diabetes. *Nature*, 409:307-312

Stressman, J.; Cohen, A.; Hammerman-Rozenberg, R.; Bursztyn, M.; Azoulay, D.; Maaravi, Y.; Jacobs, J.M. (2008). Holocaust survivors in old age: The Jeruzalem Longitudinal Study. *J.Am.Geriatr.Soc*. 56:470-477.

Snieder, H.; Doornen van den, J.L.; and Boomsma, D.I. (1999). Dissecting the genetic architecture of lipids, lipoproteins, and apolipoproteins: lessons from twin studies. *Arterioscler Thromb Vasc Biol,* 19:2826-2834

Tan, M.H.; Johns, D.; Strand, J.; Halse, J.; Madsbad, S.; Eriksson, J.W.; Clausen, J.; Konkoy, C.S. and Herz, M. for the GLAC Study Group (2004). Sustained effects of pioglitazone vs. glibenclamide on insulin sensitivity, glycaemic control, and lipid profiles in patients with Type 2 diabetes. *Diabet.Med.* 21: 859-866.

Tejero, M.E.; Cole, S.A.; Cai, G.; Peebles, K.W.; Freeland-Graves, J.H.; Cox, L.A.; Mahaney, M.C.; Rogers, J.; VandeBerg, J.L.; Blangero, J.; Comuzzie, A.G. (2005). Genome-wide scan of resistin mRNA expression in omental adipose tissue of baboons. *Int. J. Obes. Relat. Metab. Disord.,* 29:406-412

Tilg, H. and Hotamisligil, G.S. (2006). Nonalcoholic Fatty Liver Disease: Cytokine-Adipokine interplay and regulation of Insulin Resistance. *Gastroenterology* 131:934-945.

Valdez, R.; Athens, M.A.; Thompson, G.H.; Bradshaw, B.S.; Stern, M.P. (1994). Birth weight and adult health outcomes in a biethnic population on the USA. *Diabetologia,* 37:624-631

Wamala, S.P.; Lynch, J.; Horsten, M.; Mittleman, M.A.; Schenk-Gustafsson, K. and Orth-Gomér, K. (1999). Education and the metabolic syndrome. *Diabetes Care* 22:1999-2003

Wenger, N.K. (2002). Coronary heart disease: The female heart is vulnerable. *Prog.Cardiovasc.Dis*. 46: 199-229.

Winkleby, M.A. and Cubbin, C. (2003). Influence of individual and neighbourhood socioeconomic status on mortality among black, Mexican-American and white women and men in the United States. *J.Epidemiol.Community Health* 57: 444-452.

World Health Organisation (1999). Report of a WHO consultation: definition of metabolic syndrome in definition, diagnosis, and classification of diabetes mellitus and its complications. *Chapter I: Diagnosis and classification of diabetes mellitus, World Health Organization, Department of Non-communicable Disease Surveillance, Geneva. Through http://www.who.int.*

Yanjnik, C.S. (2004). Symposium on "Adipose tissue development and the programming of adult obesity"; Obesity epidemic in India: intrauterine origins ? *Proceedings of the Nutrition Society* 63: 387-396.

In: Advances in Medicine and Biology Research, Vol. 14 ISBN: 978-1-61761-930-4
Editor: Leon V. Berhardt

Chapter IV

Microbial Diversity of Biofilm Technologies used for Water Treatment

***J. Purswani*[1], *C. Pozo*[1], *R. Vílchez*[2], *B. Juárez*[1], *M.V. Martínez-Toledo*[1], *J. González-López*[1] *and B. Rodelas*[1*]**

Grupo de Microbiología Ambiental, Departamento de Microbiología e Instituto del Agua, Universidad de Granada, Spain[1]

Microbial Interactions and Processes Research Group, Department of Medical Microbiology, Helmholtz Center for Infection Research (HZI), Braunschweig, Germany[2]

Abstract

Wastewater comprises of many different sources and contaminants, and the combinations of these require specific types of remediation strategies. Biological approaches based on the use of microbial biomass against physic-chemical processes are preferred as the low-cost alternative. Contaminated groundwater, urban and industrial wastewater may contain contaminants such as nitrates, heavy metals, crude oil derivatives or even high levels of nutrients. Microbial populations continually described for degradation of recalcitrant xenobiotics, accumulation or adhesion of heavy metals and for their contribution in geochemical cycles such as nitrogen, exceed expectations when presented with new water contaminants. Thus, their highly rate of adaptation make them ideal for wastewater treatments from different sources. Biofilm technology has been proved to be one of the most advanced methods to remove in a selective way, contaminants from water resources. Biofilms are very complex habitats where the microbial cells responsible for the treating process are embedded in a polymer matrix. Water treatment by submerged filter technology requires a biofilm formation around an inert substrate. The water to be treated passes through a biofilter consisting of the inert substrate and the biofilm established. The main components of the biofilm are different physiological and morphological bacterial cells and their composition depends on the environmental conditions. The relationships among the microbial populations established, whether mutualistic or competitive, along with the combined enzymatic activity,

[*]Corresponding author (mrodelas@ugr.es)

determine the efficiency of the bioremediation process, thus the importance in the exploration of the biodiversity. Different culture dependent and culture independent techniques have been used to describe the constant and shifting biodiversity present in wastewater biotechnologies. Trends in community profiling using the 16S ribosomal RNA as a taxonomic marker have allowed the establishment of a large taxonomic database available for microbiologists, but most importantly, it has created a general consensus to describe microbial populations to date. A review of the methods used for biofilm biodiversity determination and the corresponding microbial structures in submerged biofilm technologies is presented, along with the future trends in microbial population classification.

Introduction

The bacterial life style most commonly occurring in natural and engineered environments, is that in which the bacteria live attached to a surface forming a structure known as a biofilm (Costerton et al., 1995; Danese et al., 2000; Decho, 2000; Watnick & Kolter, 2000), where they find the necessary requirements for their development. We must consider that bacteria that are not part of a biofilm, will be at the initial stage of the formation of another, in a new surface (Watnick & Kolter, 2000). Biofilms may form on solid substrates in contact with moisture, on soft tissue surfaces in living organisms and at air-liquid interfaces, and their structure is morphological and physiologically different than that exhibited by planktonic bacterial cells.

A biofilm can be defined as "a complex structure composed by communities of microorganisms irreversibly attached to a surface (material or interface, biotic or abiotic) which produce extracellular polymeric substances (EPS) and exhibit an altered metabolic state when compared to planktonic cells", especially with regard to the transcription signals and the interactions between cells (Davies et al., 1998; Geesey & Jang, 1989; Costerton, 1999; Decho, 2000; Watnick & Kolter, 2000). Biofilm EPS, which is also referred to as *slime*, is a polymeric conglomeration generally composed of extracellular DNA, proteins, and polysaccharides under various configurations.

The study of biofilms has enhanced in recent years due to increased knowledge of their impact on natural and industrial systems, as well as human health. However, biofilms also offer enormous helpful potential for bioremediation; in this sense, they structure the basis of diverse aerobic and anaerobic reactors (i.e. trickling filters, submerged filters, rotating biological contactors) for treating municipal and industrial wastewaters, and form *in situ* biobarriers to protect soil and groundwater from contamination (Cao et al., 1992; Hunter & Shaner, 2009; Ross et al., 2001). The complexity of biofilm activity and performance requires continuous research contributions from many disciplines such as microbiology, biochemistry, engineering or physics. Numerous research papers are being published frequently on biofilms and the results are resolving aspects such as structure, composition and characteristics of these special ecosystems. Consequently, the implementation of independent culture techniques for the study of microbiology of water treatment systems has supposed an important advance in the understanding of microbial ecology.

Microbial Diversity Methodology

Microbial Community Profiling Methods

Classic microbiological taxonomy based on the cultivation of isolated microorganisms and their morphological and biochemical tests, have been substituted or complemented with the analysis of macromolecules (DNA, RNA and proteins) of these microorganisms through the use of molecular biological tools. Microbial species are currently defined by polyphasic approaches based on clear rules for both genotypic and phenotypic properties seeing that the use of classical or molecular techniques alone may confer bias (Martin-Platero et al., 2009).

Genomic DNA-DNA hybridisation is the primary criterion for distinguishing species, however whole-genome comparisons for species definition may seem premature, considering that only approximately 1500 bacterial and archaeal completed genomes are available (NCBI database, 2010), and that the diversity of uncultured microbes in the environment is high. Trends in community profiling using the 16S ribosomal RNA as a taxonomic marker have allowed the establishment of a large taxonomic database creating general consensus to describe microbial populations at large. Previously, only morphological and biochemical tests were performed on isolated cultivable microorganisms, however, the sequencing of the 16S rRNA genes has been proven useful for the identification of microorganisms by both culture-dependent and culture independent approaches. Thus the identification of microbes within a biofilm structure can be assessed by culture dependent and culture independent techniques.

Culture Dependent Techniques

The techniques used to identify cultivable and isolated microorganisms include the basic and classical microbiological techniques. Among others, it is noteworthy to mention API, BIO-LOG$_{TM}$ or Phene Plate$_{TM}$ systems based on phenotypic characteristics (Dahllof, 2002; Giraffa, 2004). Tests based on the phenotypic protein (SDS-page gel analysis) and lipidic profiles (FAME analysis - Slabbinck et al., 2010) are also used to classify microbial strains. The methods mentioned are tedious, costly, time consuming and not always reproducible.

Identification methods based on DNA or RNA are quick and easily reproducible. Genomic DNA-DNA hybridisation as mentioned above is primordial for strain identification accuracy, however this method is tedious and costly, as is the genome sequencing of each strain. RISA-PCR (rRNA gene Internal Spacer Analysis), 16S rRNA gene sequencing, ARDRA-PCR (Amplification Ribosomal DNA Restriction Analysis), ITS (Internal Transcribed Spacer) and species-specific genes are all PCR based methods, and the growing online database of bacterial ribosomal genes makes these techniques the most commonly used for identification of large number of isolates. Microbial biofilms can be highly diverse or limited to only few numbers of strains, thus the method selected to determine microbial identification of strains should be chosen to the level of confidence the research requires.

Fingerprinting type methods are also used due to their simplicity and feasibility in a short time. The aim of these types of techniques is to create "genetic fingerprints" of species or even interpreted as a *barcode*, thus allowing identification through microbial typification (Vandamme et al., 1996). An example of this is the random amplification of polymorphic DNA (RAPD-PCR "Randomly Amplified Polymorphic DNA"; AP-PCR "Arbitrarily Primed

PCR"). These techniques do not require previous knowledge of the targeted DNA sequences. However, if the targeted sequence is known, the primers used are defined and the annealing conditions are more restrictive e.g. REP-PCR (Repetitive Extragenic Palindromic) which is based on the RAPD-PCR technique, but only requires a single oligonucleotide.

Other *barcode* techniques, such as ARDRA, require the amplification of DNA (mainly 16S rRNA gene or the intergenic region between 16S and 23S rRNA genes) jointly with digestion with restriction enzymes (Vaneechoutte et al., 1992). AFLP (Amplified Fragment Length Polymorphism) could also be used, however all typifying methods can only be compared against reference strains contained within each laboratory, since the magnitude of combinations (primers and strains) are indefinite, and there are no known online databases to facilitate barcode identification.

Ultimately, the exclusive use of culture dependent techniques on microbial diversity in any habitat will be highly biased, given that the estimation of cultivable bacteria is within 1% of total bacteria and only 1–5 % of the bacterial population of ground and drinking water is culturable on nutrient media as described in different studies (Byrd et al., 1991; Schaule et al., 1993; Kalmbach et al., 1997; Saby et al., 1997; Szewzyk et al., 2000; Leclerc & Moreau, 2002). Consequently, culture independent techniques are highly sought.

Culture Independent Techniques

A great advantage of the use of culture independent techniques is that biodiversity of different ecosystems can be determined spatially and throughout time, without the need to isolate microbes. The limitations are usually associated with technical aspects such as DNA/RNA extraction, DNA amplification errors (Giraffa & Neviani, 2001; Stach et al., 2001; Ogier et al., 2002; Ercolini, 2004; Forney et al., 2004;), among others. As with culture dependent techniques for biofilm diversity determination, the 16S rRNA gene is widely sequenced in culture independent techniques. Besides the numerous advantages over the use of this gene as a taxonomic marker, there are also several inconveniences, for instance the scarce discrimination observed among closely related species. Other essential functional genes such as RNA polimerase B (*rpoB*), *recA* gene family, gyrase B (*gyrB*), elongation factor Tu (*tuf*), and the *dnaK* gene, may help solve the problems, however the lack of available sequences and in some cases the lack of conserved regions, limits their use as taxonomic markers (Dahllof, 2002).

The strategies proposed for biofilm community profiling are shown in Table 1. The most important methods are fingerprinting techniques, *in situ* techniques and others such as microarrays and flow cytometry. As described above, fingerprinting techniques generate barcode-like results, however, the profile observed in DNA/RNA environmental samples are as a consequence of multiple microorganisms. The TGGE (temperature-gradient gel electrophoresis) technique is based on the amplification and partial sequencing of the 16S rRNA coding gene and generates bands representing the dominant populations within a complex bacterial community profile. This technique, together with its similar DGGE (denaturing gradient gel electrophoresis) has been widely used in recent years to examine the bacterial biodiversity in the biofilms developed in artificial ecosystems, including sludge treatment plants or wastewater treatment systems (Wagner et al., 2002; Gómez-Villalba et al., 2006; Pozo et al., 2007; Vílchez et al., 2007). For the study of microbial community using TGGE, one generally assumes that each band corresponds to a single population (species) and its density corresponds to the relative abundance. However we have to take into consideration that this statement may not be quite right, because the TGGE technique is based on the reaction of PCR amplification, and the reaction can create certain artefacts (Wang and Wang, 1996). This technique is not quantitative in ultimate terms for it will detect relative percentage populations which are at least 1% of the total bacterial community (Muyzer & Smalla, 1998). The analysis of SSCP (Single-Stranded Conformational Polymorphism) is also based on the amplification of the 16S rRNA gene, and its electrophoretic separation based on the difference in the mobility of single stranded DNA (Schwieger & Tebbe, 1998).

Table 1. Summary of culture independent methods for microbial ecological studies.

	Taxonomical level	Applications
PCR based methods:		
PCR-DGGE/TTGE/TGGE	Members of a community (genus/species)	Community fingerprinting; population dynamics; analysis of mutations.
SSCP	Members of a community (genus/species)	Community fingerprinting; population dynamics; analysis of mutations.
T-RFLP	Members of a community (genus/species)	Community fingerprinting; population dynamics.
LH-PCR	Members of a community (genus/species)	Community fingerprinting; population dynamics.
PCR-ARDRA	Members of a community (species)	Community fingerprinting; population dynamics.
RISA/ITS-PCR	Members of a community (species)	Community fingerprinting; population dynamics.
in situ methods:		
FISH	Members of a community (species)	Detection of viable cells (cultivable and non-cultivable) of microbial communities. Spatial distribution of microorganisms in different habitats.
FISH Multiple	Members of a community (species)	Same as FISH
PCR in situ	Members of a community (species)	Detection of viable cells and cells growing slowly. Detection of target sequences usually present in low copy number.
Other methods:		
Flow cytometry	Members of a community (species)	Selective count of microbial population and sub-populations in a community. Physiological analysis of cellular state.
Competitive PCR	Members of a community (species)	Detection of viable but not culturable cells (VNC)
Quantitative hybridisation	Members of a community (species)	Semi-quantitative dynamics of microbial groups within a microbial community.

Acronyms: DGGE, Denaturing Gradient Gel Electrophoresis; TGGE, Temperature Gradient Gel Electrophoresis; TTGE, Temporal Temperature Gradient Gel Electrophoresis; SSCP, Single-Stranded Conformational Polymorphism; LH-PCR, Length Heterogeneity-PCR; ARDRA, Amplification Ribosomal DNA Restriction Analysis; ITS, Internal Transcribed Spacer; RISA, rRNA gene Internal Spacer Analysis; RFLP, Restriction Fragment Length Polymorphism; FISH (Fluorescent *in situ* Hybridisation.

Other techniques used fluorescently labelled DNA through a capillary electrophoresis such as T-RFLP (Terminal Restriction Fragment Length Polymorphism), and LH-PCR (Length hetorogeneity-PCR). ARDRA-PCR (Amplification Ribosomal DNA Restriction Analysis) and RISA-PCR (rRNA gene Internal Spacer Analysis) can also be used to determine uncultured populations (Vaneechoutte et al., 1992; Liu et al., 1997; Suzuki et al., 1998).

Techniques such as microarrays (based on dot blot methodology) have been applied to determine presence and relative number of bacterial species, such as the commercialized PhyloChip. Miniaturization and automation make this technique attractive; however unknown 16S rDNA sequences will be overlooked.

Functional Diversity within Microbial Communities

The functions carried out within a biofilm will be determined by the physicochemical conditions and the genetic make-up of the microbes. Specific gene searches are also performed to describe microbial activity. Enzymatic activities in wastewater treatments such as phosphatases, glucosidases, proteases (Goel et al., 1998), or esterases (Boczar et al., 2001) and dehydrogenases (Awong et al., 1985) are useful in evaluating biological oxidation of organic material (Nybroe et al., 1992). Specific functions such as ammonia oxidizing and denitrifying activities are determined genetically though DNA amplification of *amoA* and *nosZ* genes, respectively (Nakano et al., 2008; Gomez-Villalba et al., 2006). One of the main disadvantages of the use of degenerated primers for the amplification of metabolic genes is the overestimation of gene diversity through band profiling as a cause of the frequent finding of redundant sequences.

A proteomic approach can also be used to observe functional gene profile changes within time, for it offers a unique dynamic view into cellular function. Most investigations have focused on microbial communities in surface waters. Initially, SDS-PAGE (1D electrophoresis) was used to generate protein fingerprints of communities, with no attempt made to identify proteins (Macrae & Smit 1991; Francisco et al., 2002). Later, Kan et al. (2005) used a 2-Dimensional Electrophoresis Mass Spectrometry workflow to study Chesapeake Bay microbial communities and identified eight proteins.

Microbial Structure Visualization

The increasing resolution supplied by electron microscopy has allowed for the visualization of detailed interaction between microbes, for instance EPS formation, microtubules between cells which may confer conjugation and surface appendages to attach to solid matrices. Conformational and structural changes can also be observed though scanning electron microscopy (SEM), allowing for EDX point analysis which give us some insight to e.g. metal location in varying studies (Gonzalez-Muñoz et al., 1997; Vilchez et al., 2007).

One of the inconveniences of using DNA/RNA PCR fingerprinting based methods is that they hardly give any information as to microbial number and spatial distribution. The latter has been solved through the use of techniques such as fluorescent *in situ* hybridisation

(FISH). The spatial distribution of individual bacteria can be observed in biofilm formations using fluorescent microscopes. Confocal fluorescent microscopy allow for 3D images compared to the 2D images obtained from conventional fluorescent microscopes, giving us exact positioning of bacterial groups against others, thus being able to determine mutualism, avoidance or random positioning of the biofilm studied. Different taxonomic groups can be differentiated through the use of fluorescent probes corresponding to conserved regions of order/genus/species specific rRNA. Online platforms such as proBase (www.microbial-ecology.net/probase) contain a large number of different taxonomical group rRNA probes/primers (Loy et al., 2003) and the necessary *in situ* hybridisation conditions, making it an essential tool for microbial ecologists.

Biofilm Diversity in Submerged Biofilter Technologies for Water Treatment

Nitrogen Based Compounds and Wastewaters

The implementation of continuous and increasing inorganic nitrogenous fertilizer doses and pesticides has created problems among agriculture and water consuming sectors. High concentrations of ammonium or nitrate compounds end up through lixiviation in groundwater: globally the primordial source of drinking water. The World Health Organization (W.H.O., 1985) acknowledges nitrosamines as the cause of metahaemoglobinemia and stomach cancer, through high nitrate consumption. The main ecological impact as a consequence of high nitrate concentrations in lakes and rivers is the high possibility of eutrophication systems, causing cascades of environmental imbalances, including migration and death of many living organisms. Thus the need for bioremediation technologies to counteract and reduce nitrate concentrations in both groundwater and wastewaters is essential.

Fixed film technology is broadly used for the removal of organic matter and nitrogen from urban wastewaters through the biological process of nitrification-denitrification (Galvez et al., 2003). In this type of systems, the wastewater generally first enters an anoxic column, where a major fraction of the organic matter is degraded by heterotrophic populations. The water then enters an aerated zone where nitrification occurs, and nitrate is finally supplied into the anoxic zone by partial recirculation of the nitrified effluent. A pilot scale biofilter plant containing clayey schists as support material within two columns (aerobic and anaerobic) was used successfully for the treatment of urban wastewater (Gómez-Villalba et al., 2006). Different genes were amplified, separated (TGGE) and sequenced to define diversity in phylogenic terms (16S rRNA gene), and functional terms using ammonia-oxidizing bacterial activity (*amoA* gene) and denitrifying activity (*nosZ* gene). Samples taken during three consecutive weeks and a year later presented significant variations in band patterns among samples from the aerated column (Fig.1). Diversification was more stable (less changes were observed) within the biofilm in the anoxic column. The predominant band in the aerated column was found to be evolutively close to the nitrite oxidizer genus *Nitrospira*, suggesting an important role of populations related to this genus in the conversion of nitrite to nitrate in the system. TGGE profiles based on the partial amplification of the *amoA* and *nosZ* functional genes demonstrated the presence of nitrifying and denitrifying bacteria in both the anoxic and aerated biofilters. The community profiles of denitrifying bacteria remained stable over space and time, while profiles of ammonia oxidizers greatly varied at different sampling times and locations in the reactor. Phylogeny inferred from *amoA*-TGGE analysis revealed the prevalence of *Nitrosomonas* species, with 5 sequences affiliated to *N. oligotropha*, 6 sequences affiliated to *N. europaea*, and 3 sequences that showed only a 75.7-76.1% similarity of the DNA sequence with the closest described species (*N. nitrosa*). This value is indicative of previously undiscovered species (Purkhold et al., 2000). A total of 18 *nosZ* sequences were obtained from TGGE bands, which were mostly related to *nosZ* of Gamma-proteobacteria (*Pseudomonas*), while the rest clustered in the periphery of previously known denitrifying Alpha-proteobacteria (*Bradyrhizobium* and *Azospirillum*).

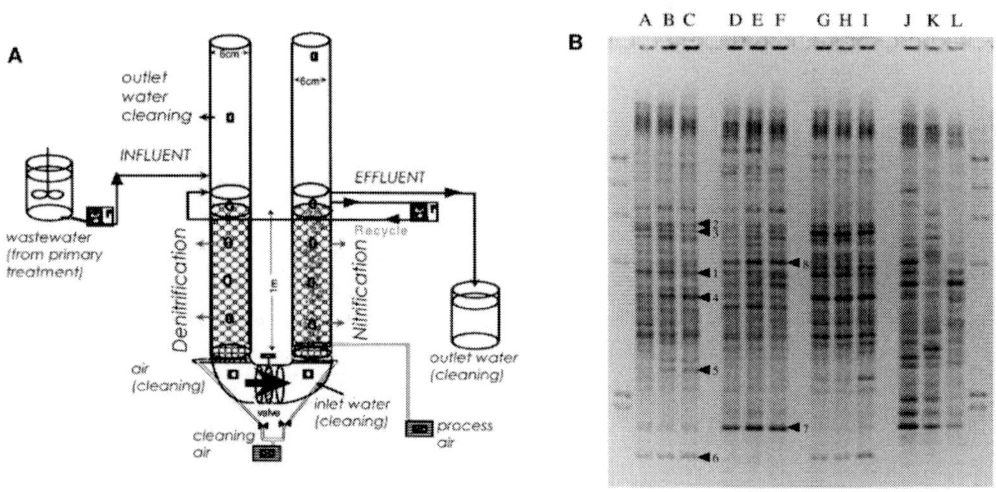

C

Band #	Closest relative	% Identity
1	Beta-proteobacteria: *Ralstonia* sp. strain MSB2080	91%
2	Beta-proteobacteria: *Aquaspirillum* sp. strain TG27	95%
3	Beta-proteobacteria: *Aquaspirillum* sp. strain TG27	100%
4	Alpha-proteobacteria: *Albidovulum inexpectatum*	96%
5	Beta-proteobacteria: *Aquaspirillum delicatum*	96%
6	Acidobacteria: *Desulfovibrio fairfieldensis*	83%
7	Nitrospira: *Nitrospira moscoviensis*	93%
8	Beta-proteobacteria: *Acidovorax delafieldii*	93%

Figure 1. Diversity of bacteria in biofilms of a submerged biofilter system (anoxic-aerated) treating urban wastewater (Gómez-Villaba et al., 2006). A) Diagram of the biofilter system in the study showing the disposition of anoxic and aerated columns. B) TGGE profiles based on the V3 region of the 16S rRNA gene. A,B,C: samples taken during three consecutive weeks from the anoxic column of the biofilter; D,E,F: samples taken during three consecutive weeks from the aerated column of the biofilter; G,H,I: samples taken from the anoxic column of the biofilter after one year of continuous operation; J,K,L: samples taken from the aerated column of the biofilter after one year of continuous operation. C) Identification of populations corresponding to bands marked (1-8) in the TGGE profiles.

Technologies based on biofilm development and the denitrification-nitrification processes are also suitable for the treatment of industrial wastewaters containing recalcitrant organic compounds. A pilot scale biofilter was successfully applied for the removal of ammonia and phenol from an industrial wastewater also carrying high salt concentrations (Cortes-Lorenzo et al., 2006; Ramos et al., 2007). The observed biofilm diversity using the 16S rRNA gene-based TGGE method resulted in mainly marine Alphaproteobacteria, grouping around the genus *Roseobacter*. Closely related species of *Roseobacter* are ubiquitous in nature, and are primary colonizers in coastal environmental surfaces (Gonzalez & Moran 1997; Dang & Lovell, 2000). They have also been described as predominant strains in initial hydrocarbon degradation in marine habitats (Harayama et al., 2004). A *Desulfobacteriaceae*-like population was detected only in the anaerobic column, related to a lineage of bacteria which have also been described as phenol and toluene degraders under anaerobic conditions (Harayama et al., 2004). Nitrite oxidizers related to a marine strain of *Nitrospira* were only present in the aerated column.

Archaea in Industrial and Domestic Wastewaters

The structure of the archaeal communities in aerobic wastewater treatment processes was investigated in conventional activated sludge by Gray et al. (2002), but in the aerobically operated WWTPs based on membrane biofilter or submerged biofilter technologies to treat industrial or urban wastewaters, Archaea occurrence has been scarcely explored.

In this sense, Gomez-Silvan et al. (2010) studied the Archaea community developed in a two-column submerged biofilter (SB) treating urban wastewater (USB treatment); a two-column SB treating industrial wastewater (ISB treatment); and a membrane bioreactor under two different conditions (MBR1 and MBR2). 16S rRNA gene amplification by PCR was used in conjunction with TGGE methods to determine community profiles. Using the Pearson coefficient-based analysis on 16S rRNA fingerprinting, clusters were separated according to wastewater origin (urban or industrial) and type of technology used (SB or MBR). All successfully sequenced archaeal bands corresponded to methanogenic Archaea from the classes *Methanobacteria* and *Methanomicrobia*. A 60% of the sequences showed a ≥98% identity to *Methanosaeta concilii*. Similar results were observed by Gray et al. (2002) from flocks in conventional activated sludge from wastewater treating urban and industrial influents, during non-aerated periods. Communities in the ISB treatment displayed the greatest divergence, due to the influence of water composition. Similar observations were previously reported (Akarsubasi et al., 2005; Gray et al., 2002; Stephanie et al., 2008). Attempts to amplify archaeal *amoA* genes in this study in search of ammonia oxidizing Archaea (AOA) were unsuccessful. Low representation of *Crenarchaeota* in WWTPs was previously detected (Gray et al 2002) and absence of AOA has also been reported in a survey of 52 nitrifying WWTPs across Europe (Daims et al., 2009). AOA were only found in five WWTPs in the USA, operated under low oxygen concentrations and high hydraulic retention times (HRT >24h) (Park et al., 2006).

Nitrate Removal from Groundwater

The biofilter technology is also suitable for the removal of toxic nitrogen compounds from groundwater (Moreno et al., 2005). A common practice consists on using activate sludge from wastewater treatment plants as inoculant for the formation of biofilms. This type of inoculants allow for the development of high-channeled biofilms, whose biological composition is heterogeneous and varies during operation of the system (Gómez et al., 2003). The development of the biofilm and its performance for nitrogen removal are influenced by the availability and concentration of C sources, as well as the temperature (Welander & Mattiason, 2003; Gómez et al., 2009). In this sense, the presence of psychrotolerant organisms in the biofilm contributes to its better performance within a wide range of temperatures (Gómez et al., 2009). In a study carried out by de la Rúa et al. (2007), the microbial diversity obtained from a groundwater nitrate remediating biofilter was analysed by using a TGGE approach, based on the analysis of the hypervariable V3 region of the 16S rRNA gene. Similarity matrix of the TGGE fingerprints showed that community profiles of biofilms formed from the autochthonous groundwater microbiota at 5 and 10°C clustered away (58% similarity) from those of biofilms formed at 20 and 30°C. Biofilms formed at all temperatures were mainly composed of species typically inhabiting freshwaters, and band sequencing allowed for the identification of representatives of denitrifying genera (*Aquaspirillum*, *Hydrogenophaga*). Psychrotolerant abilities have been described for certain species in the *Hydrogenophaga* genus (Nedwell & Rutter, 1994), and their intense presence has been reported in freshwater biofilms at winter periods (Brummer et al., 2003). Although nitrifying species have been characterised by TGGE in biofilms formed by microorganisms naturally inhabiting groundwater, the biofilms may not develop the required denitrifying activity for complete nitrate removal (de la Rúa et al., 2007). The efficiency of the process can be further improved by inoculation of the biofilm with previously selected strains displaying a high denitrifying activity (de la Rúa et al., 2007; Gómez et al., 2009). Cultured denitrifying strains isolated from submerged fixed film bioreactors have been identified through sequencing of 16S rRNA genes (Gómez et al., 2009) or classical morphological and biochemical tests (Gómez et al., 2003). Different strains belonging to *Pseudomonas alcaligenes*, *P. picketti*, *P. fluorescens, P. mandelii, Alcaligenes denitrificans*, *Hydrogenophaga* spp., *Moraxella* spp. and *Xanthobacter* spp. have been isolated. The highest denitrifying activity (2,100 nmol h. 10^9 CFU) was achieved by *Xanthobacter* strains, suggesting the importance of the denitrifying function of these populations in the biofilms (Gomez et al., 2005).

Heavy Metal Containing Waters

Heavy metals exert inhibitory effects on the growth of microorganisms by blocking essential functional groups, displacing essential metal ions, or modifying the active conformations of biological molecules (Wood & Wong, 1983: Doleman et al., 1994; Gadd & Griffiths, 1978; Li & Tan, 1994a, 1994b). However, some metals such as Co, Cu, Zn and Ni, are essential at low concentrations, since they are vital co-factors for metallo-proteins and enzymes (Eiland, 1981). Nevertheless, bacteria are frequently isolated from heavy metal polluted terrestrial systems and some have been found to resist very high concentrations of these, regardless of whether the site was previously contaminated or not. Unlike organic xenobiotic compounds, heavy metals cannot be "destroyed", thus the only mechanisms for removal of heavy metals by microorganisms are accumulation in the form of particular protein-metal associations, blockage of the metal at the cell wall and the systems of membrane transportation (Ow, 1993; Tomioka et al., 1994; Wehrhein & Wettern, 1994), precipitation in crystalline forms such as calcite (Stipp et al., 1992; Xu et al., 1996; Sturchio et al., 1997; Cheng et al., 1998) or through the interaction of extracellular precipitation of metabolic compounds (Beveridge & Doyle, 1989), biosorption or bioadsorption by live and dead cells (Ben Omar et al., 1997; Gonzalez-Muñoz et al. 1997; Zouboulis et al., 1997), and the reduction of the metal oxidation state (Mullen et al., 1992; Wang & Shen, 1995).

The release of extracellular polysaccharides (EPS) by bacteria in biofilms, favours the microorganisms in environments where metal contamination occurs. The EPS captures large quantities of metal ions such as Cd^{2+}, Cu^{2+}, Cr^{3+} and Pb^{2+} (Geesey & Jang, 1989). This occurs primarily by the formation of complex compounds between the cations and the anionic groups of the EPS surface (Geesey & Jang, 1989). The affinity for fixation depends much on the size, charge of the ion, and other factors like the composition of the EPS and pH. The biofilm reduces the concentration of the heavy metal ions that reach the bacteria, therefore enabling the microorganism to survive through higher concentrations of these contaminants, compared to the tolerance displayed by planktonic cells.

Remediation technologies applied to the removal of metal ions from contaminated water are mostly based on physicochemical processes of separation or treatment, such as activated carbon adsorption, membrane filtration, electrolysis, cation exchange, precipitation, chemical reduction, or solvent extraction (Eccles, 1999; Mulligan et al., 2001). To reduce costs on wastewater treatment, removal of heavy metals by sorption on mineral surfaces was also suggested (Garcia-Sanchez et al., 1999; Vengris et al., 2001). Heavy metals like Cd, Co, Pb, Zn can be adsorbed on the surface of calcite (Lorens, 1981; Stipp et al., 1992; Xu et al., 1996; Sturchio et al., 1997; Cheng et al., 1998). In recent years, the removal of toxic metal ions by biological methods based on the use of microbial biomass has been suggested as an attractive, low cost alternative to these conventional physicochemical processes, with important applications in remediation strategies (von Canstein et al., 1999; Costley & Wallis, 2001; Eccles, 1999; Malik, 2004; Mulligan et al., 2001). These approaches also include biofilm-based technologies. Vilchez et al. (2007) tested the suitability of a lab-scale biofilter previously inoculated with sludge from an urban waster treatment plant for the removal of copper ions from groundwater spiked with 15 mg/ml Cu^{2+}. Prokaryotic biodiversity suffered great changes in the presence of the copper cations present in the influent, compared to biofilms formed in the presence of unspiked groundwater. By using the TGGE approach, *Sphingomonas* and surrounding taxa were predominantly detected in the Cu^{2+} spiked groundwater biofilter. *Sphingomonas* spp. and closely related genera are frequently isolated from fresh or marine, pristine or contaminated aquatic environments (White et al., 1995, Balkwill et al., 2003). *Sphingomonads* spp. have been previously described for their Cu^{2+} accumulation abilities on Cu^{2+} amended plates, and have been frequently isolated in conjunction with *Acidovorax* and *Pseudomonas* strains from corroding tube pipes (Balkwill et al., 2003, Critchley et al., 2003, 2004). The same biofilter system used for Cu^{2+} removal was further proven efficient for the removal of Pb^{2+} and Cr^{3+} ions (Vílchez, 2005). The biological activities and features of the biofilms were characteristic for each of the ions tested (Fig. 2). As (III) (arsenite) could not be removed from water, although a biofilm of arsenic-resistant microorganisms was developed in the system (Fig. 3).

Figure 2. SEM analysis of the structure of microbial biofilms developed in a submerged biofilter treating heavy metal-spiked groundwater (Vílchez, 2005). A) Biofilm developed in the absence of heavy metals. B) Biofilm developed in the presence of $Cu2^{+}$ C) Biofilm developed in the absence of heavy metals. B) Biofilm developed in the presence of Pb^{2+} D) Biofilm developed in the absence of heavy metals. B) Biofilm developed in the presence of $Cr^{3+.}$

A

B

Band #	Closest relative	% Identity
1	Alphaproteobacteria: *Brevundimonas* sp.	100
2	Betaproteobacteria: *Rhodocyclus* sp.	98
3	Alphaproteobacteria: *Methylopila capsulata*	97
4	Alphaproteobacteria: *Agrobacterium tumefaciens/Rhizobium tropici*	100
5	Alphaproteobacteria: *Agrobacterium tumefaciens*	98
6	Alphaproteobacteria: *Agrobacterium tumefaciens*	98
7	Alphaproteobacteria: *Agrobacterium* sp.	97

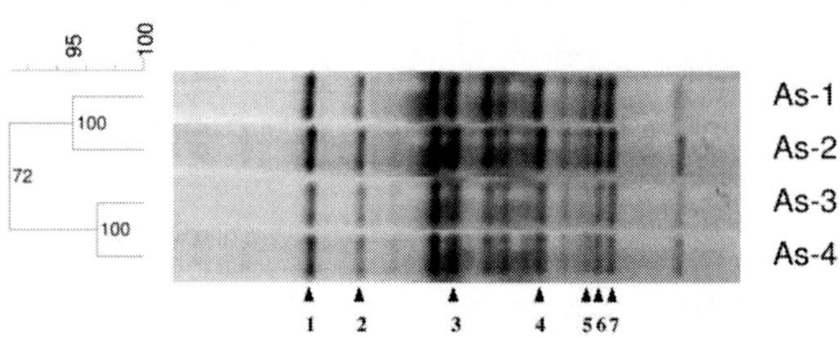

Figure 3. SEM analysis of the structure of microbial biofilms developed in a submerged in a submerged biofilter treating As (III)-spiked groundwater (Vílchez, 2005). A) Biofilm developed in the presence of As (III) (sodium arsenite) at a concentration of 70 mg/ml groundwater. B) TGGE profiling of As (III)-resistant bacteria in the biofilm sampled at four depths in the biofilm column, and identification of mayor TGGE bands by DNA sequencing and database comparison.

Current Problems and Future Trends in Microbial Population Assessment

Microbial species are currently defined by polyphasic approaches based on clear rules for both genotypic and phenotypic properties seeing that the use of classical or molecular techniques alone may confer bias, but this work is laborious and time consuming when multiple parameters need to be studied in biofilm populations.

The exclusive use of culture dependent techniques only confers a small percentage of the microbial species within complex biofilm structures, and even then, those isolated may not even be the predominant species within that community. Culture independent techniques used for microbial community identification in the biofilms have varying degrees of errors and different stages: a) Errors produced at DNA/RNA extraction; b) PCR amplification and c) resolution of separation. There are multiple DNA/RNA extraction methods, yet the continuing high number of extracted Gram negative bacteria in wastewater samples may be due to either natural selection in wastewater biofilms, or a Gram negative-biased DNA/RNA extraction. One of the main solutions to this is the mechanical rupture of samples. During PCR amplification, many errors can occur. For example, the design of universal primers is based only on currently known sequences and will miss yet unknown genera or orders. The annealing conditions are not always optimum for all microorganisms present, thus a touch-down PCR protocol may best be used. TGGE quantification will be biased due to the nature of primer annealing: if annealing is good, higher amplification will be observed, and will not imply higher representation in the biofilm. Amplification of nonsense regions and chimerical artefacts are also frequently produced in 16S rRNA gene amplification. Profiling and clustering through electrophoretic fingerprints (e.g. TGGE, DGGE) may give erroneous statistical information due to the number of artefact or nonsense bands produced (Cowell et al., 2010).

Recommendation of whole-genome comparisons for species definition may seem premature, considering that ~1500 completed bacterial and archaeal genomes are currently available and that the diversity of uncultured microbes in the environment is high. In spite of the latter, more and more bacterial genomes are being sequenced, and the rapid development of re-sequencing technologies will soon increase the number of available genome sequences by several orders of magnitude (Bentley, 2006).

Multilocus sequence analyses (MLSA) which depend on the phylogentic analyses of multiple protein-coding core genes, have been suggested to provide a criterion for microbial species boundaries (Hanage et al., 2006).

Future trends to define microbial functional groups in biofilms could be defined by the use of a shotgun proteomics approach, such as the study of a microbial biofilm community from a natural acid mine drainage (Ram et al., 2005). Approximately 2000 proteins were identified through the use of a database created from the sequencing of a microbial community sampled from the same mine but at a different location and time.

Environmental wastewater samples contain high number of transposable elements, and these may incline strain classification one way or the other, nonetheless the biofilm functionality is unique to the conditions surrounding it, and adaptation to the resources available is the key to microbial success in biotechnologies. The manipulation of microbial strains through the use of genetic engineering for human purposes, such as wastewater treatments, can only be efficient when most parameters affecting the microbial population in the biofilms are known, including the accumulation of the global biofilm gene expression, for an accomplished functioning wastewater biofilter/bioreactor.

References

Akarsubasi, A. T., Ince, O., Kirdar, B., Oz, N. A., Orhon, D., Curtis, T. P., Head, I. M. & Ince, B. K. (2005). Effect of wastewater composition on archaeal population diversity. *Water Research*, 39, 1576-1584.

Awong, J., Bitton, G. & Koopman, B. (1985). ATP, oxygen uptake rate and INT dehydrogenase activity of actinomycete foams. *Water Research*, 7, 917-921.

Balkwill, D. L., Fredrickson, J. K., & Romine, M. F. (2003). *Sphingomonas* and related genera. *In M. Dworkin (Editor) (Eds.), The Prokaryotes: an Evolving Electronic Resource for the Microbiological Community* (3rd edition, pp. 14-25). New York: Springer.

Bentley, D. R. (2006). Whole-genome re-sequencing. *Current Opinion in Genetics & Development*, 16, 545-552.

Beveridge, T. J., & Doyle, R. J. (1989). Metals ions and bacteria. In Beveridge, T. J. (Eds), *Metal ions and bacteria,* (1st edition, pp. 1-29). New York: John Wiley.

Ben Omar, N., Gonzalez Muñoz, M. T. & Arias, J. M. (1998). Struvite crystallisation on *Myxococcus xanthus*. *Chemosphere*, 36, 475-481.

Boczar, B. A., Forney, L. J., Begley, W. M., Larson, R. J. & Federle, T. W. (2001). Characterization and distribution of esterase activity in activated sludge. *Water Research*, 35, 4208-4216.

Brümmer, I. H. M., Felske, A. & Wagner-Döbler, I. (2003). Diversity and seasonal variability of beta-Proteobacteria in biofilms of polluted rivers: analysis by temperature gradient gel electrophoresis and cloning. *Applied and Environmental Microbiology*, 69, 4463-4473.

Byrd, J. J., Xu, H. & Colwell, R. (1991). Viable but nonculturable bacteria in drinking water. *Applied and Environmental Microbiology*, 57, 875–878.

Cheng, L., Sturchio, N. C., Woicik, J. C. Kemmer, K. M., Lyman, P. F. & Bedzyk, M. J., (1998). High resolution structural study of zinc ion incorporation at the calcite cleavage surface. *Surface Science*, 415, 976-982.

Cao, Y. S., Alaerts, G. J., Hooijmans, C. M. & Lubberding, H. J. (1992). Oxygen consumption and organics decomposition in drainage systems with attached biofilm. *Water Science and Technology*, 26, 683-692.

Costerton, J. W., Lewandoski, D. E., Cadwell, D. R., Corber, H. M. & Lappin-Scott, H. (1995). Microbial biofilms *Annual Review of Microbiology*, 49, 711-745.

Costerton J. W. (1999). Introduction to biofilm. *International Journal of Antimicrobial Agents*, 11, 217- 221.

Cortés-Lorenzo, C. Molina-Muñoz, M. L., Gómez-Villalba, B., Vilchez, R., Ramos, A., Rodelas, B., Hontoria, E. & González-López, J. (2006). Analysis of community composition of biofilms in a submerged filter system for the removal of ammonia and phenol from industrial wastewater. *Biochemical Society Transactions*, 34, 165-168.

Costley, S. C. & Wallis, F. M. (2001). Bioremediation of heavy metals in a synthetic wastewater using a rotating biological contactor. *Water Research*, 35, 3715-3723.

Cowell, R., Lauritzen, S. & Mortera, J. (2010). Probabilistic expert systems for handling artifacts in complex DNA mixtures. *Forensic Science International* In Press (doi:10.1016/j.fsigen.2010.03.008).

Critchley, M. M., Cromar, N. J., McClure, N. C. & Fallowfield, H. J. (2003). The influence of the chemical composition of drinking water on cuprosolvency by biofilm bacteria. *Journal of Applied Microbiology*, 94, 501-507.

Critchley, M. M., Pasetto, R. & O'Halloran, R. J. (2004). Microbiological influences in 'blue water' copper corrosion. *Journal of Applied Microbiology,* 97, 590-597.

Daims, H., Lücker, S., Mussmann, M., Brito, I., Spieck, E., Head, I., Paslier, D.L., & Wagner, M., (2009). Ammonia-oxidizing Archaea and nitrite-oxidizing *Nitrospira* in wastewater treatment plants: new insights based on molecular tools and environmental genomic. In S. J. Hall (Editor), L. L. Blackall (Editor) (Eds.), Microbial population dynamics in biological wastewater treatment, Proceeding of the ASPD5 Specialised Conference (IWA). Aalborg, Denmark: International Water Association, IWA; 2009; 80–84.

Danese, P., Pratt, L., & Kolter, R. (2000). Exopolysaccharide production is required for development of *Escherichia coli* K-12 biofilm Architecture. *Journal of Baceriology*, 182, 3593-3596.

de la Rúa, A., Rodelas, B., González-López J. & Gómez, M.A. (2007). Influence of temperature on microbial colonisation of clayey schist as a support media of a submerged filter for groundwater denitrification. *Water Science and Technology*, 55, 165-172.

Dang, H. & Lovell, C. H. (2000). Bacterial primary colonization and early succession on surfaces in marine waters as determined by amplified rRNA gene restriction analysis and sequence analysis of 16S rRNA genes. *Applied and Environmental Microbiology*, 66, 467–475.

Dahllof, I. (2002). Molecular community analysis of microbial diversity. *Current Opinion in Biotechnology,* 13, 213-217.

Decho, A. (2000). Microbial biofilms in intertidal systems: an overview. *Continental Shelf Research,* 20, 1257-1273.

Davies, D., Parsek, M., Pearson, J., Iglewski, B., Costerton, J. & Greenberg, E. (1998). The involvement of cell-to-cell signals in the development of bacterial biofilm. *Science*, 280, 295-298.

Doleman, P., Jansen, E., Michels, M., & Van Til, M. (1994). Effects of heavy metals in soil on microbial diversity and activity as shown by the sensitivity-resistance index, an ecologically relevant parameter. *Biology and Fertiity of Soils*, 17, 177-184.

Eiland, F. (1981). The effects of application of sewage sludge on microorganisms in soil. *Tidsskrift för Planteavl*, 85, 39-46.

Eccles, H. (1999). Treatment of metal-contaminated wastes: why select a biological process? *Trends in Biotechnology*, 17, 462-465.

Ercolini, D. (2004). PCR-DGGE fingerprinting: novel strategies for detection of microbes in food. *Journal of Microbiological Methods*, 56, 297-314.

Forney, L. J., Zhou, X. & Brown, C. J. (2004).Molecular microbial ecology: land of the one-eyed king. *Current Opinion in Microbiology*, 7, 210-220.

Francisco, R., Alpoim, M. C. & Morais, P. V. (2002). Diversity of chromium-resistant and -reducing bacteria in a chromium-contaminated activated sludge. *Journal of Applied Microbiology*, 92, 837-843.

Gálvez, J. M., Gómez, M. A., Hontoria, E. & González-López, J. (2003). Influence of hydraulic loading and air flowrate on urban wastewater nitrogen removal with a submerged fixed-film reactor. *Journal of Hazardous Materials*, 101, 219-229.

Giraffa, G. (2004). Studying the dynamics of microbial populations during food fermentation. *FEMS Microbiology Reviews*, 28, 251-260.

Giraffa, G. & Neviani, E. (2001). DNA-based, culture-independent strategies for evaluating microbial communities in food-associated ecosystems. *International Journal of Food Microbiology,* 67, 19-34.

Gomez, M. A., Galvez, J. M., Hontoria, E. & González-López, J. (2003). Influence of ethanol concentration on biofilm bacterial composition from a denitrifying submerged filter used for contaminated groundwater. *Journal of Bioscience and Bioengineering*, 95, 245-251.

Gómez, M. A. Rodelas, B., Sáez, F., Pozo, C., Martínez-Toledo, M. V., Hontoria, E. & González-López, J. (2005). Denitrifying activity of *Xanthobacter autotrophicus* strains isolated from a submerged fixed-film reactor. *Applied Microbiology and Biotechnology,* 68, 680-685.

Gómez, M. A., González-López, J., de la Rúa, A. (2009). Effect of temperature over start-up of a groundwater-denitrifying submerged filter inoculated with psychrotolerant bacteria. *Journal of Environmental Science and Health part A,* 44, 1298-1305.

Gómez–Villalba, B., Calvo, C., Vílchez, R., González-López, J., Rodelas, B., (2006). TGGE analysis of the diversity of ammonia-oxidizing and denitrifying bacteria in submerged filter biofilms for the treatment of urban wastewater. *Applied Microbiology and Biotechnology,* 72, 393-400.

Gómez-Silván, C., Molina-Muñoz, M., Poyatos, J. M., Ramos, A., Hontoria, E., Rodelas, B. & González-López, J. (2010). Structure of archaeal communities in membrane-bioreactor and submerged-biofilter wastewater treatment plants. *Bioresource Technology*, 101, 2096-2105.

Gray, N. D., Miskin, I. P., Kornilova, O., Curtis, T. P. & Head, I. M. (2002). Occurrence and activity of Archaea in aerated activated sludge wastewater treatment plants. *Environmental Microbiology*, 4, 158-168.

Gadd G. M., & Griffiths A. J, (1978). Microorganisms and heavy metals. *Microbial Ecology*, 4, 303-317.

Garcia-Sanchez, A., Ayuso, E. & Blas O. (1999). Sorption of heavy metals from industrial waste water by low-cost mineral silicates. *Clay Minerals*, 34, 469-477.

Geesey, G.G., & Jang L. (1989). Interactions between metal ions and capsular polymers. In Beveridge T.J., Doyle R.J., (Eds.) *Metal ions and bacteria*. (pp.325-357). New York: John Wiley.

Goel, R., Takashi, M., Hiroyasu, S. & Tomonori, M. (1998). Enzyme activities under anaerobic conditions in activated sludge sequencing batch reactor. *Water Research.,* 32, 2081-2088.

Gonzalez-Muñoz, M. T., Merroun, M. L., Ben Omar, N., Arias, J. M., (1997). Biosorption of uranium by *Myxococcus xanthus*. International *Journal of Biodeterioration and Biodegradation,* 40, 104-114.

González, J. M. & Moran, M. A. (1997). Numerical dominance of a group of marine bacteria in the alpha-subclass of the class Proteobacteria in coastal seawater *Applied and Environmental Microbiology,* 63, 4237–4242.

Harayama, S., Kasai, Y. & Hara, A. (2004). Microbial communities in oil-contaminated seawater. *Current Opinion in Biotechnology*, 15, 205–214.

Hanage, W. P., Fraser, C. & Spratt, B. G. (2006). Sequences, sequence clusters and bacterial species. *Philosophical Transactions of the Royal Society of London, part B, Biological Sciences*, 361, 1917-1927.

Hunter, W. J. & Shaner, D. L. (*2009*) Nitrogen limited *biobarriers* remove atrazine from contaminated water: laboratory studies. *Journal of Contaminant Hydrology*, 103,29–37.

Kalmbach, S., Manz, W. and Szewzyk, U. (1997) Isolation of new bacterial species from drinking water biofilms and proof of their in situ dominance with highly specific 16S rDNA probes. *Applied and Environmental Microbiology*, 63, 4164–4170.

Kan, J., Hanson, T.E., Ginter, J.M., Wang, K. & Chen, F. (2005). Metaproteomic analysis of Chesapeake Bay microbial communities. *Saline Systems*, 1, 7.

Leclerc, H. & Moreau, A. (2002) Microbiological safety of natural mineral water. *FEMS Microbiology Reviews,* 26, 207–222.

Li, F. & Tan T.C. (1994a). Effect heavy metal ions on the efficacy of a mixed bacilli BOD sensor. *Biosensors and Bioelectronics*, 9, 315-324.

Li, F. & Tan, T.C. (1994b). Monitoring BOD in the presence of heavy metal ions using poly (4-vinylpyridine) coated microbial sensor. *Biosensors and Bioelectronics,* 9, 445-455.

Liu, W.T., T.L. Marsh, H. Cheng y L.J. Forney. 1997. Characterization of Microbial Diversity by Determining Terminal Restriction Fragment Length Polymorphisms of Genes Encoding 16S rRNA. *Applied and Environmental Microbiology*, 63: 4516-4522.

Lorens, R. B. (1981). Sr, Cd, Mn and Co distribution coefficients in calcite as a function of calcite precipitation rate. *Geochimica et Cosmochimica Acta*, 45, 553-561.

Loy, A., Horn, M. & Wagner, M. (2003). probeBase: an online resource for rRNA-targeted oligonucleotide probes. *Nucleic Acids Research*, 31, 514-516.

Macrae, J. D. & Smit, J. (1991). Characterization of caulobacters isolated from waste-water treatment systems. *Applied and Environmental Microbiology*, 57, 751–758.

Mullen, M. D., Wolf, D. C., Beveridge, T. J. & Bailey, G. W. (1992). Sorption of heavy metals by the soil fungi *Aspergillus niger* and *Mucor* rouxii. *Soil Biology and Biochemistry,* 24, 129-135.

Malik, A. (2004). Metal bioremediation through growing cells. *Environ. Int.*, 30, 261-278.

Martín-Platero, A. M., Maqueda, M., Valdivia, E., Purswani, J. & Martínez-Bueno, M. (2009). Polyphasic study of microbial communities of two Spanish farmhouse goats' milk cheeses from Sierra de Aracena. *Food Microbiology*, 26, 294-304.

Moreno, B., Gomez, M. A., González-López, J. & Hontoria, E. (2005). Inoculation of a submerged filter for biological denitrification of nitrate polluted groundwater: a comparative study. *Journal of Hazardous Materials*, 117, 141–147.

Mulligan, C. N., Yong, R. N. & Gibbs, B.F. (2001). Remediation technologies for metal-contaminated soils and groundwater: an evaluation. *Engineering Geology,* 60, 193-207.

Muyzer, G. & Smalla, K. (1998). Application of denaturing gradient gel electrophoresis (DGGE) and temperature gradient gel electrophoresis (TGGE) in microbial ecology, *Antonie van Leeuwenhoek*, 73, 127–141.

Nakano, M., Shimizu, Y., Okumura, H., Sugahara, I. & Maeda, H. (2008). Construction of a consortium comprising ammonia-oxidizing bacteria and denitrifying bacteria isolated from marine sediment. *Biocontrol Science,* 13, 73-89.

Nedwell, D.B. & Rutter, M. (1994). Influence of temperature on growth rate and competition between two psychrotolerant Antarctic bacteria: low temperature diminishes affinity for substrate uptake. *Applied and Environmental Microbiology*, 60, 1984-1992.

NCBI (Nacional center for biotechnology information). Genome database [cited 2010 May 20]. *Available from: http://www.ncbi.nlm.nih.gov/sites/genome*

Nybroe, O., Jørgensen, P.E. & Henze, M. (1992). Enzyme activities in waste water and activated sludge. *Water Research*, 26, 579-584.

Ogier, J., Son, O., Gruss, A., Tailliez, P. & Delacroix-Buchet, A. (2002). Identification of the bacterial microflora in dairy products by temporal temperature gradient gel electrophoresis. *Applied and Environmental Microbiology*, 68, 3691-3701.

Ow, D. (1993). Phytochelatin-mediated cadmium tolerance in *Schizosaccharomyces pombe*. *In Vitro Cellular & Developmental Biology*, 29, 213-219.

Park, H. D., Wells, G. F., Bae, H., Criddle, C.S., Francis, C. A., 2006. Occurrence of ammonia-oxidizing Archaea in wastewater treatment plants. *Applied and Environmental Microbiology*, 72, 5643–5647.

Pozo, C., Rodelas, B., Martínez-Toledo, M.V., Vílchez, R. & González-López, J. (2007). Removal of organic load from olive washing water by an aerated submerged biofilter and profiling of the bacterial community involved in the process. *Journal of Microbiology and Biotechnology,* 17, 784-791.

Purkhold, U. Pommerening-Röser, A., Juretschko, S., Schmid, M.C., Koops, H.P. & Wagner, M. (2000). Phylogeny of all recognized species of ammonia oxidizers based on comparative 16S rRNA and *amoA* sequence analysis: implications for molecular diversity surveys. *Applied and Environmental Microbiology*, 66, 5368-5382.

Ram, R.J. Verberkmoes, N. C., Thelen, M. P., Tyson, G. W., Baker, B. J., Blake, R.C. 2nd, Shah, M., Hettich, R. L. & Banfield, J. F. (2005). Community proteomics of a natural microbial biofilm. *Science*, 308, 1915-1920.

Ramos, A.F., Gómez, M.A., Hontoria, E. & González-López, J. (2007). Biological nitrogen and phenol removal from saline industrial wastewater by submerged fixed-film reactor. *Journal of Hazardous Materials*, 142, 175-183.

Ross, N., Villemur, R., Marcandella, E. & Deschnes, L. (2001). Assessment of changes in biodiversity when a community of ultramicrobacteria isolated from groundwater is stimulated to form a biofilm. *Microbial Ecology*, 42, 56-68

Saby, S., Sibille, I., Mathieu, L., Paquin, J. L. & Block, J.C. (1997). Influence of water chlorination on the counting of bacteria with DAPI (4',6-diamidion-2-phenylindole). *Applied and Environmental Microbiology*, 63, 1564–1569.

Schaule, G., Flemming, H. C. & Ridgway, H.F. (1993). Use of 5- cyano-2,3-ditolyl tetrazolium chloride for quantifying planktonic and sessile respiring bacteria in drinking water. *Applied and Environmental Microbiology*, 59, 3850–3857.

Schwieger, F. & Tebbe, C. C. (1998). A new approach to utilize PCR single-strand-conformation polymorphism for 16S rRNA gene-based microbial community analysis. *Applied and Environmental Microbiology*, 64, 4870-4876.

Slabbinck, B., Waegeman, W., Dawyndt, P., De Vos, P. & De Baets, B. (2010). From learning taxonomies to phylogenetic learning: integration of 16S rRNA gene data into FAME-based bacterial classification. *BMC Bioinformatics*, 11, 69.

Stach, J. E., Bathe, S., Clapp, J. P. & Burns, R. G. (2001). PCR-SSCP comparison of 16S rDNA sequence diversity in soil DNA obtained using different isolation and purification methods. *FEMS Microbiology Ecology*, 36, 139-151.

Stephanie, A. F., Sierra-Alvarez, R., Altinbas, M., Hollingsworth, J., Stams, A. J. M., Smidt, H., (2008). Molecular characterization of mesophilic and thermophilic sulfate reducing microbial communities in expanded granular sludge bed (EGSB) reactor. *Biodegradation,* 19, 161–177.

Stipp, S. L., Hochella, M. F., Parks, G. A. & Leckie, J. O. (1992). Cd^{2+} uptake by calcite, solid-state diffusion and the formation od solid-solution: interface processes observed with near-surface sensitive techniques (XPS, LEED and AES). *Geochimica et Cosmochimica Acta*, 56, 1941-1954.

Sturchio N. C., Chiarello R. P., Cheng L., Lyman P. F., Bedzyk M. J., Qian Y. & Baer D. R. (1997). Lead adsorption at the calcite-water interface: synchroton X-ray standing wave and X-ray reflectivity studies. *Geochimica et Cosmochimica Acta,* 61, 251-263.

Suzuki, M.,. Rappe, M. S., & Giovannoni, S. J. (1998). Kinetic bias in estimates of coastal picoplankton community structure obtained by measurements of small- subunit rRNA gene PCR amplicon length heterogeneity. *Applied and Environmental Microbiology*, 64: 4522-4529.

Szewzyk, U., Szewzyk, R., Manz, W. & Schleifer, K. H. (2000) Microbiological safety of drinking water. *Annual Review of Microbiology*, 54, 81–127.

Tomioka, N., Uchiyama, H. & Yagi, O. (1994). Cesium accumulation and growth characteristics of *Rhodococcus erythropolis* CS98 and *Rhodococcus* sp. strain CS402. *Applied and Environmental Microbiology*, 60, 2227-2231.

Vandamme, P., Pot, B., Gillis, M., de Vos, P., Kersters, K. & Swings, J. (1996). Polyphasic taxonomy, a consensus approach to bacterial systematics. *Microbiology Reviews*, 60, 407-438.

Vaneechoutte, M., Rossau, R., De Vos, P., Gillis, M., Janssens, D., Paepe, N., De Rouck, A., Fiers, T., Claeys, G. & Kersters, K. (1992). Rapid identification of bacteria of the *Comamonadaceae* with amplified ribosomal DNA-restriction analysis (ARDRA). *FEMS Microbiology Letters*, 72, 227-233.

Vílchez, R. (2005) Eliminación de metales pesados de aguas subterráneas mediante sistemas de lechos sumergidos: estudio microbiológico de las biopelículas. *PhD Thesis, U. de Granada, Spain.*

Vílchez, R., Pozo, C., Gómez, M.A., Rodelas, B. & González-López, J. (2007). Dominance of sphingomonads in a copper-exposed biofilm community for groundwater treatment. *Microbiology*, 153, 325-337.

von Canstein, H., Li, Y., Timmis, K. N., Deckwer, W. D. & Wagner-Döbler, I. (1999). Removal of mercury from chloralkali electrolysis wastewater by a mercury-resistant *Pseudomonas putida* strain. *Applied and Environmental Microbiology*, 65, 5279-5284.

Wagner, M., Loy, A., Nogueira, R., Purkhold, U., Lee, N. & Daims, H. (2002). Microbial community composition and function in wastewater treatment plants. *Antonie Van Leeuwenhoek,* 81, 665-680.

Wang, Y. T. & Shen, H. (1995). Bacterial reduction of hexavalent chromium. *Journal of Industrial Microbiology*, 14, 159-163.

Wang, G. G. Y., & Wang, Y. (1996). The frequency of chimeric molecules as a consequence of PCR co-amplification of 16S rRNA genes from different bacterial species. *Microbiology,* 142, 1107-1114.

Watnick, P. & Kolter, R. (2000). Biofilm, city of microbes. *Journal of Bacteoriology*, 182, 2675-267.

Wehrheim, B. & Wettern, M. (1994). Comparative studies of the heavy metal uptake of whole cells and different types of cell walls from *Chlorella fusca. Biotechnology Techniques*, 8, 227-232.

Wood, J. M., & Wang, H. K. (1983). Microbial resistance to heavy metals. *Environmental Science and Technology*, 17, 582-590.

Welander, U. & Mattiasson, B. (2003). Denitrification at low temperatures using a suspended carrier biofilm process. *Water Research*, 37, 2394–2398.

White, D. C., Sutton, S. D. & Ringelberg, D. B. (1996). The genus *Sphingomonas*: physiology and ecology. *Current Opinion in Biotechnology*, 7, 301-306.

W.H.O. (World Health Organization) Report W.H.O: Health hazards from nitrates in drinking water [cited 2010 May 20]. Available from: http://whqlibdoc.who.int/euro/ehs/EURO_EHS_1.pdf (1985). *Meeting of Copenhagen 5-9* March 1984.

Xu N., Hochella M. F., Brown G. E. & Parks G. A. (1996). Co (II) sorption at the calcite-water interface: I. X-ray photoelectron spectroscopic study. *Geochimica et Cosmochimica Acta,* 60, 2801-2815.

Zouboulis, A.I., Matis, K.A. & Hancock I.C. (1997). Biosorption of metals from dilute aqueous solutions. *Separation and Purification Methods*, 26, 255-295.

In: Advances in Medicine and Biology Research, Vol. 14 ISBN: 978-1-61761-930-4
Editor: Leon V. Berhardt

Chapter V

Applying Project Management to Evidence-Based Medicine Research

Patrick Letouze
Professor at the Department of Computer Science, Universidade Federal do Tocantins – UFT (Federal University of Tocantins), Brazil e-mail: patrick.letouze@gmail.com or letouze@uft.edu.br

Abstract

Project management (PM) may be used as a guideline for Evidence-Based Medicine (EBM) research. Such kind of research consists in the use of statistically sound estimates derived from scientific studies on population samples for supporting clinical decision making in diagnosis, investigation or management of individual patients. They are put in hierarchy in the following order: systematic review and meta-analysis randomized controlled trials with definitive results, randomized controlled trials with non-definitive results, cohort studies, case-controlled studies, cross-sectional surveys, and case reports. In such studies medical protocols are of major importance and a PM approach may be applied. For instance, a patient transplantation process starts with the acknowledgment of transplant necessity, where restrictions and hypothesis are determined defining medical procedures, psycho-social attendance strategies and performance criteria (initiation). Then, transplantation protocols and plans are formulated, that is, its schedules, activities, risk identification, staff assignment, medication, exams, evaluations and surgical intervention planning are elaborated (planning). Therefore, the preparation of the surgical transplant act requires that medical procedures are performed while the patient progression is evaluated constantly (executing). Moreover, to get the patient physiologically ready for it, several exams are performed and some parameters are monitored, because some specific limits or values must be kept or achieved (controlling). After the successful surgical intervention, the patient is still monitored for a while, usually one year, and then he returns from time to time for a follow up. Basically, after the surgery until a year his medical records should be reviewed and finalized to avert potential problems on him and on future transplants (closing). Consequently, as a methodology for scientific research, PM concepts and techniques may be used for treating a patient. Furthermore, regarding medical studies' hierarchy, PM may be simultaneously applied from bottom-up, that is, if it is used in every kind of research then it would be easier to perform systematic reviews and meta-analysis. In addition, a computational data-base could be created for EBM using PM, it could have functions to statistically support primary studies focused on treatment, diagnoses, tracking, prognoses and causality. Therefore, a PM approach allied to a computational data-base could speed up medical research while facilitating the physicians work, because there would be more and new kind of ready to use statistical data for supporting EBM researches. Also, due to that standardized approach, information retrieval techniques would be more effective in discovering new knowledge.

Introduction

Medical research has been using Project Management techniques for a while, since a research with funding is seen as a project with a budget, resources and other items that demand administration (1)(2). Nonetheless, it has been proposed to consider the transplantation process as a project (3); in this case, it is a methodology to control and optimize the transplantation in terms of medical results, including a statistical support to Evidence-Based Medicine (EBM) research. Actually, the latter approach could be applied to any complex medical treatment or intervention, and to any medical research. This work is about integrating both approaches to increase the quality of evidence in a medical research.

Evidence-Based Medicine is the judicious use of the best current evidence in making decisions about the care of the individual patient. However, according to MacNutt and Livingston in (4):

> Evidence-based medicine must evolve to include clinically and contextually explicit estimates for outcomes as guides to care. Clinical trials and studies evaluating diagnostic tests should be designed and reported to enable clinicians to maximize the care of individual patients, so they can avoid doing the sometimes right things for the sometimes wrong patients.

Consequently, to apply Project Management techniques may satisfy the above issues. In order to see that, let us briefly review the Project Management phases of the Project Management Body of Knowledge (PMBOK) (5):

1) Initiation: to determine project goals, deliverables and process outputs, to document project constraints and assumptions, to define strategy, to identify performance criteria, to determine resource requirements, to define the budget and to produce a formal documentation.
2) Planning: to refine project, to create a work breakdown structure (WBS), to develop the resource management plan, to refine time and cost estimates, to establish project controls, to develop the project plan and to obtain the plan approval.
3) Execution: to commit resources, to implement resources, to manage progress, to communicate progress and to implement quality assurance procedures.

4) Control: to measure performance, to refine control limits, to take corrective action, to evaluate effectiveness of corrective action, to ensure plan compliance, to reassess control plans, to respond to risk event triggers and to monitor project activity.
5) Closing: to obtain acceptance of deliverables, to document lessons learned, to facilitate closure, to preserve product records and tools, and to release resources.

Hence, "to include clinically and contextually explicit estimates for outcomes as guides to care" is satisfied by determining the project goals, deliverables and process outputs, in the initiation phase.

Table 1. Project Management Knowledge Areas, its descriptions and processes.

AREA	Description	Processes
INTEGRATION	It is concerned with coordinating all aspects of the project plan and is highly interactive	Project plan development
		Project plan execution
		Integrated change control
SCOPE	It is concerned with the work of the project, all the processes involved with it and only the work that is required to complete the project	Initiation
		Scope planning
		Scope definition
		Scope verification
		Scope change control
TIME	It is concerned with estimating the duration of the project plan activities, devising a project schedule, and monitoring and controlling deviations from schedule	Activity definition
		Activity sequencing
		Activity duration estimating
		Schedule development
		Schedule control
COST	It is concerned to establish estimates for costs and resources and keep watch over those costs to ensure that the project stays within the approved budget	Resources planning
		Cost estimating
		Cost budgeting
		Cost control
QUALITY	It is concerned with the assurance that the project meets the requirements that the project was undertaken to produce	Quality planning
		Quality assurance
		Quality control
HUMAN RESOURCE	It is concerned with all aspects of people management and personal interaction	Organizational planning
		Staff acquisition
		Team development
COMMUNICATIONS	It is concerned to ensure that all project information including project plans, risk assessments, meeting notes, and more is collected and documented	Communications planning
		Information distribution
		Performance reporting
		Administrative closure
RISK	It is concerned with identifying and planning for potential risks that may impact the project	Risk management planning
		Risk identification
		Quality risk analysis
		Quantitative risk analysis
		Risk response planning
		Risk monitoring and control
PROCUREMENT	It is concerned with the processes involved with purchasing goods or services from external vendors, contractors and suppliers	Procurement planning
		Solicitation planning
		Solicitation
		Source selection
		Contract administration
		Contract closeout

Moreover, to satisfy the assertion "Clinical trials and studies evaluating diagnostic tests should be designed and reported to enable clinicians to maximize the care of individual patients", it is helpful to identify performance criteria – initiation phase, and to measure performance, refine control limits, take corrective action and evaluate effectiveness of corrective action – control phase. That is, from the medical research point of view, two key areas of PM knowledge would be Integration and Quality, see Table 1.

Also in (4), they argue that without a context consideration a meta-analysis may lead to inappropriate treatments. As a consequence, they propose that "One helpful approach may be for reports of meta-analyses to include, in detail, the inclusion and exclusion criteria for patients enrolled in the original studies." Again, PM satisfies this issue through project constraints and assumptions in the Initiation phase, more than that, to document project selection criteria is a part of PM, so to report, in detail, the selection criteria for patients would be analogous.

Clearly, PM techniques may be applied for administrating medical researches as a whole, but before starting the research, to correctly apply these techniques regarding the medical context may satisfy the issues highlighted in (4). Therefore, this work starts presenting some basic concepts of EBM and some medical research standards. Then it approaches the application of PM to EBM as a medical research methodology. Finally, it concludes by discussing some practical issues in the application of PM to EBM research.

Medical Research

Another definition for EBM in (6) is:

> Evidence-based medicine is the use of mathematical estimates of the risk of benefit and harm, derived from high-quality research on population samples, to inform clinical decision making in the diagnosis, investigation or management of individual patients.

In this definition three important aspects may be improved by PM: mathematical estimates, high-quality research and management of individual patients. The first aspect is a usual activity in project management, like cost estimation, activity estimation and risk qualitative analysis - see Table 1. Therefore, it is enough to incorporate the medical parameters to be estimated in the Initiation and Planning phases. The second aspect, high-quality research, is the core of this work, while the last, management of individual patients, is naturally accomplished considering the treatment as a project, as in (3).

The quality of a medical research depends on the kind of study performed and for that reason they are classified hierarchically. A non-exhaustive list of hierarchy of study designs to assess the effects of interventions from (7) is presented below and another approach to EBM hierarchy is presented in Table 2:

1) Randomized Controlled Trials (RTCs)

They are a type of scientific experiment commonly used in testing the efficacy or effectiveness of medical interventions (treatments or conditions), involving its random allocation to subjects. An important advantage is the elimination of selection bias, balancing both known and unknown prognostic factors, in the assignment of interventions. For instance, still according to (7) "The simplest form of RCT is known as the parallel group trial which randomizes eligible participants to two or more groups, treats according to assignment, and compares the groups with respect to outcomes of interest. Participants are allocated to groups using both randomization (allocation involves the play of chance) and concealment (ensures that the intervention that will be allocated cannot be known in advance)."

a) Randomized cross-over trials.

All participants receive all interventions; it is the sequence of interventions that is randomized.

b) Cluster randomized trials.

Clusters of people rather individuals are randomized to different interventions.

2) Quasi-Experimental Studies.

They differ from RCT in terms of participants allocations. They do not use random assignment to create the comparison group.

a) Non-randomized controlled studies.

Participants are allocated to concurrent comparison groups.

b) Before-and-after study.

Comparisons of outcomes occur before and after an intervention in the same or different sample of participants.

c) Interrupted time series

They are multiple observations over time that are "interrupted" by an intervention.

3) Observational Studies

They are studies to explore the effect of the interventions or exposures on health outcomes, where participants are not allocated by an investigator.

a) Cohort study

The comparison of who did or did not receive an intervention is made over time.

b) Case-control study

Groups with a specific outcome of interest are cases and groups without are controls. They are compared to evaluate the association between exposure to an intervention and the outcome.

c) Case series

They are descriptions of a number of cases of an intervention and the outcome without a comparison to a control group.

Table 2. The CEBM "levels of Evidence". Produced by Bob Phillips, Chris Ball, Dave Sackett, Dough Badenoch, Sharon Straus, Brian Haynes, martin Dawes since November 1998. Updated by Jeremy Howick March 2009. Taken from http://www.cebm.net/index.aspx?o=1025, in June 18 of 2010.

Level	Therapy/ Prevention, Aetiology/ Harm	Prognosis	Diagnosis	Differential diagnosis/ symptom prevalence study	Economic and decision analyses
1a	SR(with homogeneity*)of RCTs	SR (with homogeneity*) of inception cohort studies; CDR" validated in different populations	SR (with homogeneity*) of Level 1 diagnostic studies; CDR" with 1b studies from different clinical centres	SR (with homogeneity*) of prospective cohort studies	SR (with homogeneity*) of Level 1 economic studies
1b	Individual RCT (with narrow Confidence Interval"¡)	Individual inception cohort study with > 80% follow-up; CDR" validated in a single population	Validating** cohort study with good" " " reference standards; or CDR" tested within one clinical centre	Prospective cohort study with good follow-up****	Analysis based on clinically sensible costs or alternatives; systematic review(s) of the evidence; and including multi-way sensitivity analyses
1c	All or none§	All or none case-series	Absolute SpPins and SnNouts" "	All or none case-series	Absolute better-value or worse-value analyses " " " "
2a	SR (with homogeneity*) of cohort studies	SR (with homogeneity*) of either retrospective cohort studies or untreated control groups in RCTs	SR (with homogeneity*) of Level >2 diagnostic studies	SR (with homogeneity*) of 2b and better studies	SR (with homogeneity*) of Level >2 economic studies
2b	Individual cohort study (including low quality RCT; e.g., <80% follow-up)	Retrospective cohort study or follow-up of untreated control patients in an RCT; Derivation of CDR" or validated on split-sample§§§ only	Exploratory** cohort study with good" " " reference standards; CDR" after derivation, or validated only on split-sample§§§ or databases	Retrospective cohort study, or poor follow-up	Analysis based on clinically sensible costs or alternatives; limited review(s) of the evidence, or single studies; and including multi-way sensitivity analyses
2c	"Outcomes" Research; Ecological studies	"Outcomes" Research		Ecological studies	Audit or outcomes research

Table 2. (Continued).

3a	SR (with homogeneity*) of case-control studies		SR (with homogeneity*) of 3b and better studies	SR (with homogeneity*) of 3b and better studies	SR (with homogeneity*) of 3b and better studies
3b	Individual Case-Control Study		Non-consecutive study; or without consistently applied reference standards	Non-consecutive cohort study, or very limited population	Analysis based on limited alternatives or costs, poor quality estimates of data, but including sensitivity analyses incorporating clinically sensible variations.
4	Case-series (and poor quality cohort and case-control studies§§)	Case-series (and poor quality prognostic cohort studies***)	Case-control study, poor or non-independent reference standard	Case-series or superseded reference standards	Analysis with no sensitivity analysis
5	Expert opinion without explicit critical appraisal, or based on physiology, bench research or "first principles"	Expert opinion without explicit critical appraisal, or based on physiology, bench research or "first principles"	Expert opinion without explicit critical appraisal, or based on physiology, bench research or "first principles"	Expert opinion without explicit critical appraisal, or based on physiology, bench research or "first principles"	Expert opinion without explicit critical appraisal, or based on economic theory or "first principles"

These studies are called primary studies and more details can be found in (6) and (7). Secondary studies rank higher in the EBM hierarchy, they are Systematic Reviews (SR) and Meta-Analysis (MA). The former is a literature review focused on a question that seeks to identify, appraise, select and synthesize relevant evidence to that question, and the latter combines the results of several studies that address a set of related research hypotheses.

Consequently, before starting a medical research, it is helpful to take into account the type of study and the related standard. For instance, some medical journals even require that some studies follow the corresponding standard, as the British Medical Journal (BMJ) and the Journal of the American Medical Association (JAMA).

The high-quality research issue in medicine motivated many initiatives to propose some guidelines ("standards") for each type of medical study. Some of these standards, or similar systems, are:

- EBM - evidence-based medicine (6) (8).
- The COCHRANE Collaboration (9) (www.cochrane.org): "it is an international network of people helping healthcare providers, policy makers, patients, their advocates and carers, make well-informed decisions about human health care by preparing, updating and promoting the accessibility of Cochrane Reviews – over 4,000 so far, published online in *The Cochrane Library.*"
- STARD – Standards for the Reporting of Diagnostic accuracy studies (10) (www.stard-statement.org): "The objective of the STARD initiative is to improve the accuracy and completeness of reporting of studies of diagnostic accuracy, to allow readers to assess the potential for bias in the study (internal validity) and to evaluate its generalisability (external validity)."

- STROBE – STrengthening the Reporting of OBservational studies in Epidemiology (11) (www.strobe-statement.org): "STROBE stands for an international, collaborative initiative of epidemiologists, methodologists, statisticians, researchers and journal editors involved in the conduct and dissemination of observational studies, with the common aim of strengthening the reporting of observational studies in epidemiology."
- MOOSE – reports of Meta-analysis Of Observational Studies in Epidemiology (12).
- CONSORT 2010[1] – Consolidated Standards of Reporting Trials (13) (14) (www.consort-statement.org): "The main product of CONSORT is the CONSORT Statement, which is an evidence-based, minimum set of recommendations for reporting RCTs. It offers a standard way for authors to prepare reports of trial findings, facilitating their complete and transparent reporting, and aiding their critical appraisal and interpretation."
- PRISMA – Preferred Reporting Items for Systematic Reviews and Meta-Analysis (15) (16) (www.prisma-statement.org): "The aim of the PRISMA Statement is to help authors improve the reporting of systematic reviews and meta-analyses. We have focused on randomized trials, but PRISMA can also be used as a basis for reporting systematic reviews of other types of research, particularly evaluations of interventions. PRISMA may also be useful for critical appraisal of published systematic reviews, although it is not a quality assessment instrument to gauge the quality of a systematic review."
- GRADE[2] – the Grading of Recommendations Assessment, Development and Evaluation (8) (17) (18) (19) (20) (21) (www.gradeworkinggroup.org): "The Grading of Recommendations Assessment, Development and Evaluation (short GRADE) Working Group began in the year 2000 as an informal collaboration of people with an interest in addressing the shortcomings of present grading systems in health care. The working group has developed a common, sensible and transparent approach to grading quality of evidence and strength of recommendations. Many international organizations have provided input into the development of the approach and have started using it."
- URM – Uniform Requirements for Manuscripts submitted to biomedical journals (22) (www.icmje.org): "A small group of editors of general medical journals met informally in Vancouver, British Columbia, in 1978 to establish guidelines for the format of manuscripts submitted to their journals. This group became known as the Vancouver Group. Its requirements for manuscripts, including formats for bibliographic references developed by the National Library of Medicine (NLM), were first published in 1979. The Vancouver Group expanded and evolved into the International Committee of Medical Journal Editors (ICMJE), which meets annually. The ICMJE has gradually broadened its concerns to include ethical principles related to publication in biomedical journals."

---

1 From now on CONSORT 2010 is referred simply as CONSORT.

2 GRADE is not the only grading system for medical evidence.

- SPIRIT – Standard Protocol Items for Randomized Trials (23) (24): "It aims to produce evidence-based recommendations for key information to include in a trial protocol."
- MIBBI – Minimum Information for Biological and Biomedical Investigation (25) (www.mibbi.org): "Throughout the biological and biomedical sciences, prescriptive checklists specifying the key information to include when reporting experimental results are beginning to find favor with experimentalists, analysts, publishers and funders alike. However, such 'minimum information' (MI) checklists are usually developed independently, from within particular biologically- or technologically-delineated domains. Consequently, the full range of checklists can be difficult to establish without intensive searching, and tracking their evolution is non-trivial; they are also inevitably partially-redundant one against another, and where they overlap arbitrary decisions on wording and substructuring make integration difficult. This presents significant difficulties for the users of checklists; for example, in the area of systems biology, where data from multiple biological domains and technology platforms are routinely combined. We offer a common portal to such MI checklists; to act as a 'one-stop shop' for those exploring the range of extant projects, foster collaborative development and ultimately promote gradual integration."

From a bottom-up approach to evidence hierarchy perspective, the Minimum Information for Biological and Biomedical Investigations (MIBBI) project provides and maintains a web-based, freely accessible resource for checklist projects (and to complementary data formats, controlled vocabularies, tools and databases), for those exploring the range of extant minimum information checklists and fosters coordinated development of such checklists (25), thereby enhancing both transparency and accessibility. Clearly, MIBBI is a candidate to assist primary studies as listed above complementing the initiatives STARD, STROBE and CONSORT.

The STARD initiative seeks the improvement of accuracy and completeness for the report of studies of diagnostic accuracy, making assessable to readers the potential for bias in the study and to evaluate its generality. As diagnostic accuracy may be important to quality of evidence in observational studies and randomized controlled trials, STARD may be seen as a support to STROBE and CONSORT. While, the former wishes to provide guidance on how to report observational research well, the latter is an evidence-based, minimum set of recommendations for reporting RCTs (Randomized Controlled Trials).

Meta-analysis of Observational Studies in Epidemiology – MOOSE, is the next step to observational research, especially when "randomized controlled trials are not feasible and only data from observational studies are available." (12) Consequently, observational studies applying STROBE have greater chance to meet MOOSE requirements and the greater the number of evaluated studies in a meta-analysis its results are statistically sounder.

Also, PRISMA may be applied to observational studies and the meta-analysis from MOOSE may be used in a systematic review using it. Although, PRISMA is an evidence-based minimum set of items for reporting in systematic reviews and meta-analysis, if it is considered in the research's design, Liberati et al. in (15) recognize that:

> We developed the PRISMA statement and this explanatory document to help authors report a wide array of systematic reviews to assess the benefits and harms of a healthcare intervention. We consider most of the checklist items relevant when reporting systematic reviews of non-randomized studies assessing the benefits and harms of interventions. However, we recognize that authors who address questions relating to aetiology, diagnosis, or prognosis, for example, and who review epidemiological or diagnostic accuracy studies may need to modify or incorporate additional items for their systematic reviews.

In this quotation it is explicit the idea of considering one standard in a higher level when designing a lower level medical research, in terms of evidence hierarchy. In addition to that, SPIRIT "has finalized an evidence-based checklist that defines the key items to be addressed in trial protocols, leading to improved quality of protocols and enabling accurate interpretation of trial results." (23)

Finally, GRADE attempts to develop a common, sensible approach to grading quality of evidence and strength of recommendation. In principle, to grade a research is an *a posteriori* activity, that is, it occurs after a research has been done. Nevertheless, the design of a research may take it into account and that might improve the quality of the results (18).

The importance of measuring the quality of evidence in making healthcare management decisions is that patients and clinicians must compare different alternatives strategies that will be influenced not only by the expected advantages and disadvantages, but by their reliance on these evidences (17). The GRADE system fulfills that, and furthermore, it establishes that consideration of the context is a requirement for judging the quality of evidence, as Guyatt et al. in (18) states:

> GRADE provides a definition for the quality of evidence in the context of making recommendations. The quality of evidence reflects the extent to which confidence in an estimate of the effect is adequate to support recommendations. This definition has two important implications. Firstly, guideline panels must make judgments about the quality of evidence relative to the specific context in which they are using the evidence. Secondly, because systematic reviews do not—or at least should not—make recommendations, they require a different definition. For systematic reviews, the quality of evidence reflects the extent of confidence that an estimate of effect is correct.

Also in (18), Guyatt et al. point out five limitations that can reduce the quality of evidence: study limitations, inconsistent results, indirectness of evidence, imprecision and publication bias. Moreover, Guyatt et al. in (19) shows that GRADE is an approach suited for developing and presenting recommendations for management of patients and that the quality of the evidence is a key factor to determine the strength of a recommendation.

Besides that, Schünemann et al. in (20) explain how "the GRADE system can be used to grade the quality of evidence and strength of recommendations for diagnostic tests or strategies." And in (21), Guyatt et al. present how to incorporate considerations of resource into grading recommendations by: regarding cost as an outcome; using a balance sheet to assess benefits versus costs; making evidence profiles presents resource use, not just monetary values; specifying context as much as possible; broadening perspective if possible; judging quality of evidence for resource use; and using a formal economic modeling if it is appropriate.

Hence, as the GRADE's aim is to develop a common, sensible approach to grading quality of evidence and strength of recommendation, if quality of evidence and recommendation is desired it should permeate the design of every medical research. And for publishing purposes, every study should regard the Uniform Requirements for Manuscripts Submitted to Biomedical Journals (URM).

Project Management Applied to EBM

Medical research is going through a standardization phase. Every attempt to establish minimum requirements for doing a certain type of research, or related activity, intends to improve and ensure the quality of medical evidence or recommendation. The Project Management Body of Knowledge – PMBOK (5) may be seen as a project management standard, hence it may be a framework for medical research standards.

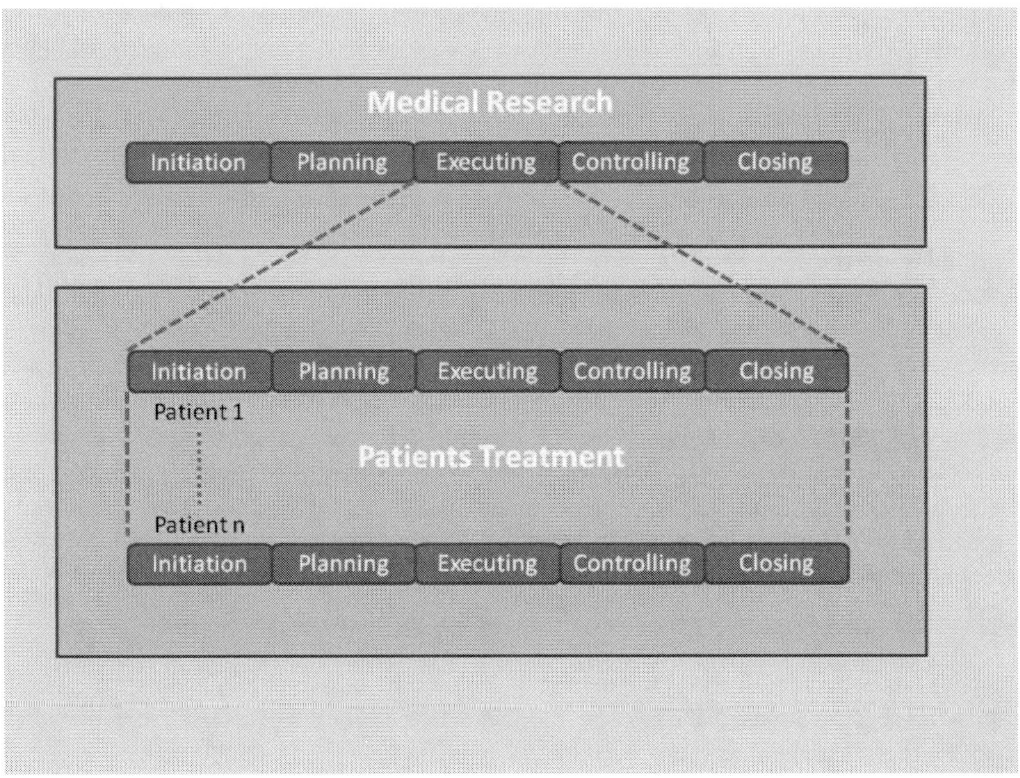

Figure 1. Application of Project Management in different levels of a medical research.

Figure 1 shows that PM may be applied in two levels in a medical research. The first and immediate level is to consider the research itself as a project; some examples are presented in (1), which is a training course, and in (2), which focus on randomized clinical trials. And the second level, the medical treatment or intervention is considered a project; this approach is justified when the treatment or intervention is complex enough, that is, when it is worth the trouble to approach the medical treatment or intervention as a project. The transplantation process is an example of the second level and it was first proposed in (3). Nonetheless, PM may be applied simultaneously to both levels and to show that a hypothetical example is proposed.

Let us say that a transplantation center decides to do a medical research with evidence-based orientation, but it also decides to apply project management in both levels. It starts with an Initiation phase followed by the Planning phase in the medical research level.

The Initiation process would have four inputs: the research description; a strategic plan, which specifies how the research inserts itself into the promoting institution; the research selection criteria, that is, which patient or research to accept or reject; and historical information, which approaches previous researches of a similar nature or in the promoting institution. And it would have three output documents: the research definition or overview, the constraints and the assumptions.

The Initiation phase would determine the research objectives, its requirements, deliverables, and process outputs, to document constraints and to document assumptions. The research overview document should have the objectives, requirements and deliverables, not only from a project perspective, but also and mainly in terms of medical objectives, requirements and deliverables. The objectives would be the purpose for undertaking the research, they should describe the final result of the research and they should describe what it is intended to do, accomplish, or produce. For instance, the purpose of the research could be to evaluate the effect of a medicine for immunosuppression in kidney transplantation preparation, that is, the administration of this drug in the peri-transplantation phase. The determination of the research objectives should follow the SMART rule (5): Specific, Measurable, Accurate, Realistic and tangible, and Time bound. For evaluating the immunosuppression drug (specific) for preparing a patient for transplantation, with a daily constant dosage (measurable and accurate) during ten days (time bound) before the transplant intervention (realistic – the transplantation is authorized, confirmed and scheduled), and some blood test, or a more convenient exam, would be performed before the first administration of the drug and before the last administration of the drug in the tenth day.

Usually, immunosuppression drugs are used after the transplantation for avoiding organ rejection and in this example it would be out of scope, but if the patient treatment is approached as a project, then any unexpected result after the transplantation would be recorded giving new directions to the research or providing new evidences to start new researches.

The requirements would be medical specifications or prerequisites that make up the objectives or deliverables, while the deliverables could be measurable outcomes, measurable results, or specific items that must be produced to consider the research completed. To elaborate the constraint document would be to evaluate time, budget and quality – especially medical evidence. Constraints would be anything that either restricts or dictates the actions of the research team. In addition, assumptions would be one of the outputs of the Initiation process and shall be used as inputs to other processes later in the research.

After approval, the research charter is the official, written acknowledgement and recognition that a research exists. It should gather the other three documents (overview, constraints and assumptions) and it should describe the preliminary roles and responsibilities of the entire staff, sponsor and others. Furthermore, it should include resource and cost estimates, and a feasibility study if possible. Finally, this phase ends with the research charter sign-off: sign-off from the sponsor or the authority.

In the Planning phase it should be considered the appropriate standard and it should be decided if the treatment or intervention should have a PM approach. One could ask why not consider the appropriate standard in the Initiation phase and the answer would be it may, but this phase is supposed to be brief, it intends to grant the approval to commit the organization's resources to working on the project. Therefore, it might be too much an effort to take into account the standard before the approval of the Initiation of the project, though it is important to highlight the need of it in the final document of this phase. For instance, the Initiation final document may be seen as a research proposal to a Research Committee or hospital's board of directors where the research is supposed to take place. By applying the concepts of PMBOK's Initiation phase, it should help the researcher to produce a solid research proposal.

The first process in the Planning phase is scope planning. The primary purpose of scope planning is to produce the scope statement and the scope management plan. The Scope Statement is about research's justification, deliverables and objectives. Here, the research charter is an input, if it was well written then it is enough to transfer the objectives and deliverables information from the charter to the scope statement. The Scope Statement should contain a comprehensive list of all research requirements.

From the Scope Statement it is produced the Scope Definition, where the research deliverables are broken down into smaller, manageable components so that research tasks and activities are planned. The decomposition aims estimation improvement, because it easier to estimate costs, time, and resources needed for individual work components than it is to estimate them for a whole body of work or deliverable. Also, it makes easier performance measures and controls assignment assuring that the responsibility for that assignment goes to the proper parties. The tools and techniques of Scope Definition are the Work Breakdown Structure (WBS) templates and decomposition. The number of levels in the WBS is based on the complexity of the research, that is, the researcher or manager is free to determine it, but it must include enough levels to accurately estimate the research time and costs. The lowest level in a WBS is called a work package, which are tasks that can be easily assigned to one person, or a team of people, with clear accountability and responsibility for completing the assignment. An interesting concept would be a milestone, which is a major accomplishment in a project, for example, in the hypothetical transplantation research each transplant could be a milestone, or a group of successful transplants. Milestones are checkpoints to help determine progress. After creating the WBS, deliverables might need further definition or refining, some new deliverables may appear which were not thought of during the Scope Planning process, then the Scope Statement must be updated.

At this point, if it was chosen to approach each patient treatment or intervention as a project, then a general Initiation and Planning phase for patient treatment should be performed. Hence, each Initiation Phase would start with the acknowledgement of the transplant necessity and ends with the elaboration of the Patient Treatment Initiation Document. This document should be approved by the designated doctor for the transplant or by the leading doctor of the transplant unit according to the chosen strategy adopted by each center. The Initiation phase should determine medical and psycho-social restrictions and hypotheses from which the medical protocols and psycho-social attendance strategies would be defined. In order to provide new instances for research on the field, performance criteria should be defined. Usually, the criteria would be measurable variables in exams and evaluations. In the Planning Phase, the transplant staff should elaborate planning documents to be used throughout the transplant process. From the Initiation Document other documents should be created such as a resource management plan and a transplant control. Transplant protocols and plans should contain schedules, activities, risk identification, staff assignment, medication, exams, evaluations and surgical intervention planning. Also, it is in here that transplant requirements and stakeholders are identified. For instance, if the kidney donor is alive then he or she is a stakeholder. Also, other members of the recipient and donor families could be stakeholders, since they may influence the transplantation.

After the elaboration of the general Patient Treatment Planning Phase, it would be possible to create the resource planning and the estimation of the medical research, which are the process of developing the Resource Management Plan (RMP), where time and cost estimates are refined. This plan should encompass several processes including resource planning, organizational planning, staff acquisition, solicitation planning, and procurement planning. The combined outputs from these processes would comprise the RMP. Here, patients should be seen as resources, and then patient selection and management should be included in the RMP. Usually, resources inputs are the WBS, resource pool description, organizational policies and activity duration. The Resource Requirements Document is a detailed description of the kinds of resources needed for research and the quantity needed for each resource. Here, to identify the activities and to sequencing them in the correct order are fundamental to estimate the duration of the involved research activities. Then time estimates are assigned as baseline measurements and used to track project activities to ensure the research is completed on time.

Then establishing the Research Planning Controls involves identifying quality standards, for example GRADE. Therefore quality inputs are quality policy, standards and regulations, and cost of quality. A Quality Management Plan should have operational definitions and checklists – the corresponding initiative checklist.

In addition, a Risk Management Plan should be created by identifying potential risks to the research and to the patient treatment or intervention, such as, budgets/ funding, schedules, scope or requirements change, medical/technical issues, patient/personnel issues, etc.

Finally, a Research Plan should be created, which should contain a developed schedule, mathematical analysis for project and medical purposes, and a cost budget baseline.

The next phase is the Executing Phase for both, research and patient levels. It simultaneously puts the research plans and each patient transplant plans into action. For each specific patient it is created an individual Patient Treatment Planning phase, where resources are designated, committed and coordinated. The actual implementation of the transplant should occur here, that is, what was planned in the Patient Treatment Planning Phase should be executed. Moreover, at this Phase the patient's progression should be evaluated constantly in relation to quality assurance procedures according to the established performance criteria.

The fourth phase is about controlling the research and each transplant, and therefore, it is called the Controlling Phase. Every parameter or performance measure used in the transplant process shall have its limits defined in the Planning Phase that are under control here. Corrective action, which shall be subject to effectiveness evaluations, must be taken when a parameter or performance measure gets out of limits. Here transplant activities are monitored. And simultaneously, all patient data should be monitored from the research control perspective.

The last phase for each patient, which is called Closing Phase, reviews the collected documentation to avert potential problems on future transplants. It releases the committed resources and it closes the records for each patient. After a sufficient number of transplants are accomplished then the research closing phase happens and ends with the elaboration of scientific paper using URM (22).

Discussion

Project Management may be applied in medical research in two levels. The first is to treat the medical research as a project. The second is to approach the patient's treatment or intervention as a project, which may provide new kind of data for analysis. This work proposed to ally these two approaches in order to improve the quality of medical evidence obtained by the medical research.

Besides, it was shown that PM may answer three major issues posed by MacNutt and Livingston in (4):

- "Evidence-based medicine must evolve to include clinically and contextually explicit estimates for outcomes as guides to care."
- "Clinical trials and studies evaluating diagnostic tests should be designed and reported to enable clinicians to maximize the care of individual patients."
- "One helpful approach may be for reports of meta-analyses to include, in detail, the inclusion and exclusion criteria for patients enrolled in the original studies."

Furthermore, from the patient's point of view, to be treated as a project may seem at first as an inhuman approach, but in practice is to give a more attentive support and a more thorough attendance. Nevertheless, the patient does not need to know the adopted approach, but if a qualified professional, such as a psychologist, explains the benefits of the PM approach, probably it will be convincing enough because his/her case would have a more detailed reporting and attention. Actually, to include a patient in a research requires legally its consent in a proper signed form, hence to explain and adopt the PM approach may be an additional reason to convince the patient to enroll in the research.

However, another problem was reported by Chan et al. in (26): "we found that the reporting of trial outcomes in journals is frequently inadequate to provide sufficient data for interpretation and meta-analysis, is biased to favor statistical significance, and is inconsistent with primary outcomes specified in trial protocols. These deficiencies in outcome reporting pose a threat to the reliability of the randomized trial literature." Consequently, they highlight three lines of action: "First, protocols should be made publicly available— not only to enable the identification of unreported outcomes and post hoc amendments but also to deter bias. Ideally, protocols should be published online after initial trial registration and prior to trial completion. Although journals constitute one obvious modality for protocol publication, academic and funding institutions should also take responsibility in providing further venues for disseminating research information. Second, deviations from trial protocols must be described in the published articles so that readers can assess the potential for bias. Third, journal editors should not only consider routinely demanding that original protocols and any amendments be submitted with the trial manuscript but that this material should also be provided to peer reviewers and preferably be made available at the journal's Web site.

Finally, trialists and journal editors should bear in mind that most individual trials may well be incorporated into subsequent reviews. Outcomes that are mentioned in published articles, but are reported with insufficient data, may not always matter when interpreting a single trial report, but they can have an important impact on meta-analyses. Unreported outcomes are even more problematic for both trials and reviews. It is therefore crucial that adequate data be reported for pre-specified outcomes independent of their results. The increasing use of the Internet by journals may help to provide the space needed to accommodate such data."

In regards to the three lines of action proposed by Chan et al. in (26), if PM is simultaneously applied as a framework from bottom-up, that is, if it is used in every kind of research then it would be easier to perform systematic reviews and meta-analysis due to a standardized approach. More than that, because of the intrinsic requirement to correct and updated the project Plan, the first line of action would be satisfied by publishing all documents of the Initiation and Planning phases, which should include all adopted medical protocols. The second and third lines of action would be accomplished by making available all versions of all documents of the Initiation and Planning phases. If some information on those documents is classified, then it should be easy to simplify them to satisfy Chan et al.'s proposition.

Finally, if a significant part of the medical research community decides to adopt PM as a framework for doing medical research, then the development of an appropriate WEB-based software would provide a computational data-base, which could have functions to statistically support primary studies focused on treatment, diagnoses, tracking, prognoses and causality. Therefore, this system could speed up medical research while facilitating the physicians work, because it would be a unified computational framework. Additionally, there would be more and new kind of ready to use statistical data for supporting EBM researches. Also, due to a PM standardized approach, information retrieval techniques would be more effective in discovering new knowledge.

References

[1] (TDR), UNICEF/ UNDP/ World Bank/ WHO Special Programme for Research and Training in Tropical Diseases. Effective project planning and evaluation in biomedical research. *World Health Organization on behalf of the Special Programme for Research and Training in Tropical*, 2005.

[2] Feine JS, Wollin SD and Rashid F. Chapter 5: How to carry out succesful clinical studies: lessons from project management. *Statistical and Methodological Aspects of Oral Health Research.* Editors: Lesaffre E, Feine JS, Leroux B and Declerck D. John Wiley and Sons Ltd., 2009.

[3] Letouze P, and Teixeira MM. Project-based transplant management as a research statistical support. *Transplant Proc.* 2009, 41:852-854.

[4] MacNutt RA, Livingston EH. Evidence-Based Medicine Requires Appropriate Clinical Context. *JAMA.* 2010, 303:454-455.

[5] Heldman, K. PMP Project Management Professional Exam Study Guide. *Sybex,* 2009; 5 edition. ISBN-10: 0470455586.

[6] T, Greenhalgh. *How to Read a Paper: The Basics of Evidence-based Medicine.* Wiley-Blackwell, 2006; 3 edition. ISBN-10: 1405139765.
[7] Centre for Reviews and Dissemination, University of York. *Systematic Reviews - CRD's guidance for undertaking reviews in health care.* Layerthorpe : York Publishing Services Ltd, 2009. ISBN 978-1-900640-47-3.
[8] Montori VM, and GuyaTT GH. Progress in Evidence-Based Medicine. *JAMA.* 2008, 300:1814-1816.
[9] Higgins JPT, Green S (editors). *Cochrane Handbook for Systematic Reviews of Interventions Version 5.0.2 [updated September 2009].* The Cochrane Collaboration, 2009. Available from www.cochrane-handbook.org., 2009.
[10] Bossuyt PM, and Reitsma JB. The STARD Iniciative. *Lancet.* 2003, 361:71.
[11] Vandenbroucke JP, von Elm E, Altman DG, Gøtzsche PC, Mulrow CD, Pocock SJ, Poole P, Schlesselman JJ and Egger M. Strengthening the Reporting of Observational Studies in Epidemiology (STROBE): Explanation and Elaboration. *Ann Intern Med.* 2007, 147:W-163-94.
[12] Stroup JD, Berlin JA, Morton SC, Olkin I, Williamson GD, Rennie D, Moher D, Becker BJ, Sipe TA and Thacker SB. Meta-analysis of Observational Studies in Epidemiology - A Proposal for Reporting. *JAMA.* 2000, 283:2008-2012.
[13] Schulz KF, Altman DG and Moher D. CONSORT 2010 Statement: updated guidelines for reporting parallel group randomised trials. *BMJ.* 2010, 340:c332.
[14] Moher D, Hopewell S, Schulz KF, Montori V, Gøtzsche PC, Devereaux PJ, Elbourne D, Egger M and Altman DG. CONSORT 2010 Explanation and Elaboration: updated guidelines for reporting parallel group randomised trials. *BMJ.* 2010, 340:c869.
[15] Liberati A, Altman DG, Tetzlaff J, Mulrow C, Gotzsche PC, Ioannidis JPA, Clarke M, Devereaux PJ, Kleijnen J and Moher D. The PRISMA statement for reporting systematic reviews and meta-analyses of studies that evaluate healthcare interventions: explanation and elaboration. *BMJ.* 2009, 339:b2700.
[16] Moher D, LIberati A, Tetzlaff J and Altman DG. Preferred reporting items for systematic reviews and meta-analyses: the PRISMA statement. *BMJ.* 2009, 339:b2535.
[17] Guyatt GH, Oxman AD, Vist GE, Kunz R, Falck-Ytter Y, Alonso-Coello P and Schünemann HJ. GRADE: an emerging consensus on rating quality of evidence and strength of recommendations. *BMJ.* 2008, 336:924-926.
[18] Guyatt GH, Oxman AD, Vist GE, Kunz R, Falck-Ytter Y and Schünemann HJ. GRADE: what is "quality of evidence" and why is it important to clinicians? *BMJ.* 2008, 336:995-998.
[19] Guyatt GH, Oxman AD, Vist GE, Kunz R, Falck-Ytter Y, Liberati A and Schunemann HJ. GRADE: going from evidence to recommendations. *BMJ.* 2008, 336:1049-1051.
[20] Schünemann HJ, Oxman AD, Brozek J, Glasziou P, Jaeschke R, Vist GE, Williams Jr JW, Kunz R and Craig J. GRADE: grading quality of evidence and strength of recommendations for diagnostic tests and strategies. *BMJ.* 2008, 336:1106-1110.
[21] Guyatt GH, Oxman AD, Kunz R, Jaeschke R, Helfand M, Liberati A, Vist GE and Schünemann HJ. GRADE: Incorporating considerations of resources use into grading recommendations. *BMJ.* 2008, 336:1170-1173.
[22] Uniform Requirements for Manuscripts. *ICMJ.* [Online] October, 2008. www.icmj.org. [Cited May 6, 2010.]

[23] SPIRIT initiative. Reporting guidelines under development. *Enhancing the QUAlity and Transparency Of helath Research.* [Online] April, 2010. [Cited in: May 6, 2010.] http://www.equator-network.org/resource-centre/library-of-health-research-reporting/reporting-guidelines-under-development/reporting-guidelines-under-development/.

[24] Editorial. Strengthening the credibility of clinical research. *Lancet.* 2010, 375:1225.

[25] Taylor CF, Field D, Sansone SA, Aerts J, Apweiler R, Ashburner M, Ball CA, Binz PA, Bogue M, Booth T, Brazma A, Brinkman RR, Clark AM, Deutsch EW, Fiehn O, Fostel J, Ghazal P, Gibson F, Gray T, Grimes G, Hancock JM, Hardy NW, Hermjakob H et al. Promoting coherent minimum reporting guidelines for biological and biomedical investigations: the MIBBI project. *Nature Biotechnology.* 2008, Vol. 26, 8:889-896.

[26] Chan A-W, Hróbjartsson A, Haahr MT, Gøtzsche PC and Altman DG. Empirical Evidence for Selective Reporting of Outcomes in Randomized Trials. *JAMA.* 2004, 291:2457-2465.

[27] Hrynaszkiewicz I, Norton NL, Vickers AJ and Altman DG. Preparing raw clinical data for publication: guidance for journal editors, authors, and peer reviewers. *BMJ.* 2010, 340:c181.

[28] Graf C, Battisti WP, Bridges D, Bruce-Winkler V, Conaty JM, Ellison JM, Field EA, Gurr JA, Marx ME, Patel M, Sanes-Miller C and Yarker YE. Good publication practice for communicating company sponsored medical research: the GPP2 guidelines. *BMJ.* 2009, 339:b4330.

[29] Hopewell S, Dutton S, Yu LM, Chan AW and Altman DG. The quality of reports of randomised trials in 2000 and 2006: comparative study of articles indexed in PubMed. *BMJ.* 2010, 340:c723.

[30] von Elm E, Altman DG, Egger M, Pocock SJ, Gøtzsche PC and Vandenbroucke JP. The Strengthening the Reporting of Observational Studies in Epidemiology (STROBE) statement: guidelines for reporting observational studies. *Journal of Clinical Epidemiology.* 2008, 61:344-349.

In: Advances in Medicine and Biology Research, Vol. 14 ISBN: 978-1-61761-930-4
Editor: Leon V. Berhardt

Chapter VI

Genetic Counseling for Pompe Disease

***Lindsay Paull*[1, 2] *and Shawn Lipinski*[1]**
Division of Medical Genetics, University of Virginia Health System, Charlottesville, VA 22908[1]
Children's National Medical Center, Washington, DC 20010, USA[2]

Abstract

The complex nature of glycogen storage disease type II or Pompe disease requires a multi-disciplinary team approach. A genetic counselor trained in metabolic or neuromuscular disorders can be a valuable member of the health care team working with patients with Pompe disease. The roles and unique clinical skills of a genetic counselor working with Pompe disease are highlighted here. These roles include assisting with the confirmation of a diagnosis, educating the family about the disease, supporting the management of disease and disease progression, and facilitating the communication process between the patient and the health care team.

Introduction

The complex nature of glycogen storage disease type II or Pompe disease requires a multi-disciplinary team approach. A genetic counselor trained in metabolic or neuromuscular disorders can be a valuable member of the health care team working with patients with Pompe disease. The roles and unique clinical skills of a genetic counselor working with Pompe disease are highlighted here. These roles include assisting with the confirmation of a diagnosis, educating the family about the disease, supporting the management of disease and disease progression, and facilitating the communication process between the patient and the health care team.

Confirming a Diagnosis

Facilitating the diagnostic process is often the first role that a genetic counselor fulfills when working with this patient population. A genetic counselor trained in neuromuscular or metabolic diseases appreciates the clinical presentation of Pompe disease and can aid in the appropriate diagnostic testing.

Clinical Presentation

Pompe disease can be classified into infantile and late-onset forms based on the condition's severity and age of onset. "Classic" infantile Pompe disease is characterized by the rapid progression of cardiomyopathy, hypotonia, feeding problems and respiratory failure with death typically occurring at less than one year of age(1). The rapid physical decline of infantile Pompe disease necessitates a timely diagnosis with sensible and prompt intervention.

In contrast, the late-onset form has a slower disease progression characterized by respiratory insufficiency and skeletal muscle weakness. The prognosis of late-onset Pompe disease depends on the degree of enzyme deficiency, disease duration, rate of disease progression and respiratory involvement. Late-onset Pompe disease usually has little to no cardiac involvement and may present at any time from childhood into adulthood(2, 3).

For late-onset Pompe disease, a patient may initially present with difficulty breathing, difficulty rising from a seated position or difficulty climbing stairs due to progressive proximal muscle weakness. Patients may also complain of muscle pain and fatigue. Proximal muscle weakness in the lower extremities results in a characteristic abnormal gait. Dyspnea and orthopnea are initial presentations in one third of late-onset patients(2).

In comparison to late-onset, an infant often presents with poor feeding, failure to thrive, hepatomegaly and hypotonia. Severe cardiomegaly can be seen on chest x-ray. Elevated urinary hexose tetrasaccharide (Hex4) can be used to reflect muscle glycogen storage in both forms of Pompe disease and has close to 100% sensitivity for infantile Pompe cases (4, 5). Laboratory work up for both types of Pompe disease may also reveal elevated serum creatine kinase (CK), elevated levels of aspartate aminotransferase (AST), alanine aminotransferase (ALT), or lactate dehydrogenase (LDH)(6).

Testing

The testing approach for Pompe disease may vary between the infantile and late-onset forms and it is important to recognize the situational value of each type of testing. Diagnostic testing for suspected Pompe cases typically involves the measurement of alpha glucosidase (GAA) enzyme activity. Although measuring GAA enzyme in cultured skin fibroblasts has been considered the gold standard for a diagnosis of Pompe disease, the culturing of fibroblasts can take up to 6-8 weeks and there is often not enough time with the infantile form to wait for the completion of this process(7). Thus, other approaches such as enzyme testing from a dried blood spot or a peripheral blood leukocyte sample coupled with elevated Hex4 levels can provide initial GAA quantification but should not be considered confirmatory. The

Pompe Disease Diagnostic Working Group recommends that a second test such as molecular testing or the measurement of GAA in a fibroblast culture be done to support the final diagnosis(3). A suspected diagnosis can enable earlier management and the push for a more prompt diagnosis has led the majority of practitioners to seek genetic testing as a means of confirming a diagnosis.

A muscle biopsy can provide yet another way of measuring GAA enzyme activity. However, the limitations of a muscle biopsy should be appreciated before this invasive form of testing is pursued. An experienced genetic counselor is trained to recognize both the benefits and limitations of a biopsy and can therefore help to inform the patient (or family) of the test's utility. For example, not only is the procedure invasive but it also involves a serious cardio-respiratory risk for the infantile Pompe patient due to the necessary administration of anesthesia(3, 8). Secondly, inaccurate measurements of GAA activity in a muscle biopsy could result from compromised sample integrity specifically if the sample does not undergo snap freezing or is handled improperly. Finally, different biopsy sites may vary in the amount of glycogen accumulation especially in adult patients. Thus, the lack of glycogen accumulation in a biopsy sample should not formally exclude a Pompe diagnosis(7).

Conversely, the benefits of a muscle biopsy include the ability to promptly determine the amount of GAA enzyme activity and sometimes glycogen accumulation in the muscle tissue,. Furthermore, a muscle biopsy is invaluable in a diagnostic workup if a patient is presenting with myopathy of unknown etiology and a GAA level can be measured as part of the differential workup.

During the diagnostic process, a patient or family may express frustration or confusion as to why an invasive muscle biopsy procedure was needed if the diagnosis could have been discovered by a simple blood test. A genetic counselor can use his or her counseling skills to handle these types of questions and emotional responses. A counselor may first assess whether the patient's frustration is originating from a lack of information or communication surrounding the diagnosis, if it is a difficulty coping with the diagnosis, or if other emotional matters are manifesting as frustration about the biopsy. The counselor can assess for the source of frustration and can also clarify the diagnostic process. Specifically, the counselor can help to explain that Pompe disease falls into a broad category of differential diagnoses in which a muscle biopsy may be used to evaluate a vast array of dystrophies, congenital or metabolic myopathies, motor neuron disorders or neuromuscular junction disorders that overlap in clinical presentation.

Molecular analysis of the only known gene implicated in Pompe disease, the GAA gene which falls on chromosome 17, allows the medical team to both confirm the biochemical diagnosis and to provide the patient and family with valuable information for family planning and genetic counseling purposes. An explanation of the benefits and limitations of molecular testing such as the test's sensitivity and specificity or the difference between targeted mutation testing and full sequencing of the gene should be provided and is often a role of the genetic counselor. One should educate the family about the potential outcomes of genetic testing and effectively communicate positive results, negative results or variants of unknown significance. Additionally, one should be prepared to discuss the implications of genetic testing on other family members including carrier risks and recurrence risks for future pregnancies. A genetic counselor can assist the physician with this discussion including the potential outcomes and the implications of genetic testing.

Mutations in Pompe disease can vary between the infantile and late-onset forms. Mutations in the infantile form, such as nonsense mutations, commonly result in the complete absence of GAA enzyme whereas late-onset individuals often have partial enzyme activity caused by a combination of mutations allowing for some production of GAA. Polymorphisms and silent mutations have also been found in the GAA gene(9). Kroos and colleagues noted over 200 mutations in the GAA gene to date and highlight a panel of common mutations in various ethnic and age groups(10). Common mutations include the IVS 1(-13 T>G) splice mutation found in about half of the late-onset Caucasian patients, the Asp645Glu mutation found in most Taiwanese infants, the Arg854X nonsense mutation found in African or African-American infants and the del525T and del exon 18 commonly seen in Dutch infants(3, 10-12).

No matter which diagnostic test is being ordered, a unique skill of the genetic counselor during the testing process is the recognition of a laboratory's reliability. In the case of Pompe disease, a genetic counselor can assist the medical team by finding a laboratory that is properly equipped to quantify the GAA enzyme or one that provides the appropriate molecular testing. As most blood samples contain maltose glucoamylase (MGA), an enzyme that can mask the deficiency of GAA and provide false negative results, laboratories experienced in testing for GAA enzyme appreciate the use of acarbose at the appropriate pH to counteract the competing properties of MGA(7). In addition, a genetic counselor will acknowledge the type of test(s) offered by the laboratory (molecular or biochemical), the turn-around time of each test, the test sensitivity and specificity, and the laboratory's billing and submission requirements.

Interpretation of Results

Upon receipt of molecular or biochemical results, genetic counselors are trained in the proper interpretation of those results. The interpretation of GAA enzyme levels in a Pompe patient may have implications for phenotypic presentation and therapeutic approaches as the disease progression often correlates with the amount of enzyme activity present. No GAA enzyme in a patient typically results in a more severe clinical picture(3). In addition, if a patient with infantile Pompe disease has little to no residual enzyme he or she may be negative for cross-reacting immunologic material ("CRIM negative"), and enzyme replacement therapy may not be as effective without concomitant immune modulation therapy(13). Specific testing for CRIM status is currently available on a research basis only. Because of the clinical implications of the recent CRIM status data it is reasonable to expect future clinical test availability.

When interpreting results, it is important to recognize the complexities of genotype-phenotype relationships within a Pompe family. Generally, the phenotypic presentation for infantile Pompe disease is similar among affected family members(14). However, Hoefsloot and colleagues(11) noted that while a number of mutations seen in homozygosity may suggest a genotype-phenotype correlation, the existence of case reports of both infantile and late-onset Pompe disease in the same family suggests strong caution in predicting outcome based on disease severity in relatives. In late-onset cases, variability in the age of onset has been seen for individuals with the same gene mutation for both familial and non-familial cases(15, 16). Thus, caution should be taken when counseling about genotype-phenotype correlations.

Once a diagnosis is confirmed in the proband and a known familial mutation has been identified, testing other at-risk family members can occur. The testing of family members may include targeted mutation analysis in a sibling, a parent or another at-risk relative. It is important to note that enzyme testing does not accurately identify carrier status, as variation in normal individuals may occur (17) .

In addition to assessing risk, families should be provided with information about prenatal diagnostic options, including enzyme and molecular testing through amniocentesis and chorionic villus sampling. Discussing prenatal options is a common and unique role of a genetic counselor. In the case of Pompe disease, GAA enzyme activity can be measured in both amniocytes and in uncultured chorionic villus samples(7). This is especially helpful for the infantile form of Pompe disease if a familial mutation was never tested for or identified. Enzyme or mutation testing on cord blood is yet another approach for a prompt diagnosis after a potentially affected child is born, and should be considered part of the diagnostic discussion. Furthermore, alternative conception or pregnancy options including ooycyte or sperm donation, *in-vitro* fertilization with pre-implantation genetic diagnosis (IVF with PGD), and adoption, should be discussed in a sensitive manner.

Education and Psychosocial Counseling for Pompe Disease

Educating about Pompe disease necessitates an appreciation of the natural history of the condition, the genetics and inheritance, and the available treatment and management options. Baker, Schuette and Uhlmann have divided patient education into seven parts 1) information gathering, 2) diagnosis, 3) risk assessment, 4) information giving, 5) psychological assessment and counseling, 6) help with decision making, and 7) ongoing client support(18).

Information gathering often begins during the diagnostic process. One should spend time learning about the patient's story and a genetic counselor can help the busy physician start this process. The approach that a genetic counselor takes to gather the patient's information may differ in comparison to other health care providers in that a counselor may focus more on psychosocial issues. For example, a counselor may spend more time inquiring about the patient's diagnostic journey by utilizing open-ended questions and inquiring about their emotional reactions along the way. A counselor may also spend extra time addressing a patient's understanding and perception of his condition and current health state as well as his experiences with the medical team. By acknowledging a patient's perceptions and attitudes, misconceptions or concerns about the condition can be addressed. By listening to a patient's medical journey, one can gather information about the stressors that impact the patient or family and make note of the patient's support systems. This information allows the genetic counselor or other health care professional to appreciate the factors that may influence a patient's coping style and also gives a foundation for how and where to best assist the family as they learn to adapt to the condition.

The collection of a thorough family history or pedigree is a valuable way to gather information about health and psychosocial issues in the family. Additionally, it sets the stage for discussion of inheritance risks, family dynamics and attitudes, and for addressing decision-making topics.

Appendix 1. Example Questions for a Comprehensive Pompe History.

Component of History	Primary example	Supportive example
History of Presenting Illness	1. Understanding the patient's journey	a. When did the patient first notice symptoms? b. How was the diagnosis given and by whom? c. How familiar is the patient with Pompe disease?
	2. Review activities of daily living	a. How has the patient's health condition affected his daily activities?
Past Medical History	1. Has the patient been diagnosed with a neuromuscular disorder or been evaluated by a neurologist?	a. Has the patient's mobility affected his/her quality of life? b. Does the patient use mobility or assistive devices? c. Is the patient involved with physical therapy?
	2. Has the patient been diagnosed with a pulmonary disorder or been evaluated by a pulmonologist?	a. Does the patient have any respiratory distress or issues (asthma, COPD[1], etc.)? b. Has the patient had any pulmonary function testing (upright and/or supine)? c. Has the patient had a sleep study test for evaluation of sleep apnea? d. Dose the patient use BiPAP[2] or CPAP[3]?
	3. Has the patient been evaluated by a cardiologist?	a. Has the patient had an echocardiogram or electrocardiogram?
	4. Has the patient been evaluated by genetics or seen a genetic counselor?	a. Has the patient had any genetic testing? b. Has the patient had any biochemical studies (i.e. enzyme analysis, antibody testing, etc.)?
	5. Has the patient initiated enzyme replacement therapy (ERT)?	a. Does the patient know his current dose of ERT? b. Who was/is the ordering physician for the patient's ERT? c. When was the patient's last infusion?
	6. Has the patient had any prior hospitalizations?	
	7. Does the patient have any concomitant medical conditions (that may have an adverse synergist effect with Pompe disease)?	
	8. Has the patient been evaluated by other specialists?	
	9. Is the patient involved in any research trials?	

Medications	1. Is the patient taking any medications known to cause myopathy or exacerbate symptoms of neuromuscular disease (i.e. statins, narcotics, etc.)? 2. Is the patient currently on enzyme replacement therapy (ERT)?	a. See #5 a-c above under Past Medical History
Allergies	1. If the patient is on ERT has the patient had an allergic reaction to ERT (i.e. uticaria, anaphylaxis, fever)?	a. Is the patient being pre-medicated before infusions?
Family History	1. Has anyone in the family been diagnosed with Pompe disease? 2. Has anyone been diagnosed with a chronic illness or known genetic condition?	a, Has anyone in the patient's family had a history of pulmonary disease? b. Have there been any deaths at a young age in the patient's family? c. Has anyone in the patient's family had any muscular or neuromuscular issues? d. Has anyone in the family had any cardiac issues?
Social History	1. What is the patient's current living environment? 2. Where does the patient work or go to school? 3. Has the patient applied for or considered applying for disability? 4. What is the patient's support system 5.Transportation	a. What is the patient's housing situation? Does the patient live on the ground level, have stairs in the home, etc.? Does the patient own a home or rent an apartment? b. Does the patient have any children/dependents? c. What supportive aids or devices does the patient have? a. What type of work does the patient do? b. Does the patient have any physical limitations at work (i.e. lifting, bending, getting up from chair?) c. Does the patient require any special assistance at school or work? a. Is the patient in a relationship? Does the partner know about the patient's diagnosis? b. What is the patient's friend/support network? a. Who takes the patient to the appointments?

Appendix 1 (Continued)

Review of Systems	1. Constitutional	a. Fatigue?
	2. HEENT	a. Otitis Media? b. Myringotomy and tympanostomy tubes? c. Speech issues? d. Eyelid drooping? e. Feeding issues?
	3. Cardiorespiratory	a. Chest pain/pressure? b. Dyspnea on exertion? c. Orthopnea? d. Postural nocturnal dyspnea? e. Edema? f. Lightheadedness or syncope? g. Palpitations?
	4. Musculoskeletal	a. Myopathy/Myalgias? b. Scoliosis? c. Gait disturbances, tripping, falling?
	5. Neurology	a. Headaches?
	6. Psychiatric	a. Depression? b. Anxiety?

[1] COPD- Chronic Obstructive Pulmonary Disease
[2] BiPAP- Bilevel Positive Airway Pressure
[3] CPAP- *Continuous Positive Airway Pressure*

Questions relating to the clinical features of Pompe disease can be asked at the time of pedigree construction and may reveal potential at-risk family members.

Relatives of the proband who are at risk for late-onset Pompe disease may present with muscle changes or respiratory issues elicited when taking the family history. Glimpses into family dynamics are often revealed at this time and appropriate empathic responses help the counselor maintain or gain rapport with the patient and family.

Importantly, one should not overlook the potential for other conditions that may be present in the family independent of Pompe disease and should be prepared to counsel the family about these potential risks.

Appendix 1 includes a list of suggested questions one could use during a comprehensive history in which there is suspicion of Pompe disease or known diagnosis of Pompe disease.

The education process would not be complete without acknowledging the autosomal recessive inheritance pattern of Pompe disease. This discussion includes a risk assessment for future pregnancies and for potential carrier family members. Appropriate risk counseling should be provided by way of explaining genes and chromosomes, assessing what the patient believes his or her risk is and presenting numeric data in various ways, such as a fraction (1 in 4) and a percent (25%) or by making an analogy to better ensure that the patient conceptualizes and appreciates what that number means to him or her.

As is the case for many autosomal recessive conditions, the disease may not be present in other family members and the diagnosis comes as a surprise to the parents or patient. Anticipating potential reactions about a new diagnosis allows one to provide appropriate responses and necessary psychosocial counseling. A counselor or health care professional needs to appreciate that people have different coping styles to difficult news. Counseling should be adjusted based on these different styles. In addition, the environment and method in which a diagnosis is presented is an important part of the education process. A diagnosis should be provided in a private room with minimal distractions, presented in an empathetic way both verbally and nonverbally, and in a manner that reflects informational reliability and accuracy so that the patient can trust one's professional knowledge. Information on support groups or educational resources can be provided at this time to help the patient learn more about the condition and to feel that he or she is not alone.

Appendix 2. Pompe Resources.

1. Acid Maltase Deficiency Association, Inc. (AMDA)
P.O. Box 700248
San Antonio, Texas 78270-0248
(210) 494-6144 or (210) 490-7161
Fax: (210) 490-7161 or (210) 497-3810
E-mail: tianrama@aol.com
Website: http://www.amda-pompe.org/
2. Association for Glycogen Storage Disease
P.O. Box 896
Durant, Iowa, USA, 52747-9769
563-785-6038
Website: http://www.agsdus.org/
3. Global Organization for Lysosomal Diseases (GOLD)
3 Albion Rd, Chalfont St Giles, Buckinghamshire, HP8 4EW Website: http://www.goldinfo.org/
4. International Pompe Association +31 35 5480461 Website: http://worldpompe.org/
5. National Organization of Rare Disorders (NORD)
55 Kenosia Avenue PO Box 1968
Danbury, CT 06813-1968
(203) 744-0100 or TDD Number: (203) 797-9590
Fax: (203)798-2291
Website: www.raredisorders.org
6. Pompe Community
Genzyme Corporation
500 Kendall Street
Cambridge, MA 02142
Office: (617) 768-9000 or Toll Free: (800) 745-4447
Website: www.pompe.com/en
7. United Pompe Foundation
5100 N. Sixth Street #119
Fresno, CA 93710
Phone: (559) 227-1898
Fax: (559) 227-1898
Email: mailto:david@unitedpompe.com
Website: http://www.unitedpompe.com/

Support groups may also provide a sense of empowerment allowing the patient to have some control over the condition and its management. Appendix 2 includes a brief listing of Pompe patient resources.

Education about the natural history and genetics of Pompe disease is a necessary part of the genetic counseling process and may occur at the time of initial consideration or once the diagnosis is established. Information about Pompe disease should be presented in a culturally and intellectually sensitive manner. The American Society of Human Genetics defined genetic counseling as "a communication process which deals with the human problems associated with the occurrence or risk of occurrence of a genetic disorder in a family"(19). Thus the information that one provides about Pompe disease should be communicated in a way that respects the family structure and dynamics, cultural and religious beliefs as well as appreciating the educational level of those receiving the information. A genetic counselor often plays an important role in this education process.

Many patients with Pompe disease, especially those with the late-onset type, have experienced a diagnostic odyssey such that they may have gone many years without having a diagnosis to explain their symptoms of muscle fatigue, weakness and breathing issues(2, 15). Thus, confirming the diagnosis is often a turning point in their psychosocial adaptation to their disease. Adaptation to a chronic condition can be a lifelong process especially since the progression of the disease may require continual changes in lifestyle. The counseling of potential lifestyle changes and how to cope with them may be needed on a number of different occasions. For example, counseling may be needed to help the patient accept the use of assistive devices and supportive therapies. This topic is discussed in more depth later in the chapter.

Counseling a patient or family about living with a chronic genetic disease is an important skill and another unique role of the genetic counselor. Psychosocial counseling is needed in a variety of situations with Pompe disease. For example, parents who lose a child to infantile Pompe disease may need the support of a genetic counselor trained in grief counseling as they experience the different stages of loss. Genetic counseling can offer a pregnant female with Pompe disease and concomitant muscle weakness support and information about pregnancy risks and management choices. Or genetic counseling may benefit the overwhelmed parents who need assistance in managing the many doctors' appointments and services for their child with the disease.

A genetic counselor's approach to a counseling session is founded on Carl Roger's belief of the person-centered method. The person-centered method respects a patient's autonomy and his or her ability to make independent decisions. Rogers believed that a counselor can empower a patient to make autonomous decisions but to do so the counselor must possess the following three attitudes in practice: unconditional positive regard or positive view of the client, empathy, and counselor genuineness, which he describes as a safe and open relationship with the client (20). These attitudes support the idea that a genetic counselor will direct the process of decision making but is cautious not to direct the outcome (21).

Counseling about a chronic disease is challenging and may take place over multiple sessions as a family's situation and understanding evolve. Discussions often involve complex topics such as coping with stigma, the "loss" of a healthy child, explaining physical differences to others and defining the roles of care givers in the family. As rapport builds and a relationship between the patient and counselor develops, the genetic counselor may have multiple discussions with the family or patient to tackle these challenging topics. In situations where long-term therapy is necessary, a genetic counselor may choose to refer to a psychologist or another mental health counselor if the topics being addressed require more consistent and thorough mental health counseling sessions.

Treatment and Management Options

The management and treatment of Pompe disease involves both supportive therapies as well as disease-specific treatment. At the current time, both forms of Pompe disease require lifelong therapy and necessitate the continual involvement of a metabolic specialist, neurologist or medical professional experienced with the disease. A genetic counselor or nurse familiar with lysosomal storage diseases is often the primary point of contact for coordinating treatment and for providing information about management options. Enzyme replacement therapy (ERT) consisting of recombinant human GAA is currently the only disease-specific treatment and it has been shown to improve both cardiac and skeletal muscle strength and function and prolong overall survival in patients with Pompe disease(22-24). Limitations to ERT include the unsatisfactory distribution of the recombinant enzyme to the muscle cells and the formation of antibodies that may interfere with the uptake of enzyme (22). However research is ongoing for new therapies. Research studies have evaluated the effectiveness of hematopoietic stem cell transfer of α-glucosidase in mice but these studies have not yet reached human clinical trials(25). Research is also being performed in the small molecule arena, looking at the effectiveness of oral chaperone therapy. This type of therapy, which enhances the transport and efficacy of the altered GAA enzyme, hopes to eliminate the need for recurrent infusions. However, chaperone therapy would be a potential treatment option only to those patients with residual enzyme present (26).

The patient with Pompe disease should be informed about all available therapies, whether the therapy is commercially available or in research studies. Many genetic counselors who work with lysosomal storage disorders have research training and are able to coordinate the consenting of a patient into a clinical trial. Thus, many genetic counselors are able to help arrange for patients to start on commercial therapy or identify appropriate clinical trials open to enrollment. Genetic counselors may help initiate and coordinate clinical research studies at their own site and coordinate the surveillance and treatment of those involved in research as part of the clinical trial team.

Administration of Enzyme Replacement Therapy

For a genetic counselor to assist in the initiation of enzyme replacement therapy, he or she must have an understanding of the following areas: an explanation of the benefits, risks and limitations of the therapy, appreciation of dosages and drug stability, coordination of HIPAA and other consent agreements, addressing insurance companies with authorization issues, writing statements of medical necessity or intent to treat letters, coordinating infusion sites (home care v. hospital infusions) and coordinating the shipment of the drug with the drug supplier and manufacturer to the infusion site. Lastly, one should be aware of the potential adverse events or reactions to therapy and provide the necessary information to the patient and infusion team to recognize these events such that a plan with the doctor and infusion team is prepared should a reaction occur.

The initiation of ERT should occur in a specialized medical care facility equipped with experienced medical staff as well as appropriate infusion and life support equipment. Presently, all patients with Pompe disease receive their ERT in this managed care setting. In the future, patients may be allowed to transfer their infusions to a home location similar to patients receiving ERT for other lysosomal storage disorders. As is the case for other disorders treated with ERT, before a patient transitions to home care infusions he or she will likely need the drug infused in a hospital or medically appropriate setting at the recommended dose for a set number of consecutive infusions without the occurrence of an adverse event. The transition of a patient to home care will require a counselor or other health care professional to seek out a home care nursing agency that is willing to work with Pompe disease and be trained on the ERT infusions. The genetic counselor or health care professional will then regularly communicate with the home care agencies to coordinate recommended laboratory studies and inquire about the patient's health status and response to infusions.

Issues such as the implementation of dose adjustments, changes in insurance companies or nursing staff, and coordination of labs require the counselor to have frequent interactions with the patient and family. As ERT is an expensive treatment, a counselor may work with a social worker or the drug manufacturer to seek out financial assistance for the patient. Often a genetic counselor has continual interaction with the patient and family thus becoming the main point of contact for coordination of treatment, management and for overall support. In this way, a genetic counselor facilitates the communication process between the patient and the health care team.

Assessing Disease Progression

Standard of Care and Multisystem Evaluation

Pompe disease is a lifelong chronic and progressive condition. Disease progression can be assessed by the doctor coordinating the patient's care by making comparisons to the patient's baseline evaluation and by meeting management and surveillance recommendations. The genetic counselor can assist in collecting the necessary medical records and review these with the doctor and patient.

The American College of Medical Genetics (ACMG) has provided a set of guidelines for the diagnosis and management of Pompe disease(3). The ACMG guidelines include detailed recommendations on appropriate diagnostic approaches, specifics about clinical tests and assessments, and information about available support groups for the Pompe community. Additional management recommendations for Pompe disease can be found in the 2009 Brazilian Consensus on the Management of Pompe Disease(27) and/or the Management and treatment of glycogenosis type II by Bembi and colleagues published in Neurology in 2008(28).

A Registry of patients with Pompe disease was created in 2004 to track the natural history and progression of the disease and to gain a better understanding of the effects of ERT. The information provided in the Registry allows providers and patients to discuss health status and set therapeutic goals based on the patient's disease state. The compilation of multi-system evaluations provides the patient with a complete "picture" of disease status and allows the patient and health care team to appreciate where therapeutic intervention may be needed. The Registry is open to all individuals with Pompe disease regardless of treatment status. It provides health care providers with a recommended schedule of assessments for management and monitoring of disease progression. These recommendations were compiled by a panel of experts within the Pompe and lysosomal storage disorder specialties. The Registry is overseen by a Board of Physician Advisors and has the world's largest collection of Pompe disease data with approximately 30 countries represented (personal communication with Julie Glick, Genzyme Registry Manager). The Pompe Registry can be accessed at www.pomp eregistry.com.

Referring to Ancillary Specialties

As a patient's disease progression is assessed, referrals to ancillary specialties are often needed. The patient typically has one specialist provider who acts as their primary coordinator of care. Ancillary specialists most often include Genetics, Neurology, Cardiology, Pulmonology, Orthopedics, Audiology, Social Work, Physical Therapy and Psychology.

Making appropriate referrals is dependent on the ancillary provider's comfort and familiarity with the disease and/or willingness to learn about the condition. Since Pompe disease is relatively rare, many health care providers are unfamiliar with the condition and may need education from an experienced individual such as a genetic counselor. One of the professional competencies of a genetic counselor as defined by the American Board of Genetic Counseling states that a counselor can plan, organize, and conduct public and professional education programs on human genetics, patient care, and genetic counseling issues(29). Thus a genetic counselor working with Pompe disease should be competent in teaching aspects of the disease to other professionals including features of disease progression, disease-specific evaluations and management, and modes of inheritance.

Once a referring specialist is selected and the evaluation complete, it is often the role of the genetic counselor or nurse to obtain the evaluation records. These reports are critical for the coordinating specialist to assess the patient's disease course. The collection of a patient's medical information requires the genetic counselor to have both organizational and time management skills. Often a genetic counselor establishes a familiarity with the specialist and his or her team through the process of education, requesting information and coordinating evaluations.

By acting as a central point of contact for the patient the genetic counselor can help to provide relief and alleviate the pressure and time required of the patient in managing his or her medical care. This assistance may lessen the stress related to a patient's condition and provides the patient with an opportunity to focus on other areas of self care such as spending time with friends and family, hobbies, or education and community involvement.

Managing Disease Progression and Follow up

Assistive Therapies, Aids and Devices

A well-informed patient understands the clinical course of Pompe disease and can better anticipate the potential changes in their physical state over time. Understanding the disease process and implications thereof facilitates management discussions between the patient and health care team.

The treating physician may raise the question of physical therapy for the patient as part of their disease management. Physical therapy has been recommended in certain instances to prevent the onset of contractures caused by the condition's excessive muscle weakness and chronic positioning (30). Low impact exercises such as water therapy may improve muscle strength and fatigue. Daily stretching is recommended for those at risk of contractures (30). Although some degree of physical therapy may assist with meeting developmental milestones, increasing strength and endurance, and decreasing the risk of contracture, physical therapy is met with caution. Over-exertion may result in increased leakage of glycogen from lysosomes or lysosomal rupture during muscle contraction, contributing to overall muscle degeneration and increased weakness(31). Thus, a conservative approach should be employed regarding physical activity in patients with Pompe disease until more definitive trials can be conducted. Furthermore, osteopenia and osteoporosis have been noted in infantile and late-onset Pompe disease and should be considered when applying weight bearing forces in exercise(3, 30). These clinical features can be addressed by periodic dual energy x-ray absorptiometry (DEXA) scans.

Occasionally, a patient will be reluctant in accepting disease progression and the need for supportive therapies. For example, the intervention of respiratory therapy such as CPAP or BiPAP may be difficult for a patient to accept but may be essential in decreasing difficulty breathing and fatigue and ultimately improve an overall quality of life. Depending on the severity of the patient's respiratory involvement, therapy may include the use of respiratory support only during the day, at night, or both. Patients with a severe presentation may require full time ventilator support by tracheostomy. The reluctance to agree to respiratory therapy is an active form of denial being exhibited by the patient. A counselor can identify this denial, label it for the patient, and make an appropriate psychosocial intervention. In a situation such as this, it is the goal of the counselor to help the patient understand that while agreeing to the assistive devices is an active acknowledgement of their disease state, it is also allowing the patient to take a pro-active role in their health outcome.

Other therapies commonly involved in the care of Pompe disease include occupational therapy to assist the patient with daily tasks and psychosocial therapy to help manage the emotional stressors associated with a chronic condition. Finally, feeding and communication difficulties caused by a decline in oral motor function and/or macroglossia may require speech and language services and the aid of a nutritional specialist especially for children with Pompe disease.

Progressive muscle weakness often necessitates the use of assistive mobility devices and aids such as walking devices, canes, wheelchairs and supportive chairs for driving or transportation. Adjustments are also often needed in the home. The installation of support bars or grips, raised toilet seats, incorporation of assistive technology, and/or moving to a ground level living environment have been helpful adjustments for many patients with progressive muscle disease. Other assistive aids include orthotic devices such as casts, splints or adaptive equipment to correct the risk of contractures(30). Hearing loss has been documented in some cases of infantile Pompe disease and thus hearing aids may be needed for some individuals(3).

The numerous assistive therapies, assistive aids and clinical appointments that are part of the management of Pompe disease may be overwhelming for the families or patient. A genetic counselor can offer assistance in the coordination of appointments to specialty clinics in a timely manner. In addition, a genetic counselor can help provide informational resources about assistive support. A genetic counselor's support in the coordination of these efforts can help the patient to acknowledge the lifestyle changes that occur with the disease and help facilitate the adaptations that are needed in response to those changes.

The genetic counselor can also reflect with the patient on the emotional impact that the disease progression has made on him or her. Addressing impact and adaptation issues of a chronic disease is a unique role of a genetic counselor and one that often requires ongoing communication and follow-up discussions. The experienced genetic counselor can explain the benefits of ancillary specialist appointments and through the collection of assessments acknowledge evidence of improvement or decline in a patient's health and/or success in reaching therapeutic endpoints. Having a better understanding of disease course and effects of therapy can encourage compliance and improve the quality of life of a patient. A counselor may also provide a patient with a connection to other families with Pompe disease who are willing to address questions, share experiences, and provide a supportive ear. The counselor can facilitate these connections in a medically appropriate way that respects HIPAA and patient confidentiality.

Pediatric to Adult Transition

For those patients diagnosed with Pompe disease as young adolescents, transitioning from pediatric to adult health care is a critical juncture for both them and the health care team. Freed and Hudson suggested that a patient's cognitive function and the progressive nature of the health condition are two primary factors affecting the pediatric to adult health care transition(32). One should appreciate the cognitive level of the patient and help to educate and guide the patient toward adult care responsibilities at an appropriate intellectual level. A discussion about health care transition and taking responsibility for one's Pompe disease should promote the patient to reflect on his or her physical health state, encourage the patient to appreciate the infusion procedures, recognize adverse events to therapy, and identify and communicate between adult providers including a primary care physician and ancillary specialists(33). Emphasis on consistency with appointments and infusions, being an advocate (for oneself and others with the disease), and recognizing one's capabilities and limitations are especially important for the patient as he or she prepares to become more independent. A genetic counselor can assist with the successful transition of care from pediatrics to adult by discussing the transition process with the family and helping to seek out adult specialists familiar with the disease.

Conclusion

The diagnosis and management of Pompe disease requires a large multi-disciplinary team approach. Advances are ongoing to understand the complex course of Pompe disease in an effort to identify and refine the appropriate therapeutic approaches. The future promises an increased understanding of the disease as well as advancements in therapies and assistive technology for patients. Each of these areas of advancement has implications for genetic counselors. Having a choice of treatments available will require counselors to offer appropriate education and help facilitate the patient's decision-making process about his or her therapy. Perhaps one of the most important future roles of a genetic counselor is the continuation of educating health care professionals. Specifically a genetic counselor will continue to educate about the disease and how best to address a new diagnosis in a way that lessens the emotional impact and heightens communication between the patient and the health care provider.

The roles highlighted in this chapter reflect the many clinical skills that a genetic counselor can contribute when working with the Pompe population. Genetic counselors experienced with lysosomal storage disorders bring a unique combination of genetic knowledge and counseling expertise. This combination provides the patient with access to genetic and medical education, assistance with treatment and management as well as psychosocial services all in one setting. Their unique combination of skills makes a genetic counselor an essential and valuable contributor to the health care team.

Reviewed by

Jennifer A. Sullivan, MS, CGC Duke University Medical Center

Supported by

The Univeristy of Virginia's Genetic Counseling Fellowship in Lysosomal Storage Disorders is supported by Genzyme Corporation.

References

[1] Kishnani PS, Hwu WL, Mandel H, Nicolino M, Yong F, Corzo D. A retrospective, multinational, multicenter study on the natural history of infantile-onset Pompe disease. *J Pediatr* 2006;148(5):671-676.

[2] Diagnostic criteria for late-onset (childhood and adult) Pompe disease. *Muscle Nerve* 2009;40(1):149-60.

[3] Kishnani PS, Steiner RD, Bali D, Berger K, Byrne BJ, Case LE, et al. Pompe disease diagnosis and management guideline. *Genet Med* 2006;8(5):267-88.

[4] An Y, Young SP, Kishnani PS, Millington DS, Amalfitano A, Corz D, et al. Glucose tetrasaccharide as a biomarker for monitoring the therapeutic response to enzyme replacement therapy for Pompe disease. *Mol Genet Metab* 2005;85(4):247-54.

[5] Young SP, Stevens RD, An Y, Chen YT, Millington DS. Analysis of a glucose tetrasaccharide elevated in Pompe disease by stable isotope dilution-electrospray ionization tandem mass spectrometry. *Anal Biochem* 2003;316(2):175-80.

[6] Winkel LP, Hagemans ML, van Doorn PA, Loonen MC, Hop WJ, Reuser AJ, et al. The natural course of non-classic Pompe's disease; a review of 225 published cases. *J Neurol* 2005;252(8):875-84.

[7] Winchester B, Bali D, Bodamer OA, Caillaud C, Christensen E, Cooper A, et al. Methods for a prompt and reliable laboratory diagnosis of Pompe disease: report from an international consensus meeting. *Mol Genet Metab* 2008;93(3):275-81.

[8] Ing RJ, Cook DR, Bengur RA, Williams EA, Eck J, Dear Gde L, et al. Anaesthetic management of infants with glycogen storage disease type II: a physiological approach. *Paediatr Anaesth* 2004;14(6):514-9.

[9] Martiniuk F, Mehler M, Tzall S, Meredith G, Hirschhorn R. Sequence of the cDNA and 5'-flanking region for human acid alpha-glucosidase, detection of an intron in the 5' untranslated leader sequence, definition of 18-bp polymorphisms, and differences with previous cDNA and amino acid sequences. *DNA Cell Biol* 1990;9(2):85-94.

[10] Kroos M, Pomponio RJ, van Vliet L, Palmer RE, Phipps M, Van der Helm R, et al. Update of the Pompe disease mutation database with 107 sequence variants and a format for severity rating. *Hum Mutat* 2008;29(6):E13-26.

[11] Hoefsloot LH, van der Ploeg AT, Kroos MA, Hoogeveen-Westerveld M, Oostra BA, Reuser AJ. Adult and infantile glycogenosis type II in one family, explained by allelic diversity. *Am J Hum Genet* 1990;46(1):45-52.

[12] Labrousse P, Chien YH, Pomponio RJ, Keutzer J, Lee NC, Akmaev VR, et al. Genetic heterozygosity and pseudodeficiency in the Pompe disease newborn screening pilot program. *Mol Genet Metab* 2010;99(4):379-83.

[13] Kishnani PS, Goldenberg PC, DeArmey SL, Heller J, Benjamin D, Young S, et al. Cross-reactive immunologic material status affects treatment outcomes in Pompe disease infants. *Mol Genet Metab* 2010;99(1):26-33.

[14] Smith WE, Sullivan-Saarela JA, Li JS, Cox GF, Corzo D, Chen YT, et al. Sibling phenotype concordance in classical infantile Pompe disease. *Am J Med Genet A* 2007;143A(21):2493-501.

[15] Hagemans ML, Winkel LP, Van Doorn PA, Hop WJ, Loonen MC, Reuser AJ, et al. Clinical manifestation and natural course of late-onset Pompe's disease in 54 Dutch patients. *Brain* 2005;128(Pt 3):671-7.

[16] Kroos MA, Pomponio RJ, Hagemans ML, Keulemans JL, Phipps M, DeRiso M, et al. Broad spectrum of Pompe disease in patients with the same c.-32-13T->G haplotype. *Neurology* 2007;68(2):110-5.

[17] Tinkle B, Leslie N. Glycogen Storage Disease Type II (Pompe Disease). *In. Seattle: University of Washington;* 2008.

[18] Baker D, Schuette J, Uhlmann W. A guide to genetic counseling. New York: Wiley-*Liss Inc*.; 1998.

[19] Genetic counseling. *Am J Hum Genet* 1975;27(2):240-2.

[20] Veach P, LeRoy B, Bartels D. Facilitating the Genetic Counseling Process. New York: *Springer-Verlag*, Inc.; 2003.

[21] Bartels DM, LeRoy BS, McCarthy P, Caplan AL. Nondirectiveness in genetic counseling: a survey of practitioners. *Am J Med Genet* 1997;72(2):172-9.

[22] Beck M. Therapy for lysosomal storage disorders. *IUBMB Life* 2010;62(1):33-40

[23] Klinge L, Straub V, Neudorf U, Schaper J, Bosbach T, Gorlinger K, et al. Safety and efficacy of recombinant acid alpha-glucosidase (rhGAA) in patients with classical infantile Pompe disease: results of a phase II clinical trial. *Neuromuscul Disord* 2005;15(1):24-31.

[24] Kishnani PS, Nicolino M, Voit T, Rogers RC, Tsai AC, Waterson J, et al. Chinese hamster ovary cell-derived recombinant human acid alpha-glucosidase in infantile-onset Pompe disease. *J Pediatr* 2006;149(1):89-97.

[25] Douillard-Guilloux G, Richard E, Batista L, Caillaud C. Partial phenotypic correction and immune tolerance induction to enzyme replacement therapy after hematopoietic stem cell gene transfer of alpha-glucosidase in Pompe disease. *J Gene Med* 2009;11(4):279-87.

[26] Parenti G. Treating lysosomal storage diseases with pharmacological chaperones: from concept to clinics. *EMBO Mol Med* 2009;1(5):268-79.

[27] Llerena J. The Brazilian Consensus on the Management of Pompe Disease. *J Pediatr* 2009;155(4 Suppl):S47-56.

[28] Bembi B, Cerini E, Danesino C, Donati MA, Gasperini S, Morandi L, et al. Management and treatment of glycogenosis type II. *Neurology* 2008;71(23 Suppl 2):S12-36.

[29] American Board of Genetic Counseling. *Practice-Based Competencies.* In; 2006.

[30] Case LE, Kishnani PS. Physical therapy management of Pompe disease. *Genet Med* 2006;8(5):318-27.

[31] Griffin JL. Infantile acid maltase deficiency. I. Muscle fiber destruction after lysosomal rupture. *Virchows Arch B Cell Pathol Incl Mol Patho*l 1984;45(1):23-36.

[32] Freed GL, Hudson EJ. Transitioning children with chronic diseases to adult care: current knowledge, practices, and directions. *J Pediatr* 2006;148(6):824-7.

[33] Enns G. The adolescent with an inborn error of metabolism: medical issues and transition to adulthood. *Adolesc Med* 2002;13(2):315-329.

In: Vitamin E: Nutrition, Side Effects and Supplements
Editor: Alice E. Lindberg
ISBN: 978-1-61728-757-2

Chapter VII

Therapeutic Values of Vitamin E Supplements in Clinical Practices

Chi Chiu Wang
Department of Obstetrics & Gynaecology, The Chinese University of Hong Kong, Prince of Wales Hospital, Shatin, New Territories, Hong Kong

Abstract

Free radical generation, redox imbalance and oxidative stress have been implied as the key pathological mechanism in the development and process of many diseases. Vitamin E is a very common dietary compound that functions as a potent antioxidant to scavenge toxic free radicals, regulate redox status and resolve further oxidative damage. Vitamin E has been comprehensively investigated in cell line experiments in vitro and animal models in vivo. Recently it has been extensively evaluated its protection potentials in large scale epidemiological and population studies. However the therapeutic values of vitamin E supplements in promoting health conditions in general population and preventing or treating disease conditions in major randomized controlled trials are still open to question. The studied diseases include inflammation, ischemia and reperfusion, degeneration, aging and cancers. In addition, many other antioxidants including vitamin E do interfere with the efficacy of other antioxidants and pharmaceutics used in clinical management and result in unwanted toxic side effects. In this chapter, we will critically review the therapeutic values of vitamin E in clinical applications from its chemistry, pharmacology, metablism, clinical significance to medical recommendation. The prospective and significant implications of Vitamin E to the clinical practices of modern medicine are discussed.

Background

In 1922, Evans and Bishop described factor X, a nutrient found in vegetable oil that cured sterility in rats maintained on a lard diet [1]. Pure vitamin E was first isolated from wheat germ oil in 1936 and its chemical structure was defined and synthesis was achieved in 1938. The first time vitamin E was recognized by the Food and Nutrition Board of the National Research Council was in 1968. Systemic vitamin E was approved by the FDA in 1941.

The majority of vitamin E's benefits stem from its antioxidant qualities. That means it combines with oxygen and destroys free radicals. Vitamin E protects polyunsaturated fats which is important to cell membranes. For example, vitamin E protects lung cells that are in constant contact with oxygen and white blood cells that help fight disease. Vitamin E also acts as an antioxidant in foods. The vitamin E in vegetable oils helps keep them from being oxidized and turning rancid. Likewise, it protects other oxygen-sensitive compounds such as vitamin A in foods from being destroyed by damaging oxidation reactions. This makes vitamin E a useful food preservative.

Chemistry

Natural vitamin E is comprised of four tocopherols, namelyα, β, γ , and δ, and four tocotrienols (α, β, γ , δ). The most studied is α-tocopherol. The name tocopherol comes from the Greek tocos (childbirth), phero (to bear), and ol (alcohol) [2]. Tocopherols and tocotrienols are lipid soluble compounds that are synthesized by plants and required by most animals, including humans. The most recognized role for α-tocopherol is as a lipid soluble antioxidant required for the preservation of cell membranes, where it reacts quickly with peroxyl radicals to preserve polyunsaturated fatty acids [3]. d-α-tocopherol is the naturally occurring form with the greatest vitamin activity. The synthetic form (SRR-α-tocopherol, dl-α-tocopherol) is dl-α-tocopherol. Commercial vitamin E preparations are formulated primarily from synthetic dl-α-tocopheryl acetate. This acetate ester confers stability to the compound, but is only half as active as the natural form. Synthetic vitamin E is an equal mixture of eight stereoisomers of α-tocopherol [4]. All isomers have an identical chromanol group and hence equal antioxidant activity [5]. Synthetic α-tocopherol contains only 12.5% pure RRR-α-tocopherol and equal amounts of the other forms. However, only the four 2R-α-tocopherol isoforms are efficiently retained in the body.

The isoforms of vitamin E differ in the degree and site of methylation in the chromanol ring and the configuration of the methyl groups in the phytyl-side chains. The degree of methylation in the chromanol ring determines the antioxidant activity of each form of vitamin E. The biological activities of the different forms of vitamin E are expressed in international units per milligram (IU/mg). The relative activities are based on an assay using a biological system in which the amount of natural vitamin E required to prevent fetal resorption in rats deficient in vitamin E compared to RRR-α-tocopherol [6]. However, using this data to compare the biological activities of the different forms of natural vitamin E can make interpretation difficult, because the human requirements and tissue concentrations of the two most important natural forms of vitamin E, namely α- and γ-tocopherol, are markedly different.

Though α-tocopherol having twice the antioxidant activity of γ-tocopherol [7], there is some evidence that γ-tocopherol is superior to α-tocopherol at detoxifying nitrogen dioxide (NO_2) and that it is more effective at inhibiting peroxynitrite-induced lipid peroxidation of phosphatidylcholine lipososmes [8]. This inhibition occurs because reactive nitrogen-oxide species (RNOS) can be trapped by γ-tocopherol, which leads to the formation of 5-nitro-γ-tocopherol. γ-tocopherol is highly reactive toward RNOS because it has one less methyl group and thus a reactive position available to trap electrophiles, whereas α-tocopherol is fully substituted in the chromanol ring [9]. On the other hand, γ-tocopherol concentrations in patients correlate well with clinical conditions [10,11].

Vitamin E Supplement

Primary vitamin E deficiency is generally found only in premature and low-birth-weight infants. Secondary causes include fat malabsorption syndromes (e.g., cystic fibrosis, chronic liver disease, abetalipoproteinemia, and intestinal resection) and some hematological disorders (e.g., β-thalassemia major, sickle-cell anemia, and glucose-6-phosphate dehydrogenase deficiency). Patients with vitamin E deficiency have abnormal erythrocyte membrane morphology due to oxidative stress, and the characteristic acanthocytosis is associated with a reduction in half life of red cell. Long-term deficiency in vitamin E can lead to neurological abnormalities, including ataxia, hyporeflexia, blindness, and dementia.

As an antioxidant with powerful punch, the benefits of antioxidant role of vitamin E may actually go much further. In general, there is significant evidence vitamin E supplement can protect against cancer, heart disease, strokes, cataracts, and may slow the deterioration associated with aging. As a dynamic cancer fighter, vitamin E protects cells and DNA from damage that can turn cancerous. It reduces the growth of tumors while enhancing immune function and preventing precancerous substances from being turned into carcinogens. Vitamin E protects artery walls and keeps the low-density lipoprotein (LDL) cholesterol from being oxidized. Oxidation of LDL cholesterol marks the beginning of clogged arteries. Vitamin E also keeps the blood thin by preventing blood platelets from clumping together. High levels of vitamin E in the body decrease the risk of a non-fatal heart attack or stroke in most people. Women who suffer from fibrocystic breast disease can often find relief with vitamin E supplementation. Fibrocystic breast disease is characterized by painful breasts, sometimes with benign lumps or swelling, starting several days before the menstrual period. Many women report that vitamin E helps reduce hot flashes and other symptoms of menopause. Vitamin E can be beneficial to people with diabetes. It enhances the action of insulin and improves blood glucose metabolism by reducing oxidative stress. This humble nutrient keeps the nervous system healthy by protecting the myelin sheaths that surround nerves. It also appears to prevent mental degeneration due to aging, possibly including Alzheimer disease. Athletes need to get adequate amounts of vitamin E. The body's own metabolism creates free radicals during excessive aerobic exercise. Vitamin E therapy also treats claudication-pains in the calf muscles that occur at night or during exercise. Premature babies receive vitamin E to reduce or prevent oxygen damage to the retina of the eye as a result of artificial ventilation. Vitamin E may limit lung damage caused by air pollution. It appears that vitamin E can reduce the activity of such common air pollutants as ozone and nitrogen dioxide. Vitamin E may increase the wound healing rate because it minimizes oxidation reactions in the wound and also keeps the wound moist. Vitamin E can slow down the oxidation of fats that occurs in aging, it may indeed delay or prevent some diseases or a loss of function related to aging.

Recent studies have reported improved short-term memory in older adults who took supplemental vitamin E.

Clinical Indications

Vitamin E is required as primary treatment approved by FDA in individuals with a rare genetic abnormality that does not allow for the appropriate absorption of vitamin E (Table 1). Patients who suffer from malabsorptive disorders such as cystic fibrosis, chronic bowel disease (e.g., Crohn's, Celiac, and Whipple's disease), or who have undergone gastrointestinal surgery (e.g., post gastrectomy or gastric bypass), may be at an increased risk for vitamin E deficiency. Premature or low birth weight infants also are particularly at high risk for vitamin E deficiency [12].

Table 1. Clinical indications and recommended dosage of Vitamin E.

Usage	Clinical Indications	Dosage
primary treatment	vitamin E deficiency*	Adults: 60-75 IU, po, QD
	malabsorption syndrome*	Children: 1 IU/kg, po, QD
	prematurity*	Neonates: 25-50 IU, po, QD
prophylaxis	total parental nutrition	Adults: 10 IU, iv, QD
	oxygen toxicity	Preterm infants: 15-100 IU/kg, po, QD
supportive	Alzheimer's disease	Adults: 1,000 IU, po, QD
	premenstrual symptoms	Adults: 400 IU, po, QD
	menopause symptoms	Adults: 800 IU, po, QD
adjunctive treatment	G6PD deficiency	Adults: 800IU po, QD
nutritional supplementation	healthy individuals	Lactating and adolescent females: RDA 19 IU, po, QD
		Pregnancy and adolescent females: RDA 15 IU, po, QD
		Adult and adolescent males: RDA 15 IU, po, QD
		Children 9-13 yr: RDA 11 IU, po, QD
		Children 4-8 yr: RDA 11 IU, po, QD
		Children 1-3 yr: RDA 11 IU, po, QD
		Infants 6-12 m: AI 5 IU, po, QD
		Infants ≤6 m: AI 4 IU, po, QD
	fat malabsorption	Adults: 100-400 IU, po, QD
	beta-thalassemia*	Adults: 750 IU, po, QD
	sickle cell anemia*	Adults: 450 IU, po, QD

IU: International units, 1 IU equal to the biological activity of 1 mg of dl-α-tocopherol acetate, 670 μg of d-α-tocopherol, 830 μg of d-α-tocopheryl acid succinate, 735 μg of d-α-tocopheryl acetate, 910 μg of dl-α-tocopherol, and 1.12 mg of dl-alpha-tocopheryl acid succinate.

*: FDA-approved indications

RDA: Recommended Dietary Allowance, revised by the IOM Food and Nutrition Board in 2000

AI: Adequate Intake

Vitamin E has been used off-label as other primary treatment for various conditions including Alzheimer's disease (AD) [13,14], glucose-6-phosphate dehydrogenase deficiency (G6PD) [15], premenstrual syndrome (PMS) [16], and hot flashes [17].

Studies linking vitamin E with AD have been promising. The combined use of vitamin E and donepezil in patients, has been reported to slow the progression of cognitive decline. Long-term combined therapy with donepezil 5 mg and vitamin E 1000 IU daily may also help [18]. Conversely, daily supplementation with vitamin E does not delay the onset or progression from mild cognitive impairment to a diagnosis of AD [19]. Previous data and guidelines suggest some benefit from a twice daily dose of vitamin E (1000 IU) [20,21]. However, this data has received criticism due to differences in baseline characteristics between study groups. Subjects receiving vitamin E were younger and had more severe dementia than control patients [22,23].

For the adjunctive treatment of G6PD deficiency, two open trials have reported that vitamin E 800 IU, alone or in combination with selenium, reduces hemolysis and reticulocytosis and increases half-life of red blood cell and vitamin E serum concentrations [24].

One double-blind randomized study has reported reduced affective and physical symptoms associated with PMS in females taking 400 IU per day of d-α-tocopherol [25]. Vitamin E has also been suggested to reduce breast symptoms of PMS [26]. Nevertheless, the data supporting the use of vitamin E in the treatment of hot flashes is marginal [27]. The North American Menopause Society recommends nonprescription therapies such as vitamin E as an option to consider in women not responding to lifestyle modifications [28].

For patients with a fat malabsorption (e.g., cystic fibrosis, chronic cholestasis, chronic bowel disease, inflammatory bowel disease, post-gastrectomy, or gastric bypass surgery), vitamin E can be used as nutritional supplementation. For oxygen toxicity, including retrolental fibroplasia prophylaxis in premature neonates receiving high concentrations of oxygen, vitamin E can be used as prophylactic treatment although the American Academy of Pediatrics considers this practice investigational and does not recommend routine use. For adjunctive treatment and prevention of photoaging, including palliation of fine facial wrinkles, and reduction the appearance of scars or stretch marks are also implicated.

Other Applications

Vitamin E is a lipid soluble vitamin can be found naturally in cellular membranes where it plays an important role in the suppression of free radical-induced lipid peroxidation, hence vitamin E supplementation is used primarily for its antioxidant properties with clinical potentials. As a free radical scavenger, vitamin E protects membrane-bound polyunsaturated fatty acids and other oxygen-sensitive substances from oxidation. Vitamin E is the predominant lipophilic antioxidant for low-density lipoproteins (LDL). But remember there is no single oxidant responsible for LDL oxidation has been identified, and it is likely that many factors, including transition metals, 15-lipoxgenase, myeloperoxidase-derived oxidants, and reactive nitrogen species, are involved [29]. Antioxidants protect LDL from oxidative modification and therefore may contribute to the reduction of cardiovascular disease (CVD). Oxidative injury has also been implicated in cancer; many carcinogens create free radicals that damage DNA and other cellular structures, initiating and promoting tumor development [30]. Therefore, it is attractive to consider vitamin E as an agent to inhibit the development of neoplasms. Vitamin E may prevent cancer by inhibiting proliferation and angiogenesis, inducing apoptosis, and enhancing immune function.

Although vitamin E is most commonly noted for its antioxidant activity, vitamin E may also act through several other mechanisms including antiplatelet and immunomodulation effects. The antiplatelet effects of vitamin E is related to inhibition of protein kinase C stimulation as opposed to its antioxidant properties [31]. Vitamin E also decreases the incidence of red blood cell hemolysis in patients with G6PD deficiency. For the immunomodulation effects, vitamin E enhances lymphocyte proliferation, decreases production of immunosuppressive prostaglandins E2, and decreases levels of immunosuppressive serum lipid peroxides.

The purported mechanisms of topically applied vitamin E involve antioxidant effects and effects against photoaging. Vitamin E is found in the skin and increases from childhood to maturity and then decreases significantly in old age. Vitamin E and coenzyme Q10 have been found to synergistically inhibit the ultraviolet (UV) deletion of squalene, cholesterol, and unsaturated fatty acids in skin surface lipids [32]. Topically applied vitamin E may result in antioxidant effects in the skin. Vitamin E absorbs UV light in the region of the solar spectrum, which is responsible for sun-induced effects on the skin [33].

Clinical Pharmacology

Vitamin E usually is administered orally but may be given parenterally as part of a multivitamin injection. Systemically absorbed vitamin E is bound to plasma beta-lipoproteins and is widely distributed, particularly in fat tissues. Vitamin E is metabolized in the liver to glucuronides of tocopheronic acid and excreted primarily via the bile. Small amounts are eliminated in the urine. Approximately 20-60% of dietary vitamin E is absorbed from the gastrointestinal tract per oral route. Absorption requires adequate pancreatic function and bile salts. Water-miscible vitamin E solutions may be more easily absorbed than vitamin E in oil preparations and are more beneficial in preventing vitamin E deficiency in patients with steatorrhea. Vitamin E can be metabolized and delivered to tissues by a variety of routes and that low serum vitamin E concentrations do not necessarily equate to vitamin E deficiency. Vitamin E deficiency may also be organ specific, serum tocopherol concentrations not necessarily indicative of those in peripheral tissues [34]. Following topical application of vitamin E, vitamin E is substantially absorbed in the skin, with no evidence of conversion within the skin to free alpha-tocopherol [35]. Data are unavailable regarding the pharmacokinetics of vitamin E when used in patients with either hepatic or renal impairment. On the other hand, vitamin E crosses the placenta and is distributed into breast milk. The amount of vitamin E that crosses the placenta appears to be less than the amount transferred to the infant via breast milk.

For vitamin E absorption, prolonged and regular use of mineral oil can theoretically interfere with absorption of fat-soluble vitamins, including vitamin E [36]. Although this interaction has not been substantiated in controlled clinical trials, it has been recommended that administration of the two agents be separated. Orlistat, also known as tetrahydrolipstatin, is a drug designed to treat obesity, has been shown to inhibit the absorption of a vitamin E acetate supplement by 60% [37]. Patients should be advised to take a daily multiple vitamin supplement that contains vitamin E while receiving orlistat. And also, cholestyramine (and colestipol), a bile acid sequestrant binds bile in the gastrointestinal tract to prevent its reabsorption in Crohn's disease, has been shown to reduce the absorption of fat soluble vitamins and potentially reduce the absorption of vitamin E [38]. Oxygen is a necessary component during porfimer [39] or verteporfin [40] for photodynamic therapy and radiation therapy and for palliative treatment of obstructing endobronchial non-small cell lung carcinoma and obstructing esophageal cancer. Therefore, vitamin E and other antioxidants would be expected to reduce the porfimer and or verteporfin treatment.

Dosage

The dietary requirement for vitamin E is often ascribed to the intake of polyunsaturated fatty acids within the diet. Vitamin E, is a fat-soluble vitamin found in many foods including vegetable oils, wheat germ, cereal grains, fruits, green vegetables, meat, eggs, and certain types of fish. Since vitamin E is widely distributed in the diet, primary deficiency in otherwise healthy humans is unlikely. Vitamin E supplementation may be necessary in patients who have a pre-existing vitamin E deficiency. Vitamin E serum concentrations below 0.5 mg/dl are suggestive of deficiency. Total body stores of vitamin E are estimated to be 3-8 g, which appear adequate for greater than 4 years of a vitamin E-deficient diet.

Clinical dosage should be individualized by prescriber based on severity of vitamin E deficiency (Table 1). Recommended tolerable upper intake levels as maximum dosage limits by the Institue of Medicine (IOM) of the National Academies, Food and Nutrition Board (2000) in healthy, non-vitamin deficient are 1,000 IU/day, p.o. for adults and elderly; 800 IU/day, p.o. for adolescents; and 200-600 IU/day, p.o. for children. Maximum dosage limits for infants is not determinable. Specific guidelines for dosage adjustments in hepatic and renal impairment are not available; use with caution in patients with liver and kidney disease. Supplementation of vitamin E specifically during pregnancy and lactation is not necessary for females with normal fat absorption. Oral supplementation of Vitamin E in amounts exceeding the recommended dietary allowance (RDA) during pregnancy and breast-feeding should be approached with caution. Topically applying vitamin E to the breasts should be avoided as it could theoretically expose the infant to additional oral vitamin E [41].

Deficiencies of vitamin E are rare and evidence of toxicity is equally rare. Prolonged use of large doses of vitamin E (400-800 IU/day) has, in rare instances; caused nausea/vomiting, diarrhea, breast enlargement, fatigue, intestinal cramps, weakness, blurred vision, headache, rash, gonadal dysfunction, increased serum cholesterol and triglycerides, increased urinary estrogens and androgens, and decreased serum thyroxine, triiodothyronine, and creatinuria. These effects are reversible following discontinuance of the drug. Prolonged use of excessive dosages of vitamin E (> 800 IU/day) can cause bleeding in patients with vitamin K deficiency and can alter thyroid, pituitary, and adrenal hormone metabolism. Large doses of vitamin E also can result in an altered immune response and sexual dysfunction. Necrotizing enterocolitis has occurred in low-birthweight infants receiving large dosages (200 IU/day) of a hyperosmolar vitamin E preparation.

A recent report suggests that mega-doses ($\geq$ 400 IU/day p.o.) of vitamin E in adults may increase all-cause mortality [42]. However, the increases were not found to be statistically significant. It remains to be seen if such data are clinically significant. The current advice is to not exceed 400 IU/day p.o. of vitamin E in adults; this is consistent with the general therapeutic use of vitamin E from other clinical data and recommendations overall. High-dosage clinical trials were often small and were performed in patients with chronic diseases (those who might be at increased risk of all-cause mortality). The generalizability of the findings to healthy adults is uncertain. Precise estimation of the dosage threshold at which risk increases is difficult. Because of conflicting data about vitamin E supplementation, it is not surprising to see disagreement about the recommended daily allowance for α-tocopherol in humans and to observe that γ-tocopherol has not yet been included [43,44,45].

Vitamin E, in large doses, can aggravate iron deficiency anemia. In a small number of patients receiving iron therapy, a poorer response was noted in those receiving vitamin E concomitantly. The underlying mechanism for this is unclear. Hypoprothrombinemia secondary to vitamin K deficiency may also be exacerbated by large doses of vitamin E because of potential vitamin K antagonistic activity. Clinically, vitamin E should be used cautiously in patients receiving warfarin. The addition of high-dose vitamin E to a stabilized regimen of warfarin has been noted to precipitate hypoprothrombinemia [46].

Metabolism and Mechanisms

The intestinal absorption of vitamin E requires the intake and digestion of dietary fat, which is enhanced by the production of bile acids from the liver. Dietary vitamin E bound in micelles formed within the intestine is absorbed, along with triglycerides and cholesterol, by a passive process into enterocytes. Chylomicrons containing vitamin E are assembled and secreted into the lymph. Inhibition of the scavenger receptor class B type I (SR-BI) blocks up to 80% of α-tocopherol uptake, and suggests a role for SR-BI in intestinal tocopherol transport [47].

The cytochrome P450 isoforms, CYP3A, are probably the most important of all the drug-metabolizing enzymes in humans as they metabolize ~50% of all drug oxidations, and high quantities of this enzyme, in particular CYP3A4, are found in both the liver (29% of total) and the intestine (70% of total). Vitamin E can interfere with drug metabolism by increasing the expression of CYP3A4 within the liver and thereby increasing the metabolism of certain drugs [48]. High-dose vitamin E supplementation may severely interfere with drug metabolism through activation of CYP3A4. The induction of CYP3A4 in humans is mainly through activation of the pregnane X receptor (PXR) [49]. Any effect of vitamin E on PXR activation may affect hormone synthesis pathways in humans which may have undesirable effects in some populations.

In the circulation, chylomicrons interact with lipoprotein lipase (LPL) to release non-esterified fatty acids and triglycerides. During LPL-mediated delipidation of lipoproteins, as chylomicron remnants are formed, they can then exchange surface components, including tocopherols, with high-density lipoprotein (HDL). HDL can then transfer tocopherols to other lipoproteins in the circulation [50]. This pathway is particularly important for individuals with abetalipoproteinemia and homozygous familial hypobetalipoproteinema because of absent or extremely low plasma levels of apolipoprotein B-containing lipoproteins.

α-tocopherol is preferentially incorporated into very low-density lipoprotein (VLDL) particles, and γ-tocopherol is excreted in the bile [51]. Chylomicron and HDL metabolism is important for tocopherol distribution to lipoproteins for patients with impaired transport of tocopherols due to rare inherited disorders of lipid and lipoprotein metabolism. However VLDL particles, even abnormal ones, are preferentially enriched with RRR-α-tocopherol [52]. Supplementation with vitamin E was not required in this group of patients, because oxidative stress was not evident, nor did they exhibit any clinical signs of tocopherol deficiency.

The effect of α-tocopherol on the oxidation of lipoproteins and its potential role as an antiatherogenic supplement have received considerable attention in the last decade. Moreover, a number of studies have described a role for α-tocopherol beyond its antioxidant function. α-tocopherol has been shown to inhibit smooth-muscle cell proliferation [53], endothelial dysfunction [54] and platelet aggregation [55] by a protein kinase-C dependant mechanism; it has also been found to inhibit monocyte adhesion to endothelial cells [56] and macrophage-mediated lipid peroxidation in vitro [57]. These diverse functions of α-tocopherol and the potential pro-oxidant effects observed in some studies [58] may account for the paradoxical results observed with human clinical trials on the prevention of recurrent atherosclerosis [59].

Although α-tocopherol is important, it is not the only determinant in the resistance of LDL to oxidation [60]. Ubiquinol-10 (CoQ10H2), the reduced form of coenzyme Q, is a well-studied co-antioxidant for α-tocopherol [61]. α-tocopherol radical (TO•) can act to transfer, rather than trap electrons. Also, CoQ10H2 can act as a better antioxidant than α-tocopherol in LDL because the semiquinone radical formed can leave the lipoprotein particle rather than promote further peroxidation [62]. Therefore, the rate of radical stress is important in determining the amounts co-antioxidants available to interact with α-tocopherol and this in turn influences the role of α-tocopherol within the LDL particle.

Vitamin E can act as a pro-oxidant, especially when given in large quantities. It is converted to an α-tocopherol oxyl radical, which if not reduced by vitamin C or glutathione, can produce harmful effects. Oxidation of α-tocopherol may occurr in the disease condition and exceed lipid peroxidation, though the lesions were not depleted of α-tocopherol. The products of α-tocopherol oxidation, tocopherylquinone, and tocopheryl epoxides, were <20% of the total tocopherols, with tocopherylquinone the major product formed. Tocopherylquinone is generated in the presence of two electron (nonradical) oxidants such as hypochlorite and peroxynitrite. The oxidation products of tocopherylquinone formed were similar to those formed when LDL is oxidized in the presence of nitrite. Because oxidized lipids co-exist with α-tocopherol, it questions the rationale for using antioxidants such as α-tocopherol to prevent atherosclerosis [63].

Urinary metabolites of vitamin E, known as the "Simon metabolites", both tocopheronic acid and the subsequent tocopheronolactone were detected in urine after high-dose supplementation with α-tocopherol [64]. These metabolites are generated following the antioxidant action of vitamin E in vivo with the chroman ring being opened after oxidation [65]. Tocopheronolactone is produced from 2,5,7,8-tetramethyl-2 (2'-carboxyethy)-6-hydroxychroman (α-CEHC, a metabolite of ∝-tocopherol) in the presence of oxygen from sample handling [66]. Plasma concentrations of vitamin E can be raised only 3-fold [67], regardless of how much α-tocopherol is given. High concentrations of vitamin E may indeed be harmful as reflected by high urinary concentrations of α-CEHC.

Clinical Trials

Vitamin E has been intensely studied over the past several decades mainly because of its ability to serve as a chain-breaking antioxidant to prevent the propagation of lipid peroxidation and reduce free radical damage. Many chronic conditions of aging are thought to be associated with this action, including cardiovascular disease, diabetes mellitus, disorders of immune function and the central nervous system, cancer, and cataracts (Figure 1).

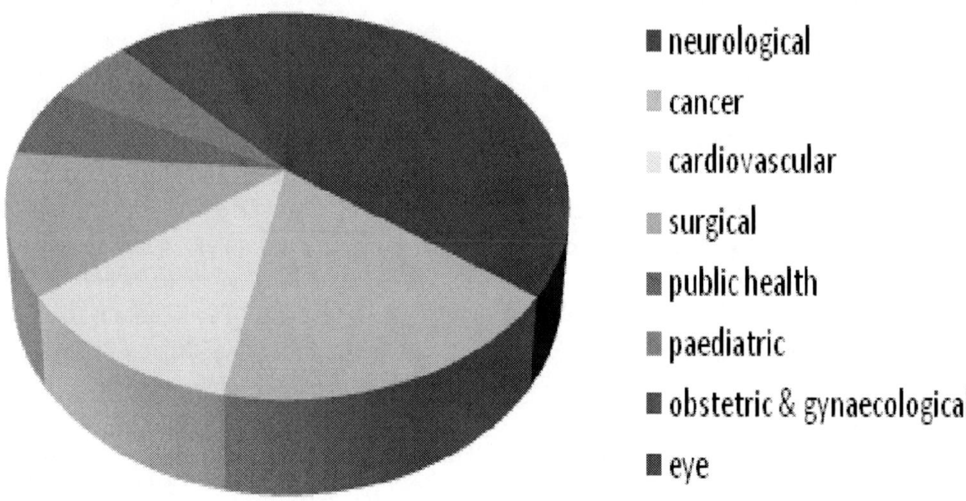

Figure 1. Meta-analysis studies of Vitamin E.

Up until the beginning of the 2000s, most studies using supplemental vitamin E were neutral or favorable with respect to its effect on these conditions. Unfortunetly, more recently, several reports were published showing that supplementation with vitamin E was associated with increased mortality and an increased risk of gastrointestinal cancer and heart failure [68]. Though meta-analyses cannot conclude such adverse outcomes of vitamin E supplement alone in general health and several major diseases conditions (Table 2), efficacy and safety of vitamin E supplement rase a major concerns to both medical community and general public [69].

The results of studies addressing the use of vitamin E in cancer have been inconsistent and inconclusive. Some studies have reported the potential benefits of vitamin E supplementation for prostate cancer. A cohort study suggested an inverse association between supplemental vitamin E use and the risk of metastatic or fatal prostate cancer among current smokers or smokers who stopped smoking [70].

Table 2. Results of meta-analysis on Vitamin E vs controls in health and disease*.

Date	Data	Authors	Disease	Outcome	Studies	Particip-ants	Quality	Main Results	Risk	95% CI	Conclusions
Jan 2010	Feb 2008	Bjelakovic et al	healthy subjects + various diseases[a]	mortality	37	123,761	>70% low risk of bias	increase mortality	RR:1.04	1.01-1.06	no evidence for prevention of mortality
Jan 2010	Aug 2001	McGrath & Soares-Weiser	neuroleptic-induced tardice dyskinesia	clinical improvement[e]	6	256	small trials with uncertain quality of randomosation	no benefit	RR:0.95	0.89-1.02	no evidence for improvement
Apr 2009	Jan 2009	Abu-Amara et al	ischemia reperfusion injury in liver resection	mortality, hospital stay, blood transfusion	1	47	adequate blinding, high risk of bias	no significant difference	-	-	-
Apr 2009	Dec 2006	Ranganathan & Ramaratnam	epilepsy	seizure control, adverse effects of antiepileptic drugs & quality of life	1	24	poor methodology, small numbers of participants, unclear allocation concealment	decrease in seizure frequency	OR:26.73	5.46-130.92	no reliable evidence for support use
Apr 2009	Jan 2003	Caraballoso et al	lung cancer	Incidence (I) & mortality (M)	1	14,573	unclear allocation concealment	no effect	I RR: 0.98 M RR: 0.93	I: 0.81-1.19 M: 0.73-1.19	no evidence for prevention of incidence & mortality
Mar 2009	Dec 2007	Mestre et al	Huntington's diseases	diseases progression[d]	1	81	unclear allocation concealment	no significant difference	-	-	none proved to be effective
Jan 2009	Aug 2007	Evans JR	age-related macular degeneration	distance visual acuity	1	582	stringent selection criteria	no significant difference	RR:1.05	0.70-1.55	no evidence for efficacy

Table 2. (Continued).

Jan 2009	Jan 2001	Kleijnen & Mackerras	intermittent claudication	physical outcomes[b]	2	113	small, poor, only 2 included patients' subjective evaluation	no effect	OR:0.35	0.17-0.74	insufficient evidence for effective treatment
Apr 2008	Dec 2007	Orrell et al	Amyotrophic lateral sclerosis	Survival	1	288	unclear allocation concealment	no significant difference	RR:1.01	0.89-1.15	insufficient evidence for efficacy
Apr 2008	Nov 2007	O'Brien & Pandit	hypertrophic and keloid scars	cosmesis, itching, pain	1	80	poor quality	had improvement	RR:0.79	0.65-0.96	weak evidence
Apr 2008	Mar 2007	Brion et al	preterm infants	Mortality (M) & morbidity (IVH-I & Sepsis-S)[g]	M: 12 I: 7 S: 4	M: 2,028 I: 1755 S: 1,009	-	no significant difference in M; significantly reduced I risk; increased S risk	M RR: 0.97 I RR: 0.85 S RR: 1.52	M: 0.83-1.14 I: 0.73-0.99 S: 1.13-2.04	no effect on mortality; reduced the risk of IVH; increased risk of sepsis
Apr 2008	Oct 2006	Albers et al	cisplatin neurotoxicity	nerve conduction parameters[h]	1	27	unclear allocation concealment	no differ in sural amplitude (SA), significantly lower in total neuropathy score (TNS)	SA 0.10 TNS 0.36	SA: -5.57-5.57 TNS: 0.15-0.83	insufficient to conclude
Mar 2008	Nov 2007	Bjelakovic et al	gastrointestinal cancers	overall occurrence (O) & mortality (M)	M: 3 O: 1	M: 26,157 O: 14,573	adequate	no effect	M RR: 1.02 O RR: 1.11	M: 0.98-1.06 O: 0.93-1.34	no evidence for prevention of ocurrence & mortality
Mar 2008	Jan 2007	Issac et al	Alzheimer's diseases (AD) + mild cognitive impairment MCI)	AD: survival time [c]; MCI: progression to AD	2	924	stringent selection criteria	AD: some benefit; MCI: no difference in progression and adverse events	OR:0.49	0.25-0.96	no evidence for efficacy

Table 2. (Continued).

Feb 2008	Oct 2007	Rumbold et al	pre-eclampsia	pregnancy outcomes & management	2	133	-	-	-	-	-
Jun 2005	Mar 2005	Etminan et al	Parkinson's disease	risk of disease	1	420	-	statistically insignificant association	OR: 0.78	0.53-1.15	do not conclusively document a beneficial effect
Apr 2004	Dec 2002	Shekelle et al	cardiovascular disease	cardiovascular deaths (D), myocardial infarction (MI)	D: 5 MI: 5	D: 16171 MI: 45200	adequate	did not demonstrate signficant difference in mortality; demonstarted significant difference in MI	D RR: 0.97 Fatal MI RR: 0.97	D: 0.80-1.19 Fatal MI: 0.74-1.27	did not generally support any positive benefit in prevention of death, no effect on risk of MI

*multivitamin supplement or in combination with other pharmaceuticals excluded;

[a]gastrointestinal, cardiovascular, neurological, ocular, dermatological, rheumatoid, renal, endocrinological, or unspecified;

[b]pain, disability, walking distance, ankle-arm pressure index, standard treadmill test, vascular surgery, amputation and death;

[c]death, institutionalisation, loss of 2 cut of 3 basic activities of daily living and severe dementia;

[d]death, neuropsychiatric symptoms, dementia, postural instability, falls, wheelchair confinement, bedridden status, retirement;

[e]symptoms of tardive dyskinesia, general mental state changes, acceptability of treatment more than 50% after at least 6 weeks;

[f]pregnancy continuation >14 days, biochemical paremeters, duration of conservative management, or indications for delivery;

[g]bronchopulmonary dysplasia, patent ductus arteriosus, sepsis, cerebral hemorrhage (IVH)

[h]sural amplitude (SA), abnormal sensory amplitude (ASA), total neuropathy score (TNS)

A prior controlled clinical trial reported a 32% decrease in the incidence of prostate cancer among subjects receiving α-tocopherol compared to those not receiving α-tocopherol [71]. The reduction was seen for clinical prostate cancer, but not for latent cancer. In this study, α-tocopherol and β-carotene supplementation, were given either separately or together, to 29,133 prostate cancer patients who were smokers for 5-8 years. In contrast, in the Selenium and Vitamin E Cancer Prevention Trial (SELECT), which was a long-term, multi-center, placebo-controlled trial, more than 35,000 men ≥ 50 years of age were randomized to receive either selenium 200 mcg and vitamin E 400 mg, selenium and placebo, vitamin E and placebo, or placebo. The trial was stopped after about 7 years when interim results found that neither drug prevented prostate cancer. Further, 2 preliminary trends of vitamin E slightly increasing the risk of prostate cancer and selenium slightly increasing the risk of diabetes were found [72]. A post-interventional followup of the The Alpha-Tocopherol, Beta-Carotene Cancer Prevention (ATBC) Study, which studied the incidence of cancer and mortality following vitamin E (50 mg daily) and beta-carotene supplementation in prostate cancer patients, reported that the initial beneficial effects of α-tocopherol disappeared during postinterventional followup [73]. Results of the Heart Outcomes Prevention Evaluation Study Extension (HOPE-TOO) trial showed that vitamin E supplementation does not prevent cancer, including prostate cancer. High-risk patients (e.g., diabetes, vascular disease) enrolled in this trial were administered a dose of 400 IU daily and were followed for a median of 7 years [74]. In the The Supplementation en Vitamines et Mineraux Antioxydants (SU.VI.MAX) study, vitamin E (30 mg) provided protection in men but not women, resulting in a significant reduction in total cancer incidence [75]. In this trial, researchers concluded that supplementation may be effective in men due to lower baseline levels of antioxidants.

Although early evidence had suggested the potential for vitamin E to reduce cardiovascular disease (CVD) [76,77,78,79,80], more recent outcome studies have not shown benefits for vitamin E. The Heart Outcomes Prevention Evaluation (HOPE) study has shown no effect of vitamin E on cardiac outcomes in high-risk patients with diabetes or vascular disease [81]. The Women's Health Study (WHS), a study of approximately 40 thousand women receiving a daily dose of 600 IU, has shown no overall benefit for major cardiovascular events [82]. Currently, the American Heart Association guidelines do not recommend vitamin E therapy for women as a preventative measure for CVD [83]. As for males, results of the SU.VI.MAX study, which included a total of 13,017 subjects (including 5,141 men aged 45-60) followed for 7.5 years, has reported that supplementation with a multivitamin (containing 30 mg of vitamin E) did not reduce the incidence of ischemic CVD incidence in men [84]. Heart failure occurred in a large randomized controlled trial assessing the effects of Vitamin E (400 IU/day) on cardiovascular events and cancer. Researchers concluded that vitamin E was an independent risk factor for the development of heart failure and hospitalization for heart failure [85].

Epidemiological data seemed to support a benefit from both dietary and supplemental vitamin E on risk reduction of CVD, colon cancer, and AD. Especially noteworthy was that no increased risk of morbidity or mortality was observed. However prospective randomised studies still cannot confrim the beneficial of vitamin E supplement in healthy and disease subjects. The conflicting results is mainly due to some of the studies had fewer participants, too many subjects dropped out or studies stopped early, other essential vitamin and nutrients, mineral capsule or multivitamins received, suboptimal supplement levels received, unevenly group matched, poor compliance, etc. Nevertheless, most meta-analyses suffer from lack of heterogeneity due to too short study period to confirm the effect of mortality, various dose of supplement, inevitable skepticism of study selection, different forms of vitamin E, healthy and co-morbidities to cause unmatched study populations, treatments, baseline parameters and protocols. The conclusions in these meta-analysis should not be accepted due to inappropriate statistical methodology, lack of control of the quality of the studies included in the meta-analysis, and study selection bias. Correction of any one of these factors could negate the marginally significant results. To date, no statistician has come forth with such an analysis.

Conclusion

It is important that the effects of vitamin E supplementation is dependent upon the degree of pre-existing deficiency. Vitamin E supplement may reduce oxidative stress in human populations with high oxidative stress, but the effect may occur only with higher doses or in combination with co-antioxidants. The question of whether an appropriate dose of one or more forms of vitamin E, given alone or with in combination with other antioxidants, has a role in the prevention and treatment of disease remains to be answered in new clinical trials examining supplementation of tocopherols in humans.

References

[1] Evans HM, Bishop KS. On the existence of a hitherto unrecognised dietary factor essential for reproduction. *Science* 1922; 56: 650-651.

[2] Roberts DCK. Vitamin E. *Sydney: Australian Professional Publications*, 1990.

[3] Young IS, Woodside JV. Antioxidants in health and disease. *J Clin Pathol* 2001; 54: 176-186.

[4] Pryor WA. Vitamin E and Carotenoids: Abstracts. Lagrange, IL: *Veris*, 1996.

[5] Blatt DH, Leonard SW, Traber MG. Vitamin E kinetics and the function of tocopherol regulatory proteins. *Nutrition* 2001; 17: 799-805.

[6] Bunyan J, McHale D, Green J, Marcinkiewicz S, Biological potencies of ε- and ζ1-tocopherol and 5-methyltocol. *Br J Nutri* 1961; 15: 253-257.

[7] Kayden HJ, Traber MG. Absorption, lipoprotein transport, and regulation of plasma concentrations of vitamin E in humans. *J Lipid Res* 1993; 34: 343-358.

[8] Cooney R, Franke A, Harwood P, Hatch-Pigott V, Custer L, Mordan L. γ -Tocopherol detoxification of nitrogen dioxide: superiority to α-tocopherol. *Proc Natl Acad Sci USA* 1993; 90: 1771-1778.

[9] Jiang Q, Christen S, Shigenaga MK, Ames BN. γ-Tocopherol, the major form of vitamin E in the US diet, deserves more attention. *Am J Clin Nutr* 2001; 74: 714-722.

[10] Ohrvall M, Sundlof G, Vessby B. γ , but not α, tocopherol levels in serum are reduced in coronary heart disease patients. *J Intern Med* 1996; 239: 111-117.

[11] Kontush A, Spranger T, Reich A, Baum K, Beisiegel U. Lipophilic antioxidants in blood plasma as markers of atherosclerosis: the role of α-carotene and γ-tocopherol. *Atherosclerosis* 1999; 144:117-122.

[12] Bieri JG, Corash L. Hubbard VS. Medical uses of vitamin E. *N Engl J Med* 1997;308:1066-1071.
[13] Sano M, Ernesto C, Thomas RG, et al. A controlled trial of selegiline, alpha-tocopherol, or both as treatment for Alzheimer's disease. *N Engl J Med* 1997;336:1216-1222.
[14] Petersen RC, Thomas RG, Grundman M, et al. Vitamin E and donepezil for the treatment of mild cognitive impairment. *N Engl J Med* 2005;352:2379-2388.
[15] Chan AC, Chow CK, Chiu D. Interaction of antioxidants and their implication in genetic anemia. *Proc Soc Exp Biol Med* 1999;222:274-282.
[16] London RS, Murphy L, Kitlowski KE, et al. Efficacy of alpha-tocopherol in the treatment of the premenstrual syndrome. *J Reprod Med* 1987;32:400-404.
[17] The North American Menopause Society. Treatment of menopause-associated vasomotor symptoms: position statement of The North American Menopause Society. *Menopause* 2004;11:11-33.
[18] Klatte ET, Scharre DW, Nagaraja HN, et al. Combination therapy of donepezil and vitamin E in Alzheimer's disease. *Alzheimer Dis Assoc Disord* 2003;17:113-116.
[19] Petersen RC, Thomas RG, Grundman M, et al. Vitamin E and donepezil for the treatment of mild cognitive impairment. *N Engl J Med* 2005;352:2379-2388.
[20] Sano M, Ernesto C, Thomas RG, et al. A controlled trial of selegiline, alpha-tocopherol, or both as treatment for Alzheimer's disease. *N Engl J Med* 1997;336:1216-1222.
[21] Doody RS, Stevens JC, Beck C, et al. Practice parameter: Management of dementia (an evidence-based review). Report of the Quality Standards Subcommittee of the American Academy of Neurology. *Neurol* 2001;56:1154-1166.
[22] Tabet N, Birks J, Grimley Evans J, et al. Vitamin E for Alzheimer's disease. *Cochrane Database Syst Rev* 2003:CD002854.
[23] Delagarza VW. Pharmacologic Treatment of Alzheimer's Disease: An Update. *Am Fam Physician* 2003;68:1365-1372.
[24] Chan AC, Chow CK, Chiu D. Interaction of antioxidants and their implication in genetic anemia. *Proc Soc Exp Biol Med* 1999;222:274-282.
[25] London RS, Murphy L, Kitlowski KE, et al. Efficacy of alpha-tocopherol in the treatment of the premenstrual syndrome. *J Reprod Med* 1987;32:400-404.
[26] Abraham GE. Nutritional factors in the etiology of the premenstrual tension syndromes. *J Reprod Med* 1983;28:446-464.
[27] Barton DL, Loprinzi CL, Quella SK, et al. Prospective evaluation of vitamin E for hot flashes in breast cancer survivors. *J Clin Oncol* 1998;16:495-500.
[28] The North American Menopause Society. Treatment of menopause-associated vasomotor symptoms: position statement of The North American Menopause Society. *Menopause* 2004;11:11-33.
[29] Mashima R, Witting PK, Stocker R. Oxidants and antioxidants in atherosclerosis. *Curr Opin Lipidol* 2001; 12: 411-418.
[30] Prasad KN, Edwards-Prasad J. Vitamin E and cancer prevention: recent advances and future potentials. *J Am Coll Nutr* 1992;11:487-500.
[31] Diaz MN, Frei B, Vita JA, et al. Antioxidants and atherosclerotic heart disease. *N Engl J Med* 1997;337:408-416.
[32] Passi S, De Pita O, Puddu P, et al. Lipophilic antioxidants in human sebum and aging. *Free Radic Res* 2002;36:471-477.

[33] Sorg O, Tran C, Saurat JH. Cutaneous vitamins A and E in the context of ultraviolet- or chemically-induced oxidative stress. *Skin Pharmacol Appl Skin Physiol* 2001:14:363-372.
[34] Minehira-Castelli K, Leonard SW,Walker QM, Traber MG, Young SG. Absence of VLDL secretion does not affect alpha-tocopherol content in peripheral tissues. *J Lipid Res* 2006; 47: 1733-1738.
[35] Alberts DS, Goldman R, Xu MJ, et al. Disposition and metabolism of topically administered alpha-tocopherol acetate: a common ingredient of commercially available sunscreens and cosmetics. *Nutr Cancer* 1996;26:193-201.
[36] Robinson DS, Benjamin DM, McCormack JJ. Interaction of warfarin and nonsystemic gastrointestinal drugs. *Clin Pharmacol Ther* 1971;12:491495.
[37] Xenical® (orlistat) package insert. Nutley, NJ: *Roche Laboratories Inc.*; 2005 Jan.
[38] Questran® and Questran® Light (cholestyramine) package insert. Spring Valley, NY: *Par Pharmaceutical Inc*; 2002 July.
[39] Photofrin® (porfimer) package insert. Birmingham, AL: *Axcan Scandipharm Inc.*; 2003 Aug.
[40] Visudyne® (verteporfin for injection) package insert. Duluth, GA: *Novartis Ophthalmics, Inc.;* 2003 Apr.
[41] Cawley MJ, Guse TM, Laroia A et al. Propofol withdrawal syndrome in an adult patient with thermal injury. *Pharmacotherapy* 2003;23:933-939.
[42] Miller ER 3rd, Pastor-Barriuso R, Dalal D, et al. Meta-analysis: high-dosage vitamin E supplementation may increase all-cause mortality. *Ann Intern Med* 2005;142:37-46.
[43] Horwitt MK. Critique of the requirement for vitamin E. *Am J Clin Nutr* 2001; 73: 1003-1005.
[44] Traber M. Vitamin E: too much or not enough? *Am J Clin Nutr* 2001; 73: 997-998.
[45] Bieri JG. Comments on the new dietary reference intake for vitamin E. *Am J Clin Nutr* 2002; 75: 781-782.
[46] Wells PS, Holbrook AM, Crowther NR et al. Interaction of warfarin with drugs and food. *Ann Intern Med* 1994;121:676-683.
[47] Reboul E, Klein A, Bietrix F, Gleize B, Malezet-Desmoulins C, Schneider M, Margotat A, Lagrost L, Collet X, Borel P. Scavenger receptor class B type I (SR-BI) is involved in vitamin E transport across the enterocyte. *J Biol Chem* 2006; 281: 4739-4745.
[48] Brigelius-Flohe R. Induction of drug metabolizing enzymes by vitamin E. *J Plant Physiol* 2005; 162:797-802.
[49] Wilkinson GR. Drug metabolism and variability among patients in drug response. *N Engl J Med* 2005; 352: 2211-2221.
[50] Kayden HJ, Traber MG. Absorption, lipoprotein transport, and regulation of plasma concentrations of vitamin E in humans. *J Lipid Res* 1993; 34: 343-358.
[51] Traber MG, Kayden HJ. Preferential incorporation of α-tocopherol vs γ -tocopherol in human lipoproteins. *Am J Clin Nutr* 1989; 49: 517-526.
[52] Traber MG, Burton GW, Hughes L, Ingold KU, Hidaka H, Malloy M, Kane J, Hyams J, Kayden HJ. Discrimination between forms of vitamin E by humans with and without genetic abnormalities of lipoprotein metabolism. *J Lipid Res* 1992; 33: 1171-1182.
[53] Tasinato A, Boscoboinik D, Bartoli G, Maroni P, Azzi A. d-α-Tocopherol inhibition of vascular smooth muscle cell proliferation occurs at physiological concentrations,

correlates with protein kinase C inhibition, and is independent of its antioxidant properties. *Proc Natl Acad Sci USA* 1995; 92: 12190-12194.

[54] Keaney JF Jr, Guo Y, Cunningham D, Shwaery GT, Xu A, Vita JA. Vascular incorporation of α-tocopherol prevents endothelial dysfunction due to oxidized LDL by inhibiting protein kinase C stimulation. *J Clin Invest* 1996;98:386-394.

[55] Freedman JE, Farhat JH, Loscalzo J, Keaney JF Jr. α-Tocopherol inhibits aggregation of human platelets by a protein kinase C-dependent mechanism. *Circulation* 1996; 94: 2434-2440.

[56] Devaraj S, Li D, Jialal I. The effects of α−tocopherol supplementation on monocyte function. Decreased lipid oxidation, interleukin-1 beta secretion, and monocyte adhesion to endothelium. *J Clin Invest* 1996;98:756-763.

[57] Devaraj S, Jialal I. The effects of α-tocopherol on critical cells in atherogenesis. *Curr Opin Lipidol* 1998;9:11-15.

[58] Thomas SR, Stocker R. Molecular action of vitamin E in lipoprotein oxidation: implications for atherosclerosis. *Free Radic Biol Med* 2000; 8:1795-1805.

[59] Steinberg D. Is there a potential therapeutic role for vitamin E or other antioxidants in atherosclerosis? *Curr Opin Lipidol* 2000;11:603-607.

[60] Esterbauer H, Dieber-Rotheneder M, Striegl G, Waeg G. Role of vitamin E in preventing the oxidation of low-density lipoprotein. *Am J Clin Nutr* 1991;53:314S-321S.

[61] Neuzil J, Weber C, Kontush A. The role of vitamin E in atherogenesis: linking the chemical, biological and clinical aspects of the disease. *Atherosclerosis* 2001; 157: 257–283.

[62] Ingold K, Bowry V, Stocker R,Walling C. Autoxidation of lipids and antioxidation by α-tocopherol and ubiquinol in homogeneous solution and in aqueous dispersions of lipids: Unrecognized consequences of lipid particle size as exemplified by oxidation of human low density lipoprotein. *Proc Natl Acad Sci USA* 1993;90:45-49.

[63] Terentis AC, Thomas SR, Burr JA, Liebler DC, Stocker R. Vitamin E oxidation in human atherosclerotic lesions. *Circ Res* 2002;90:333-339.

[64] Simon EJ, Eisengart A, Sundheim W, Milhoart AT. The metabolism of vitamin E. II. Purification characterisation of urinary metabolites of α−tocopherol. *J Biol Chem* 1956:221:807-817.

[65] Galli F, Lee R, Dunster C, Kelly FJ. Gas chromatography mass spectrometry analysis of carboxyethylhydroxychroman metabolites of α- and γ -tocopherol in human plasma. *Free Radic Biol Med* 2002;32:333-340.

[66] Schultz M, Leist M, Petrzika M, Gassmann B, Brigelius-Flohe R. Novel urinary metabolite of α-tocopherol, 2,5,7,8-tetramethyl-2(2'-carboxyethyl)-6-hydroxychroman, as an indicator of an adequate vitamin E supply? *Am J Clin Nutr* 1995;621(6 Suppl):1527S-1534S.

[67] Traber MG, Rader D, Acuff RV, Ramakrishnan R, Brewer HB, Kayden HJ. Vitamin E dose-response studies in humans with use of deuterated RRR-α-tocopherol. *Am J Clin Nutr* 1998; 68: 847–853.

[68] Clarke MW, Burnett JR, Croft KD. Vitamin E in human health and disease. *Crit Rev Clin Lab Sci* 2008;45:417-450.

[69] Bell SJ, Grochoski GT. How safe is vitamin E supplementation? *Crit Rev Food Sci Nutr.* 2008;48(8):760-774.

[70] Chan JM, Stampfer MJ, Ma J, et al. Supplemental vitamin E intake and prostate cancer risk in a large cohort of men in the United States. *Cancer Epidemiol Biomarkers Prev* 1999;8:893-899.

[71] Heinomen OP, Albanes D, Virtamo J, et al. Prostate cancer and supplementation with alpha-tocopherol and beta-carotene: incidence and mortality in a controlled trial. *J Natl Cancer Inst* 1998;18:440-446.

[72] National Cancer Institute news release. October 27, 2008. Review of prostate cancer prevention study shows no benefit for use of selenium and vitamin E supplements. *Accessed May 05, 2010. Available on the World Wide Web at www.cancer.gov/newscenter/pressreleases/SELECTresults2008*

[73] Virtamo J, Pietinen P, Huttunen JK, et al. Incidence of cancer and mortality following alpha-tocopherol and beta-carotene supplementation: a postintervention follow-up. *JAMA* 2003;290:476-485.

[74] Lonn E, Bosch J, Yusuf S, et al. Effects of long-term vitamin E supplementation on cardiovascular events and cancer: a randomized controlled trial. *JAMA* 2005;293:1338-1347.

[75] Hercberg S, Galan P, Preziosi P, et al. The SU.VI.MAX Study: a randomized, placebo controlled trial of the health effects of antioxidant vitamins and minerals. *JAMA* 2004;164:2335-2342.

[76] Stampfer MJ, Hennekens CH, Manson JE, et al. Vitamin E consumption and the risk of coronary disease in women. *N Engl J Med* 1993;328:1444-1449.

[77] Rimm EB, Stampfer MJ, Ascherio A, et al. Vitamin E consumption and the risk of coronary disease in men. *N Engl J Med* 1993;328:1450-1456.

[78] Stephens NG, Parsons A, Schofield PM, et al. Randomised controlled trial of vitamin E in patients with coronary disease: Cambridge Heart Antioxidant Study (CHAOS). *Lancet* 1996;347:781-786.

[79] Hodis HN, Mack WJ, LaBree L, et al. Serial coronary angiographic evidence that antioxidant vitamin intake reduces progression of coronary artery atherosclerosis. *JAMA* 1995;273:1849-1854.

[80] Rapola JM, Virtamo J, Haukka JK, et al. Effect of vitamin E and beta carotene on the incidence of angina pectoris. *JAMA* 1996;275:693-698.

[81] Yusuf S, Dagenais G, Pogue J, et al. The Heart Outcomes Prevention Evaluation Study Investigators. Vitamin E supplementation and cardiovascular events in high-risk patients. *N Engl J Med* 2000;342:154-160.

[82] Lee MI, Cook NR, Gaziano JM, et al. Vitamin E in the primary prevention of cardiovascular disease and cancer: the Women's Health Study: A randomized controlled trial. *JAMA* 2005; 294:56-65.

[83] Mosca L, Women's LJ, Benjamin EJ, et al. Evidence-based guidelines for cardiovascular disease prevention in women. *Circulation* 2004;109:672-693.

[84] Hercberg S, Galan P, Preziosi P, et al. The SU.VI.MAX Study: a randomized, placebo controlled trial of the health effects of antioxidant vitamins and minerals. *JAMA* 2004;164:2335-2342.

[85] Yusuf S, Dagenais G, Pogue J, et al. The Heart Outcomes Prevention Evaluation Study Investigators. Vitamin E supplementation and cardiovascular events in high-risk patients. *N Engl J Med* 2000;342:154-160.

In: Advances in Medicine and Biology Research, Vol. 14 ISBN: 978-1-61761-930-4
Editor: Leon V. Berhardt

Chapter VIII

The Influence of the Nutrient Load on the Formation of Escherichia Coli Biofilms

J.S. Teodósio, M. Simões, L.F. Melo and F.J. Mergulhão*

LEPAE, Department of Chemical Engineering, Faculty of Engineering, University of Porto, Rua Dr. Roberto Frias, s/n, 4200-465 Porto, Portugal

Abstract

The attachment of microorganisms to surfaces and the subsequent biofilm development are very complex processes affected by several variables. Among these variables, it has been shown that the nutrient/substrate concentration may influence the structure of biofilms growing in aqueous environments. The goal of this study was to assess the effect of the nutrient load on biofilm formation by *Escherichia coli* JM109(DE3) by using three substrate feed concentrations (glucose concentrations of 0.55, 0.15 and 0.050 $g.L^{-1}$) and four dilution rates (0.0042, 0.0126, 0.0800 and 0.1650 h^{-1}) on separate experiments. Biofilms were formed in a flow cell reactor under turbulent conditions, with a Reynolds number of approximately 5000, by using a recirculating system that enables independent control of the dilution rate and flow velocity. Three nutrient load ranges were obtained on the independent experiments and our results show that when high nutrient loads were used, planktonic cell concentration was reduced and glucose consumption was proportional to the nutrient load. Additionally, when moderate nutrient loads were tested, higher planktonic cell concentrations were observed when compared to the high nutrient load conditions. When the nutrient load was further reduced planktonic cell concentration was further increased. Unlike the glucose consumption values obtained for the high nutrient load conditions, glucose consumption for the moderate and low nutrient loads was not dependent on the nutrient load that was used. Furthermore, when the system was fed with the highest nutrient concentration (0.55 $g.L^{-1}$), glucose consumption was the same even when the dilution rate was increased.

*Corresponding author (filipem@fe.up.pt)

However, for the intermediate nutrient concentration (0.15 $g.L^{-1}$), glucose consumption was enhanced with the increase in dilution rate.

As a general trend, *E. coli* biofilm formation under flow conditions was favored at higher nutrient loads and this could be measured both in terms of biofilm thickness as well as wet weight. There were no significant effects on the amount of biofilm formed by moderate and low nutrient loads.

1. Introduction

In nature, medical or industrial systems, microorganisms are prone to live in communities forming aggregates that are called biofilms (Bryers, 1999; Davey and O'Toole, 2000; O'Toole et al., 2000; Stoodley et al., 2002; Watnick and Kolter, 2000). Biofilm formation requires cell attachment which is influenced by the presence of nutrients, ions and organic material that are required for cell viability and growth (Poulsen, 1999). Bacteria grow in biofilms forming microcolonies embedded in extracellular polymeric substances – EPS (Sihorkar and Vyas, 2001). EPS are composed of polysaccharides, proteins, phospholipids, teichoic and nucleic acids as well as other polymeric substances hydrated to 85 to 95%. These extracellular substances play an important role in biofilm maintenance since they protect inhabitants from biocides and other antimicrobial agents, determine their ability to degrade toxic substances and facilitate nutrient concentration necessary for bacterial survival (Chmielewski and Frank, 2003),. Mature biofilms are a microcosm in which bacteria can float sharing water channels that provide passages for the exchange of nutrients and waste products (Shi and Zhu, 2009; Sihorkar and Vyas, 2001).

In many industries biofilms are the cause of serious problems. Common examples of deleterious effects of biofilm formation are found in food processing industries as well as in water distribution systems. In fact, it is often discovered that the contamination of dairy products is due to the presence of biofilms. Previous reports by Simões (2009) indicate that biofilm cells are more resistant to antimicrobials when compared to planktonic cells and thus new control strategies applied to industry must be developed. A more detailed knowledge about biofilm formation will allow the development of more efficient cleaning and disinfection processes with positive outcomes on public health.

The first report of a foodborne associated bacterial biofilm concerned the adhesion of *Salmonella* species to food surfaces (Shi and Zhu, 2009). This microorganism is associated to poor sanitation of equipment in food processing environments (Chmielewski and Frank, 2003). Several microorganisms have been reported as food product contaminants or contact surface contaminants, some of them potentially pathogenic. In the diary industries, the most commonly found bacteria belong to the genus *Enterobacter*, *Lactobacillus*, *Listeria*, *Micrococcus*, *Streptococcus*, *Bacillus* and *Pseudomonas* (Simões et al., 2010). Bacterial pathogens like *Listeria monocytogenes*, *Yersinia enterocolitica*, *Campylobacter jejuni* and *Escherichia coli* O157:H7, were also associated to food contact surfaces (Kumar and Anand, 1998).

In this chapter we evaluate the effect of the nutrient load on *E. coli* JM109(DE3) biofilm formation. Besides being a typical indicator of the sanitary quality of water (Van Houdt and Michiels, 2005), *E. coli* forms biofilms in food-processing environments (Marouani-Gadri et al., 2009; Sharma and Anand, 2002) and on water-distribution systems (Juhna et al., 2007). *E.*

coli is a natural component of the human flora and its presence in the environment, or in foods, generally implies some history of contamination. It is considered to be a good indicator organism since it should always be present in relatively large numbers when the pathogen may be present, allowing rapid detection (Adams and Moss, 1995).

1.1 Biofilm Formation Process

Biofilm establishment and development involves a complex process that results from the combination of environmental conditions, cellular metabolism and genetic control systems. Mature biofilms are the result of cellular adaptations and growth cycles determined by the environmental nutrient diffusion and flow-dynamic conditions (Kjelleberg and Molin, 2002).

The first step on biofilm formation consists on the attachment of microorganisms to surfaces. First published evidences of bacterial attachment where described by Zobell (1943), who studied the adhesion of soil bacteria to slide surfaces. Bacterial attachment depends not only on the physical characteristics of the solid material like solid critical surface tension, the surface free energy, surface defects and humidity (Shi and Zhu, 2009), but also on the cell surface properties like the presence of extracellular appendages such as fimbriae, the cell-to-cell communication and EPS production (Simões et al., 2010). Physicochemical properties of cell surfaces are important in active bacterial adhesion. Most bacterial cells have a negatively charged surface which prevents their adhesion to other surfaces due to electrostatic repulsive forces. In order to overcome this initial repulsion, fimbriae, flagella and pili make their contribution with hydrophobic residues that promote attachment (Kumar and Anand, 1998).

The environmental factors, such as, pH, temperature, ionic concentration, and nutrient composition are important factors on the phenotypic change from the planktonic to sessile form (biofilm cell). In the planktonic state bacterial cells live as free individual floating organisms and in a sessile form they live attached to a surface, forming biofilms and functioning as a community (Poulsen, 1999).

The production of highly hydrated EPS by biofilm cells is also an important contribution to the initial adherence of the bacteria to the surface. This EPS plays an important role on the formation and maintenance of microcolonies, on biofilm structure, on the resistance to the environmental conditions and to antimicrobial agents and allows the capture of nutrients from the surrounding media (Pereira and Vieira, 2001).

Cell adhesion to a surface is a time dependent process that can be divided in the reversible and irreversible phases. The reversible phase is like an initial approximation of the cell to the surface and therefore long-distance interactions including electrostatic, hydrophobic interactions and van der Walls forces are involved. During the reversible phase microorganisms have the ability to detect the presence of nutrients on the surface. The irreversible phase occurs with the binding to the surface by short-range forces like hydrophobic, dipole-dipole forces, hydrogen interactions, or even covalent bonds (Poulsen, 1999).

In food processes, organic molecules and nutrients are accumulated on the surfaces of equipments and pipes and as a result a variety of macromolecular and colloid materials will be attracted into the surface, forming a conditioning film which, in turn, promotes the attachment of microorganisms (Hood and Zottola, 1997). This initial reversible binding will become irreversible when cell flagella and fimbriae establish connection to the surface.

From a simple microcolony, a layer of cells covering the surface will grow and divide using nutrients from the fluid environment and from the formed film, changing surface microenvironment properties (Kumar and Anand, 1998). Continuous attachment of cells to the surface, growth and EPS production will form a multilayer biofilm. As mentioned before, biofilm stability is influenced by the nutrient availability since cell death may promote biofilm detachment (Poulsen, 1999). Biofilms can reach thicknesses of a few millimeters depending on the culture conditions (Melo et al., 1992) and it has been reported for *Pseudomonas* species that an increase in flow velocity or in nutrient concentration is associated with the increase of cell attachment (Simões et al., 2010).

In the present work we studied biofilms formed by only one organism although in the food industry different microorganisms with different nutrient requirements can form a biofilm. Mixed species biofilms are normally thicker and more stable than pure biofilms since there is a higher adaptation capacity to the environmental conditions (James et al., 1995).

Cell replication and growth will occur in a biofilm together with convective and diffusive transport of oxygen and nutrients. However, as the biofilm ages, attached bacteria must be able to detach and migrate to other surfaces where they can establish new biofilms. Several factors can be involved in the detachment process, such as the presence of chemicals, associated for example with cleaning-in-place procedures (CIP) in the food industry, or shear stress and fluid dynamics effects of the media and also altered properties from the bacteria or substrate (Simões et al., 2010; Wong, 1998). When CIP procedures are applied they should guarantee cleaning of organic and inorganic contamination, disinfection from live microorganisms and elimination of chemical residues (Vlková et al., 2008).

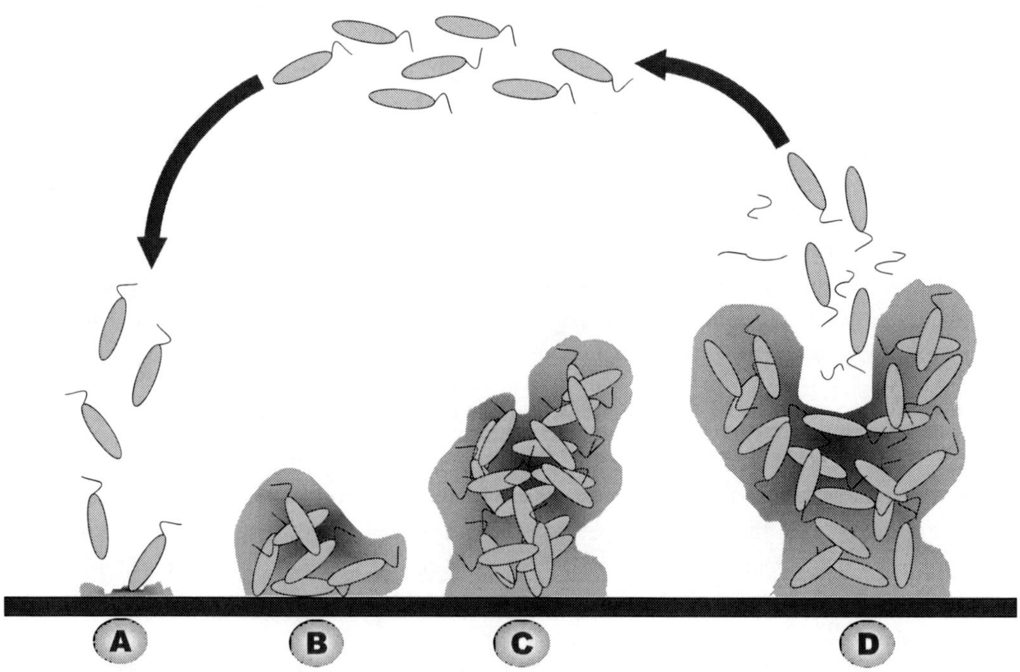

Figure 1. Schematic representation of a biofilm formation process: A – Initial reversible cell surface attachment, B – Transition from reversible to irreversible attachment with EPS matrix production, C – Mature biofilm with microcolonies formation, D – Biofilm cell detachment to planktonic stage (some daughter cells attach to surface to form new biofilm) (adapted from Van Houdt and Michiels (2005)).

The environment around the biofilm, including physical and chemical properties, determines the detachment process. Abrasion, erosion and sloughing processes can occur. Abrasion includes the scrap of biofilm by cleaning utensils or suspended particles whereas erosive detachment is usually associated to the fluid shear stress forces at biofilm-fluid interface. Sloughing is described as periodic detachment process of large particles of biomass from the biofilm (usually daughter cells) allowing these bacteria to migrate to a new surface and form new biofilms (Kumar and Anand, 1998; Moore et al., 2000; Stoodley et al., 2002). Biofilm formation steps can be observed in Figure 1.

1.2 Microbial Biofilms in Food Industry

Foodborne disease has been described by the World Health Organization (WHO) as "a disease of an infectious or toxic nature caused by, or thought to be caused by, the consumption of food or water" (Adams and Moss, 1995). Foodborne diseases, most of all with microbial origin, are considered to be the most widespread problem in the world and an important cause of reduced economic productivity (Adams and Moss, 1995).

The formation of biofilms in food equipment and pipes is a serious problem that can have consequences such as economic losses due to food spoilage, equipment damage and in some cases can be a risk factor affecting public health. There are numerous reports on the persistence of foodborne pathogens in biofilms formed on contact surfaces, affecting the quality and safety of the foods (Simões et al., 2010). Poor sanitation of food contact surfaces, equipment and processing environments have been reported as food disease outbreaks specially those involving *Listeria monocytogenes* and *Salmonella* species (Bower et al., 1996; Chmielewski and Frank, 2003). The pathogens *Yersinia enterocolitica*, *Campilobacter jejuni*, *Staphylococcus* spp. and *Escherichia coli* O157:H7 have also been reported as foodborne pathogens due to their ability to form biofilms on food contact surfaces (Marouani-Gadri et al., 2009; Shi and Zhu, 2009; Simões et al., 2010). Additionally, the growth of microorganisms and the deposit of by-products from their metabolism in food contact surfaces may cause serious problems like the increased corrosion rate, the reduced performance in heat transfer operations, and the clogging of pipes, cooling towers and water reservoirs (Chmielewski and Frank, 2003; Kumar and Anand, 1998; Poulsen, 1999; Wong, 1998). Biofilms are also a problem for example on water cooling systems or in air conditioning systems, where the human pathogen *Legionella pneumophila* can grow. Prevention by cleaning these systems is the best way to avoid the release of aerosols that constitute a serious health risk (Poulsen, 1999).

In the food industry, when pathogenic microorganisms form biofilms on food contact surfaces, problems with cross-contamination and post-processing contamination can occur. Cross-contamination can occur through direct contact or indirectly via contaminated equipment and utensils (Adams and Moss, 1995).

This kind of contamination by bacteria is thought to be a major contributing factor for epidemic or sporadic foodborne disease (Bloomfield and Scott, 1997; Chen et al., 2001).

For example, during food handling and food preparation, the microorganisms existing in raw foods can contact with the surfaces of the processing environment. Previous studies by Scott and Bloomfield (1990), reported that after the initial contact with microorganisms, bacteria like *Escherichia coli*, *Staphylococcus aureus* and *Salmonella* spp., can survive on hands, equipment surfaces or clothes for several hours or even days. Post-processing contamination refers to the persistence of microorganism in food product after cleaning procedures. This kind of contamination is normally associated to a low shelf life of the product and to aerosol formation during cleaning processes. Other potential sources of bacterial aerosol contamination are the floor, outdoor air, air conditioning systems and the personnel. In these cases air is the vehicle for pathogenic microorganism transmission (Kang and Frank, 1990).

In food processing environments, planktonic and biofilm bacteria encounter conditions of stress like temperature variations, dehydration and the presence of disinfectants or antimicrobial agents. Biofilm cells are morphologically different from planktonic cells and are able to resist in adverse environmental conditions. In fact, biofilm cells can be up to 500 times more resistant to antimicrobial agents including antibiotics and biocides (Trachoo, 2003).

In the context of food hygiene, biofilm control represents one of the most persistent challenges in food and industrial environments; since little is understood about the effectiveness of cleaning processes employed for biofilm removal. When an inappropriate cleaning process is applied, biofilm formation can be enhanced and bacterial cell resistance may build up. Also biofilm microorganisms seem to be more resistant towards antimicrobials, disinfectants or antibiotics (Chmielewski and Frank, 2003; Poulsen, 1999; Simões et al., 2010).

In most food processing industries, food contact surfaces are cleaned daily, however, places like storage tanks, walls, and pump exteriors are cleaned occasionally thus facilitating biofilm formation. Many problems associated with biofilm formation can be reduced by effective cleaning of food-processing equipment. There are some instances where mechanical cleaning is effective however the equipment design does not facilitate the access of the cleaning agent. Also stainless-steel surfaces exhibited better hygienic properties resisting to damage caused by cleaning process (Bower et al., 1996).

The judicious choice of process conditions that differ from the optimal growth conditions for microorganisms is one way to prevent biofilm formation. Also, an effective cleaning procedure should dissolve or break the biofilm EPS so that the cleaning agent can access and act upon the viable cells. Effective cleaning methods involve the use of a detergents that are specific for the microorganism found in the process and usually more than one cleaning process is applied to guarantee total removal of that microorganism (Chmielewski and Frank, 2003; Poulsen, 1999). In general, procedures for biofilm removal involve the dissolution of biofilm and organic material to which it adheres, followed by the application of disinfectants in order to inactivate and kill the microorganisms after cell detachment. Biological strategies for biofilm control using enzymes are also being studied. The enzymes to be used depend on the biofilm specific microflora (for the interaction with biofilm EPS) and given the heterogeneity of living microorganisms on most biofilms it likely that a mixture of specific enzymes will be required for the total degradation of biofilms (Kumar and Anand, 1998).

1.3 Nutrient-Load Effect on Biofilm

A variety of results on nutrient effect and biofilm formation and behavior have been reported during the last years. Most of the work is done with *Pseudomonas* species and mixed cultures. For these types of cultures, several reports by independent groups suggest that an increase in nutrient concentration enhances biofilm growth (Peyton, 1996; Rochex and Lebeault, 2007; Simões et al., 2006; Wijeyekoon et al., 2004). However, other studies have shown that an increase in nutrient concentration can promote cell detachment on biofilms (Rochex and Lebeault, 2007; Stoodley et al., 2001; Telgmann et al., 2004).

Despite the lack of information on *E. coli* biofilms, contradictory results have also been reported. As for *Pseudomonas* species and mixed cultures, some authors suggest that glucose will inhibit *E. coli* biofilm formation (Domka et al., 2006), an effect that can probably be explained by catabolite repression (Jackson et al., 2002), while others demonstrate the beneficial effect of glucose on biofilm formation (Bühler et al., 1998). In a growing *E. coli* culture, cell attachment seems to be influenced by nutrient starvation (Ito et al., 2008), although for *Pseudomonas* species starvation may promote cell detachment (Delaquis et al., 1989; Hunt et al., 2004).

Previous studies, by Tijhuis et al. (1994), reported that an increase in substrate loading will also increase biofilm thickness. The biofilm morphology was shown to be influenced by an increase in substrate loading and a decrease in detachment forces, resulted from the combination of these factors (Tijhuis et al., 1996). Another report by Wijeyekoon et al. (2004) investigated the effect of substrate loading rate on biofilm formation. Using microscopic methods, it was shown that an increase in substrate concentrations led to the formation of compact biofilms with lower porosity. Using *Pseudomonas aeruginosa*, Peyton (1996) defined biofilm thickness as the distance through which substrate and nutrients must diffuse to penetrate the biofilm and showed that a significant increase in biofilm thickness was associated to an increase in substrate loading rate, although shear stress variations did not have a direct effect on thickness.

As previously mentioned, biofilm formation is determined by the combination of attachment and detachment forces. In a food processing environment, the time needed for biofilm development can be very short. Attachment and detachment processes depend on physical and chemical factors such as shear forces and nutrient availability that will influence density and the external and internal strength of the biofilm (Chen et al., 2005; Garny et al., 2008; Klapper et al., 2002). Hydrodynamic conditions will influence biofilm formation in terms of nutrients and oxygen supply and will influence shear forces and thus the capacity of cells to attach a surface (Simões et al., 2007).

2. Experiental Set-up

2.1 Biofilm System and Sampling

An experimental apparatus including a recirculating tank, where planktonic cells grow, and a vertical flow cell for biofilm formation, was used in order to study the influence of nutrient load on the formation of biofilms by *E. coli* (Figure 2).

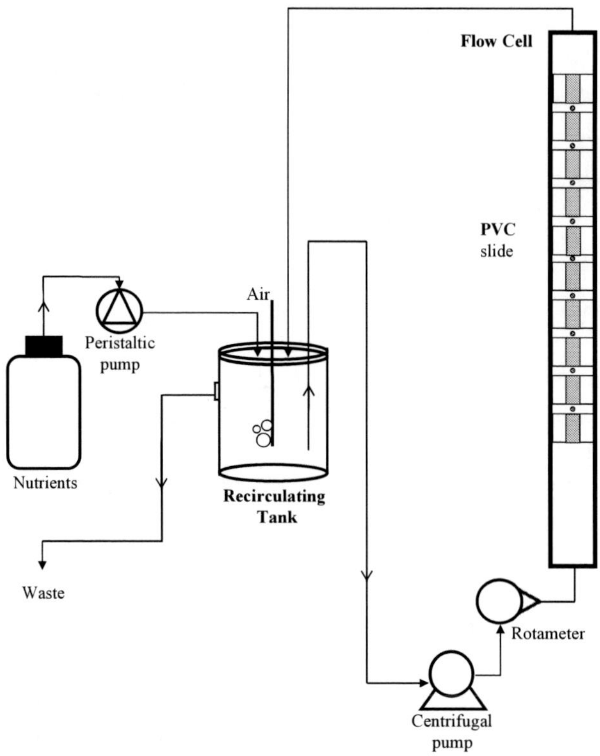

Figure 2. Schematic representation of the experimental system used for biofilm formation.

The recirculating tank was built with a cooling jacket to allow temperature control and the flow cell is composed by a semicircular Perspex duct (3.0 cm of diameter, corresponding to a hydraulic equivalent diameter of 1.8 cm and 1.09 m in length) with 10 apertures on its flat wall to fit removable rectangular pieces of Perspex (coupons). This system allows individual sampling of each coupon without disturbing the biofilm formed on the others (Simões et al., 2008). Polyvinyl chloride (PVC) slides (2 cm × 1 cm) were glued onto the coupons and were in contact with the bacterial suspension circulating in the flow cell reactor system. Biofilm was formed by recirculation of the bacterial *E. coli* suspension at 30 °C during 13 days. For each sampling day, the system was stopped to remove a coupon and carefully started again maintaining the same flow conditions. Sampling was made from the top to the bottom of the flow cell and each removed coupon was substituted with a new one that was previously cleaned and kept in ethanol. Biofilms were grown under turbulent flow with a Reynolds number of 5000.

2.2 Bacterial Growth and Culture Conditions

E. coli JM109(DE3) was used to study the influence of the nutrient load on biofilm formation. Two *E. coli* starter cultures were obtained by inoculation of 250 μL of glycerol stocks (kept at -80°C) to a volume of 0.250 L of inoculation media. This concentrated media consisted of 5.5 $g.L^{-1}$ glucose, 2.5 $g.L^{-1}$ peptone, 1.25 $g.L^{-1}$ yeast extract in phosphate buffer (1.88 $g.L^{-1}$ KH_2PO_4 and 2.60 $g.L^{-1}$ Na_2HPO_4), pH 7.0. The culture was grown on two 1 L shake-flasks (each containing 0.250 L of culture), incubated overnight at 30 ºC with orbital agitation (120 rpm). Culture growth was evaluated by measuring the optical density (O.D.) at 610 nm. The contents of both shake-flasks were pooled together and the resulting culture with an O.D. of 1 was used to continuously inoculate (at a flow-rate of 1.4 $L.h^{-1}$) the recirculation tank that had a working volume of 1L. When a higher capacity recirculation tank was used (5L) the culture was used to inoculate a 5L aerated tank containing 2 L of concentrated media and cellular growth proceeded until an OD of 1 was reached prior to inoculation of the recirculation tank.

In different experiments, recirculation tanks with different working volumes (1L or 5 L) were used and on the beginning of the inoculation these tanks contained 0.5 L or 2.5 L (respectively) of sterile, aerated water. During the experiment, these tanks were fed at different flow rates ($L.h^{-1}$) in order to obtain different dilution rates (h^{-1}).

Nutrient load, defined here as the product of the feed concentration and the dilution rate ($g.L^{-1}.h^{-1}$) was changed by varying not only the dilution rates but also the nutrient concentration on the feed.

For that purpose, a high concentration media composed of 0.55 $g.L^{-1}$ glucose, 0.25 $g.L^{-1}$ peptone, 0.125 $g.L^{-1}$ yeast extract and phosphate buffer (0.188 $g.L^{-1}$ KH_2PO_4 and 0.26 $g.L^{-1}$ Na_2HPO_4), pH 7.0 was used on some experiments whereas less concentrated media was used on other experiments.

These less concentrated feed contained 0.15 $g.L^{-1}$ or 0.05 $g.L^{-1}$ of glucose and the remaining nutrients had their concentrations proportionally reduced. All media components were purchased from Merck KGaA (Germany). Table 1 shows the summary of all the performed experiments and studied nutrient load ranges.

Table 1. Combination of three nutrient feed concentrations and four dilution rate values, resulting on three nutrient load ranges: low, intermediate and high. These ranges are represented on Figure 3. The symbols that are used on that figure are also included on the table for clarity.

Nutrient feed concentration ($g.L^{-1}$)	Dilution rates (h^{-1})				Range	Symbol
	0.0042	0.0126	0.0800	0.1650		
	Nutrient load ($g.L^{-1}.h^{-1}$)					
0.55	0.0023				Low	●
		0.0070			Intermediate	□
0.15	0.0006				Low	○
			0.0120		High	Δ
				0.0248	High	▲
0.05				0.0083	Intermediate	■

2.3 Analytical Methods

For biofilm cells analysis, flow cell coupons were removed daily and thickness and wet weight were determined. Biofilm thickness was determined using a digital micrometer. In this process, a needle connected to the digital micrometer (VS-30H, Mitsubishi Kasei Corporation) was applied on the top of the biofilm to cross the sample into the bottom of the biofilm. Ten measurements were made at random points of each coupon and the average value was determined. Wet weight of the biofilm was determined by weighing each coupon before the experience and comparing this value with the weight of the same coupon at each sampling time.

For planktonic cells analysis, optical density and glucose concentration were assayed. Optical density was measured using a Spectrophotometer at 610 nm (T80 UV/VIS Spectrometer/PG Instrument, Ltd). Glucose quantification was performed by the Dinitrosalicylic colorimetric method (DNS), as described by Miller (1959). In this method, sample and DNS reagent (25 μL each) are mixed and incubated at 80 °C for 5 min.

A volume of 250 μL of distilled water is added and the absorbance is measured at 540 nm. A calibration curve was previously constructed using glucose standards within the concentration range of 0.01-1.0 $g.L^{-1}$. Glucose consumption values were calculated by mass balance, multiplying the glucose concentration difference (feed minus concentration on the tank) by the feed flow rate.

3. Results and Discussion

We have tested six nutrient load conditions in order to assess the influence of this parameter on the formation of *E .coli* JM109(DE3) biofilms under dynamic conditions. The nutrient load was calculated by multiplying the dilution rate of the system by the feed concentration. Three feed concentrations were used (0.55, 0.15 and 0.050 $g.L^{-1}$) and four dilution rates were obtained by variation of the tank volume (and consequently the system volume) or the feed rate as described on Table 1. Three ranges of nutrient loads were obtained, the low range (0.0006 and 0.0023 $g.L^{-1}.h^{-1}$), the intermediate range (0.0070 and 0.0083 $g.L^{-1}.h^{-1}$) and the high range (0.0120 and 0.0248 $g.L^{-1}.h^{-1}$). This setup enabled not only the analysis of the nutrient load effect but also the dilution rate effect when the feed concentration was kept constant.

Figure 3 presents all the results from these experiments where the planktonic growth, glucose consumption, and biofilm thickness, wet weight and density were evaluated. Panels 1A, 2A and 3A show the influence of nutrient load on the optical density of planktonic cells. A decreasing optical density tendency can be observed on the graphics comparing low, intermediate and high nutrient load ranges. Higher optical density values are observed in panel 1A, for the conditions of 0.55 and 0.15 $g.L^{-1}$ glucose concentration and low dilution rate (0.0042 h^{-1}). Comparing panels 1A and 2A, it is possible to observe that a 3-fold increase in dilution rate using the same glucose concentration of 0.55 $g.L^{-1}$, results in lower optical density values. For the high nutrient load range, panel 3A shows even lower optical density values. In this case, when the system was fed with a nutrient solution containing 0.15 $g.L^{-1}$ of glucose, the dilution rate variation had no influence on planktonic growth.

Observing panels 1B, 2B and 3B concerning glucose consumption, we verify that on the highest nutrient range glucose consumption is higher and that on this range the increase in glucose consumption is proportional to the nutrient load increase of about 2-fold (panel 3B). For the low and intermediate nutrient loads (panels 1B and 2B), lower glucose consumptions were obtained and in fact, for these ranges glucose uptake is not dependent on the nutrient load. Additionally, a 1.7-fold increase on the nutrient load caused a 3-fold increase on glucose consumption when comparing the highest nutrient load from panel 2B with the lowest nutrient load from panel 3B. The fact that on low nutrient load ranges glucose consumption is not affected by this parameter and, after a certain threshold value is reached, glucose consumption becomes very sensitive to the nutrient load suggests a metabolic shift on the system.

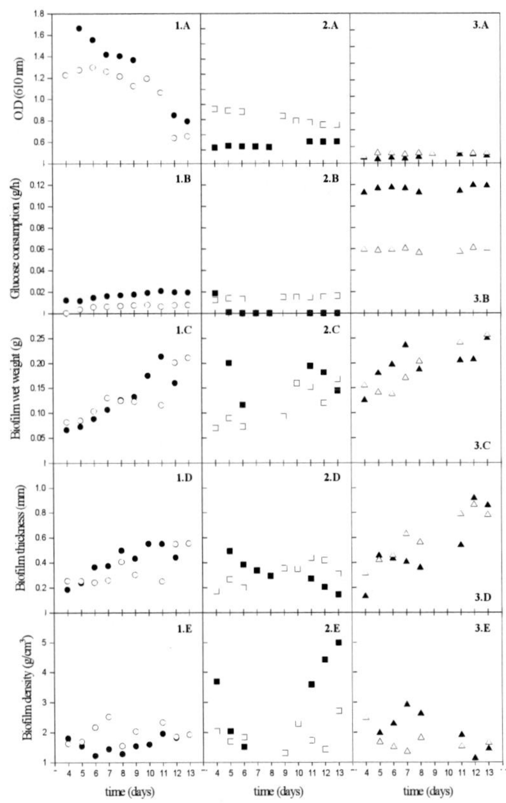

Figure 3. Experimental results for E. coli biofilms formed on the three nutrient load ranges: low (1) (○ - 0.0006 and ● - 0.0023 g.L-1.h-1), intermediate (2) (□ - 0.0070 and ■ - 0.0083 g.L-1.h-1), high (3) (△ - 0.0120 and ▲ - 0.0248 g.L-1.h-1). A - Optical density (610 nm); B – Glucose consumption (g.h-1); C – Biofilm wet weight (g); D – Biofilm thickness (mm); E – Biofilm density (g/cm3).

The effect of nutrient load on *E. coli* biofilms was quantified by measuring wet weight and thickness for each experimental day. The results of biofilm wet weight can be observed on panels 1C, 2C and 3C and the results of biofilm thickness can be observed on panels 1D, 2D and 3D. Both parameters seem to increase with increasing nutrient load ranges showing an opposite tendency when compared to planktonic cell behavior. Higher biofilm wet weight and thickness were obtained for the conditions of high nutrient load with glucose concentration of 0.15 $g.L^{-1}$ and on this range no effect of the 2-fold increase on the dilution rate was detected (panels 3C and 3D). Additionally, when the lowest dilution rate is used (0.0042 h^{-1}), a 3.7-fold increase in nutrient load has no significant effect on biofilm formation (graphics 1C and 1D). When the dilution rate is increased, for this same nutrient concentration of 0.15 $g.L^{-1}$, reaching higher nutrient load values (panels 3C and 3D), there is a visible increase in biofilm parameters. Although the effects of nutrient concentration and dilution rate are distinct when different nutrient load ranges are compared, the general trend is that thicker and heavier biofilms are obtained when using the highest nutrient load range as mentioned in literature (Peyton, 1996; Tijhuis et al., 1994).

Biofilm density graphic 2E, shows that denser biofilms are obtained in conditions of low nutrient concentration (0.05 $g.L^{-1}$) and high dilution rate (0.1650 h^{-1}). For the nutrient concentration of 0.15 $g.L^{-1}$, higher density values are also observed for the high dilution rate (0.1650 h^{-1}) (graphics 1E and 3E).

For high glucose concentration (0.55 $g.L^{-1}$) a 3-fold increase in dilution rate did not have an impact on biofilm density (graphics 1E and 2E). Our results suggest that lower glucose concentrations and high dilution rates favor an increase in biofilm density whereas higher nutrient loads favor heavier and thicker biofilms.

In conclusion, our results demonstrate the visible impact of nutrient load on *E. coli* biofilm formation. In the food industry the implementation of new strategies for biofilm control and cleaning is required. It is important to understand the interaction between bacteria and the food process surfaces and thus the mechanisms responsible for biofilm formation and development. The varying process conditions such as nutrient load, will promote the formation of biofilms with distinct phenotypes and consequently distinct susceptibilities to antimicrobials. The knowledge of the process conditions promoting the formation of susceptible biofilms will help the development of effective control strategies.

Acknowledgments

This research has been funded by the Project N° PTDC/BIO/69092/2006 supported by the Portuguese Science Foundation (FCT).

References

Adams, M. R., & Moss, M. O. (1995). *Food Microbiology*. Cambridge, UK: The Royal Society of Chemistry.

Bloomfield, S. F., & Scott, E. (1997). Cross-contamination and infection in the domestic environment and the role of chemical disinfectants. *Journal of Applied Microbiology*, *83*, 1-9.

Bower, C. K., McGuire, J., & Daeschel, M. A. (1996). The adhesion and detachment of bacteria and spores on food-contact surfaces. *Trends in Food Science & Technology*, *7*, 152-157.

Bryers, J. D. (1999). *Encyclopedia of Bioprocess Technology - Fermentation, Biocatalysis, and Bioseparation*. Flickinger, M. C., & Drew, S. W. (eds), John Wiley & Sons, NY, USA.

Bühler, T., Ballestero, S., Desai, M., & Brown, M. R. W. (1998). Generation of a reproducible nutrient-depleted biofilm of *Escherichia coli* and *Burkholderia cepacia* *Journal of Applied Microbiology*, *85(3)*, 457-462.

Chen, M. J., Zhang, Z., & Bott, T. R. (2005). Effects of operating conditions on the adhesive strength of Pseudomonas fluorescens biofilms in tubes. *Colloids and Surfaces B: Biointerfaces, 43(2)*, 61-71.

Chen, Y., Jackson, K. M., Chea, F. P., & Schaffner, D. W. (2001). Quantification and Variability Analysis of Bacterial Cross-Contamination Rates in Common Food Service Tasks. *Journal of Food Protection, 64(1).*

Chmielewski, R. A. N., & Frank, J. F. (2003). Biofilm Formation and Control in Food Processing Facilities. *Comprehensive Reviews in Food Science and Food Safety, 2*, 22-32.

Davey, M. E., & O'Toole, G. A. (2000). Microbial Biofilms: from Ecology to Molecular Genetics. *Microbiology and Molecular Biology Reviews, 64(4)*, 847-867.

Delaquis, P. J., Caldwell, D. E., Lawrence, J. R., & Mccurdy, A. R. (1989). Detachment of *Pseudomonas-Fluorescens* from Biofilms on Glass Surfaces in Response to Nutrient Stress. *Microbial Ecology, 18(3)*, 199-210.

Domka, J., Lee, J., & Wood, T. K. (2006). YliH (BssR) and YceP (BssS) regulate *Escherichia coli* K-12 biofilm formation by influencing cell signaling. *Applied and Environmental Microbiology, 72(4)*, 2449-2459.

Garny, K., Horn, H., & Neu, T. (2008). Interaction between biofilm development, structure and detachment in rotating annular reactors. *Bioprocess and Biosystems Engineering, 31(6)*, 619-629.

Hood, S. K., & Zottola, E. A. (1997). Adherence to stainless steel by foodborne microorganisms during growth in model food systems. *International Journal of Food Microbiology, 37*, 145-153.

Hunt, S. M., Werner, E. M., Huang, B., Hamilton, M. A., & Stewart, P. S. (2004). Hypothesis for the role of nutrient starvation in biofilm detachment. *Applied and Environmental Microbiology, 70(12)*, 7418-7425.

Ito, A., May, T., Kawata, K., & Okabe, S. (2008). Significance of rpoS during maturation of *Escherichia coli* biofilms. *Biotechnology and Bioengineering, 99(6)*, 1462-1471.

Jackson, D. W., Simecka, J. W., & Romeo, T. (2002). Catabolite repression of *Escherichia coli* biofilm formation. *Journal of Bacteriology, 184(12)*, 3406-3410.

James, G. A., Beaudette, L., & Costerton, J. W. (1995). Interspecies bacterial interactions in biofilms. *Journal of Industrial Microbiology, 15*, 257-262.

Juhna, T., Birzniece, D., Larsson, S., Zulenkovs, D., Sharipo, A., Azevedo, N. F., Menard-Szczebara, F., Castagnet, S., Feliers, C., & Keevil, C. W. (2007). Detection of Escherichia coli in biofilms from pipe samples and coupons in drinking water distribution networks. *Applied and Environmental Microbiology, 73(22)*, 7456-7464.

Kang, Y. J., & Frank, J. F. (1990). Characteristics of Biological Aerosols in Dairy Processing Plants. *Journal of Dairy Science, 73*, 621-626.

Kjelleberg, S., & Molin, S. (2002). Is there a role for quorum sensing signals in bacterial biofilms? *Current Opinion in Microbiology, 5*, 254-258.

Klapper, I., Rupp, C. J., Cargo, R., Purvedorj, B., & Stoodley, P. (2002). Viscoelastic fluid description of bacterial biofilm material properties. *Biotechnology and Bioengineering, 80(3)*, 289-296.

Kumar, C. G., & Anand, S. K. (1998). Significance of microbial biofilms in food industry: a review. *International Journal of Food Microbiology, 42*, 9-27.

Marouani-Gadri, N., Augier, G., & Carpentier, B. (2009). Characterization of bacterial strains isolated from a beef-processing plant following cleaning and disinfection - Influence of isolated strains on biofilm formation by Sakai and EDL 933 *E. coli* O157:H7. *International Journal of Food Microbiology, 133(1-2)*, 62-67.

Melo, L. F., Bott, T. R., Fletcher, M., & Capdeville, B. (1992). *Biofilms - Science and Technology.*

Miller, G. L. (1959). Use of dinitrosalicylic acid reagent for determination of reducing sugar. *Analytical Chemistry, 31(3)*, 426-428.

Moore, G. F., Dunsmore, B. C., Jones, S. M., Smejkal, C. W., Jass, J., Stoodley, P., & Lappin-Scott, H. M. (2000). *Community structure and co-operation in biofilms.* Cambridge, UK: Cambridge University Press.

O'Toole, G., Kaplan, H. B., & Kolter, R. (2000). Biofilm formation as microbial development. *Annual Review of Microbiology, 54*, 49-79.

Pereira, M. O., & Vieira, M. J. (2001). Effects of the interactions between glutaraldehyde and the polymeric matrix on the efficacy of the biocide against Pseudomonas fluorescens biofilms. *Biofouling, 17(2)*, 93-101.

Peyton, B. M. (1996). Effects of shear stress and substrate loading rate on *Pseudomonas aeruginosa* biofilm thickness and density. *Water Research, 30(1)*, 29-36.

Poulsen, L. V. (1999). Microbial Biofilm in Food Processing. *Lebensmittel-Wissenschaft und-Technologie, 32*, 321-326.

Rochex, A., & Lebeault, J. M. (2007). Effects of nutrients on biofilm formation and detachment of a *Pseudomonas putida* strain isolated from a paper machine. *Water Research, 41(13)*, 2885-2892.

Scott, E., & Bloomfield, S. F. (1990). The survival and transfer of microbial contamination via cloths, hands and utensils. *Journal of Applied Bacteriology, 68*, 271-278.

Sharma, M., & Anand, S. K. (2002). Biofilms evaluation as an essential component of HACCP for food/dairy processing industry - a case. *Food Control, 13(6-7)*, 469-477.

Shi, X., & Zhu, X. (2009). Biofilm formation and food safety in food industries. *Trends in Food Science & Technology, 20*, 407-413.

Sihorkar, V., & Vyas, S. P. (2001). Biofilm Consortia on Biomedical and Biological Surfaces: Delivery and Targeting Strategies. *Pharmaceutical Research, 18(9)*, 1247-1254.

Simões, L. C., Azevedo, N., Pacheco, A., Keevil, C. W., & Vieira, M. J. (2006). Drinking water biofilm assessment of total and culturable bacteria under different operating conditions. *Biofouling, 22(2)*, 91-99.

Simões, M., Simões, L. C., & Vieira, M. J. (2008). Physiology and behavior of *Pseudomonas fluorescens* single and dual strain biofilms under diverse hydrodynamics stresses. *International Journal of Food Microbiology, 128(2)*, 309-316.

Simões, M., Simões, L. C., & Vieira, M. J. (2009). Species association increases biofilm resistance to chemical and mechanical treatments. *Water Research, 43(1)*, 229-237.

Simões, M., Simões, L. C., & Vieira, M. J. (2010). A review of current and emergent biofilm control strategies. *LWT - Food Science and Technology, 43*, 573-583.

Simões, M., Pereira, M. O., Sillankorva, S., Azeredo, J., & Vieira, M. J. (2007). The effect of hydrodynamic conditions on the phenotype of *Pseudomonas fluorescens* biofilms. *Biofouling, 23(4)*, 249-258.

Stoodley, P., Sauer, K., Davies, D. G., & Costerton, J. W. (2002). Biofilms as complex differentiated communities. *Annual Review of Microbiology, 56*, 187-209.

Stoodley, P., Wilson, S., Hall-Stoodley, L., Boyle, J. D., Lappin-Scott, H. M., & Costerton, J. W. (2001). Growth and detachment of cell clusters from mature mixed-species biofilms. *Applied and Environmental Microbiology, 67(12)*, 5608-5613.

Telgmann, U., Horn, H., & Morgenroth, E. (2004). Influence of growth history on sloughing and erosion from biofilms. *Water Research, 38(17)*, 3671-3684.

Tijhuis, L., van Loosdrecht, M. C. M., & Heijnen, J. J. (1994). Formation and growth of heterotrophic aerobic biofilms on small suspended particles in airlift reactors. *Biotechnology Bioengineering, 44(5)*, 595-608.

Tijhuis, L., Hijman, B., & Van Loosdrecht, M. C. M. (1996). Influence of detachment, substrate loading and reactor scale on the formation of biofilms in airlift reactors. *Applied Microbiology and Biotechnology, 45*, 7-17.

Trachoo, N. (2003). Biofilms and the food industry. *Songklanakarin Journal of Science and Technology, 25(6)*, 807-815.

Van Houdt, R., & Michiels, C. W. (2005). Role of bacterial cell surface structures in *Escherichia coli* biofilm formation. *Research in Microbiology, 156(5-6)*, 626-633.

Vlková, H., Babák, V., Seydlová, R., Pavlík, I., & Schlegelová, J. (2008). Biofilms and Hygiene on Dairy Farms and in the Dairy Industry: Sanitation Chemical Products and their Effectiveness on Biofilms - a Review. *Czech Journal of Food Science, 26(5)*, 309-323.

Watnick, P., & Kolter, R. (2000). Biofilm, city of microbes. *Journal of Bacteriology, 182(10)*, 2675-2679.

Wijeyekoon, S., Mino, T., Satoh, H., & Matsuo, T. (2004). Effects of substrate loading rate on biofilm structure. *Water Research, 38(10)*, 2479-2488.

Wong, A. C. L. (1998). Biofilms in Food Processing Environments. *Journal of Dairy Science, 81*, 2765-2770.

Zobell, C. E. (1943). The Effect of Solid Surfaces Upon Bacterial Activity. *Scripps Institution of Oceanography, 204*, 39-56.

In: Advances in Medicine and Biology Research, Vol. 14
Editor: Leon V. Berhardt
ISBN: 978-1-61761-930-4

Chapter IX

Physiology and Pathophysiology of Megalin-Mediated Endocytosis in Renal Proximal Tubule Cells

Akihiko Saito*[*1], Hiroyoshi Sato[2], Noriaki Iino[3] and Tetsuro Takeda[4]

Department of Applied Molecular Medicine, Niigata University Graduate School of Medical and Dental Sciences, 1-757 Asahimachi-dori, Chuo-ku, Niigata, 951-8510, Japan[1]

Department of Applied Molecular Medicine, Niigata University Graduate School of Medical and Dental Sciences, 1-757 Asahimachi-dori, Chuo-ku, Niigata, 951-8510, Japan[2]

Division of Clinical Nephrology and Rheumatology, Niigata University Graduate School of Medical and Dental Sciences, 1-757 Asahimachi-dori, Chuo-ku. Niigata, 951-8510, Japan[3]

Division of Clinical Nephrology and Rheumatology, Niigata University Graduate School of Medical and Dental Sciences, 1-757 Asahimachi-dori, Chuo-ku. Niigata, 951-8510, Japan[4]

Abstract

Receptor-mediated endocytosis is a pivotal function of renal proximal tubule cells (PTC) to reabsorb and metabolize substantial amounts of proteins in glomerular filtrates. The function accounts for the essential conservation of nutrients, carrier-bound vitamins and trace elements filtered by glomeruli. Impairment of the process results in a loss of such substances and development of proteinuria, an important clinical sign of kidney disease and a risk marker for cardiovascular disease. Megalin is a multi-ligand endocytic receptor abundantly expressed primarily at clathrin-coated pits of PTC, playing a central

[*]Corresponding author: Akihiko Saito, M.D., Ph.D., Department of Applied Molecular Medicine, Niigata University Graduate School of Medical and Dental Sciences, 1-757 Asahimachi-dori, Chuo-ku, Niigata, 951-8510, Japan, Telephone: 81-25-227-0915, Fax 81-25-227-0914, e-mail: akisaito@med.niigata-u.ac.jp

role in the process. Megalin is a large (~600 kDa) glycoprotein member of the low-density lipoprotein receptor family. Megalin cooperates with other membrane molecules, such as the cubilin-amnionless complex, Na^{+}/H^{+} exchanger isoform 3, type IIa sodium phosphate cotransporter and chloride channel-5 in PTC, in order to transport various ligands into the endocytic pathway to lysosomes. Many intracellular adaptor proteins have been identified to interact with the cytoplasmic tail of megalin for intracellular trafficking. Megalin is also known to participate in signal transduction in the cells. Megalin-mediated endocytic overload of glomerular-filtered proteins leads to damage of PTC. Advanced glycation end products, generated in the circulation of patients with diabetes, are also presumed to injure PTC via megalin-mediated endocytosis. Further studies are needed to elucidate the mechanism of megalin-mediated endocytosis and to promote future development of strategies for preventing damage of PTC.

Introduction

Receptor-mediated endocytosis is a pivotal function of renal proximal tubule cells (PTC), through which the cells reabsorb and metabolize proteins and other substances from glomerular filtrates. This reabsorbing process is extremely efficient as evidenced by the virtual protein-free urine in humans, and it accounts for the essential conservation of nutrients, carrier-bound vitamins and trace elements filtered by glomeruli. Impairment of the process results in a loss of such substances and development of proteinuria. Megalin is a membrane receptor that plays a central role in the endocytic function of PTC. Megalin cooperates with various molecules in the cells, taking up ligands into the endocytic pathway to lysosomes, as well as mediating signal transduction. In this review, we focus on recent study progress on megalin and its associated molecules. We also discuss how impaired or overloaded endocytosis induces PTC damage that is associated with the onset of proteinuria and development of chronic kidney disease.

Megalin, a Major Endocytic Receptor in PTC

Megalin is a large (~600 kDa) glycoprotein member of the low-density lipoprotein (LDL) receptor family (Hjalm et al., 1996; Saito, Pietromonaco, Loo, & Farquhar, 1994) that is primarily expressed at clathrin-coated pits and partly at microvilli of PTC (Figure) (Christensen, Verroust, & Nielsen, 2009; Verroust, Kozyraki, Hammond, Moestrup, & Christensen, 2000). Megalin contains a huge extracellular domain responsible for its multispecific functions. The domain consists of 4,398 amino acids (in humans) and is made by three types of characteristic repeats of the LDL receptor family: 1) 36 cysteine-rich complement-type repeats organized in four clusters, 2) 16 growth factor repeats separated by 8 YWTD containing spacer regions involved in pH dependent release of ligands in endosomal compartments (Davis et al., 1987), and 3) a single epidermal growth factor-like repeat. The extracellular domain is followed by a single transmembrane segment and a cytoplasmic domain of 209 amino acids. The cytoplasmic tail contains two endocytic motifs (NPXY) mediating clustering into clathrin-coated pits and a NPXY-like motif (NQNY) involved in apical sorting of the receptor (Takeda, Yamazaki, & Farquhar, 2003), as well as other protein interaction motifs (SH3 and PDZ domains) and phosphorylation sites (Hjalm et al., 1996; Saito et al., 1994). The physiological potential of these regulatory motifs has not yet been fully understood.

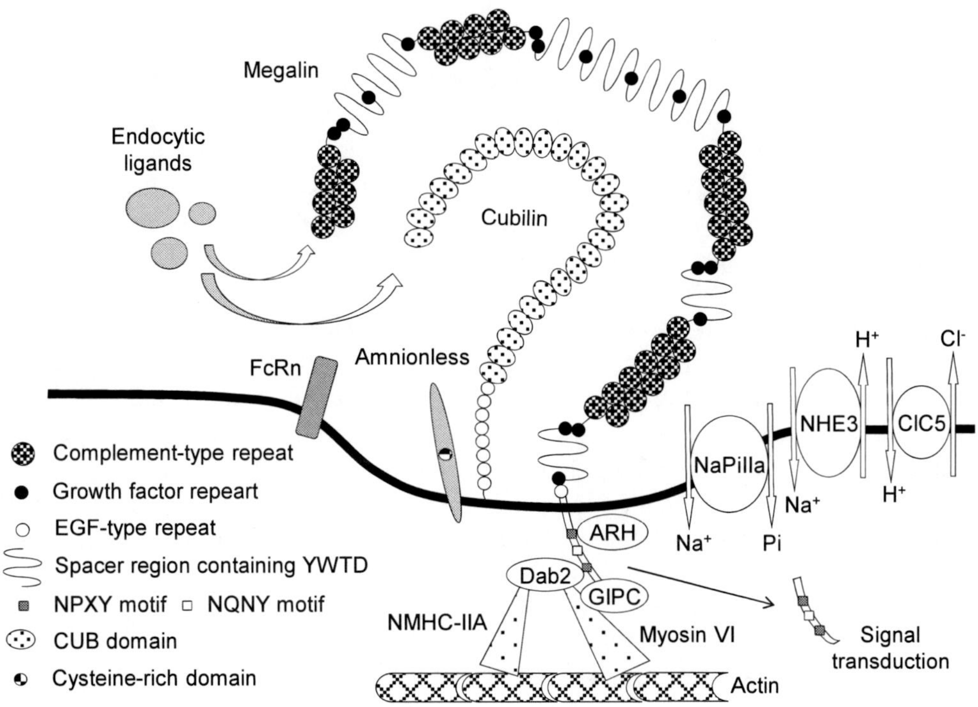

Figure 1. Megalin and its associated molecules involved in receptor-mediated endocytosis in PTC. On the apical membrane of PTC, various molecules are involved in the process of receptor-mediated endocytosis. Megalin, playing a central role in endocytosis, cooperates with other membrane proteins such as the cubilin-amnionless complex (CUBAM), NHE3, NaPiIIa and ClC5. Megalin and CUBAM directly bind a variety of ligands, whereas NHE3 and ClC5 are involved in endosomal acidification, which is important for further processing of endocytosed proteins. FcRn seems to be involved in a retrieval pathway in PTC to reclaim albumin from glomerular filtrates. Megalin also interacts with intracellular adaptor proteins such as ARH, Dab2 and GIPC. Dab2 binds to motor proteins, myosin VI and NMHC IIA, which may mediate endocytic trafficking of the molecular complexes through actin filaments. The cytoplasmic tail of megalin may be released from the membrane by γ-secretase-like activities and involved in intracellular signal transduction.

Megalin plays a critical role in the reabsorption of glomerular-filtered substances including albumin and low-molecular-weight proteins. Also, megalin may take up proteins that are released by PTC to the apical tubular space. Megalin knockout mice display low-molecular-weight proteinuria and albuminuria (Leheste et al., 1999). Furthermore, patients with Donnai-Barrow and facio-oculo-acoustico-renal syndromes, caused by mutations in the megalin gene, show increased urinary excretion of albumin and low-molecular-weight proteins (Kantarci et al., 2007). In this endocytic process, meglin mediates the conservation of carrier bound vitamins and trace elements filtered by glomeruli, including vitamin D (Nykjaer et al., 2001), vitamin A (Christensen et al., 1999), vitamin B_{12} (Moestrup et al., 1996), and iron (Kozyraki et al., 2001). Megalin cooperates with a variety of molecules at the apical membranes and also in the cytoplasm of PTC (Figure) as described in the next section.

Molecules Associated with Megalin's Functions in PTC

1) Cubilin-amnionless complex (CUBAM)

Cubilin is a 460-kDa peripheral glycoprotein, thus lacking transmembrane and intracellular segments, yet anchored to the apical membranes in PTC. It was originally identified as the receptor for intrinsic factor-vitamin B_{12} (Seetharam, Christensen, Moestrup, Hammond, & Verroust, 1997; Seetharam, Levine, Ramasamy, & Alpers, 1988), and its gene defects are the causes of hereditary megaloblastic anaemia 1 or Imerslund-Gräsbeck syndrome (selective vitamin B_{12} malabsorption with proteinuria) (Aminoff et al., 1999). Cubilin is also involved in the absorption of various protein ligands present in glomerular filtrates, including albumin, transferrin and vitamin D-binding protein (Christensen et al., 2009). Cubilin is known to interact with megalin for its endocytic functions (Kozyraki et al., 2001; Yammani, Seetharam, & Seetharam, 2001); however, it is bound more firmly by a protein called amnionless, forming a complex named CUBAM, to be translocated to the plasma membrane (Coudroy et al., 2005; Fyfe et al., 2004). Amnionless, a 38-50 kDa membrane protein with a single transmembrane domain, was initially identified as a component for normal development of the trunk mesoderm derived from the middle streak (Kalantry et al., 2001). Its gene defects also cause hereditary megaloblastic anaemia (Tanner et al., 2003). Nonetheless, the role of amnionless in PTC is not fully understood.

2) Na^+/H^+ exchanger isoform 3 (NHE3)

NHE3, the main NHE isoform in PTC, mediates isotonic reabsorption of approximately two thirds of the filtered NaCl and water, the reabsorption of bicarbonate, and the secretion of ammonium (Bobulescu & Moe, 2009). It also contributes to the reabsorption of filtered citrate, amino acids, oligopeptides by providing H^+ used for the H^+-coupled cotransporters. Enhanced NHE3 activity is assumed to be a factor for increased Na^+ reabsorption and development of hypertension in diabetes. NHE3 was reported to interact with megalin in intermicrovillar clefts of PTC (Biemesderfer, DeGray, & Aronson, 2001; Biemesderfer, Nagy, DeGray, & Aronson, 1999). After endocytosis with megalin, NHE3 is postulated to utilize the outward transvesicular Na^+ gradient of endocytic vesicles and early endosomes to drive inward movement of H^+ and endosomal acidification, which is important for dissociating reabsorbed ligand proteins off megalin both for further processing of the ligands and for recycling of megalin to the cell surface.

3) Type IIa sodium phosphate cotransporter (NaPi-IIa)

Renal reabsorption of inorganic phosphate is mediated by NaPi-IIas of PTC. Changes in renal phosphate handling are mainly attributable to altered NaPi-IIa brush border membrane expression (Hernando et al., 2005). Parathyroid hormone (PTH) induces inactivation of NaPi-IIa by endocytic retrieval and degradation. Adequate steady-state expression of NaPi-IIa and the capacity of PTC to react on PTH-mediated inactivation of NaPi-IIa by endocytosis and intracellular translocation were found to require the presence of megalin using kidney-specific megalin gene knockout mice (Bachmann et al., 2004). However, NHE3 and NaPiIIa, both expressed at apical microvilli, were reported to localize in distinct domains: NHE3 within lipid rafts and NaPiIIa in nonrafts (Riquier, Lee, & McDonough, 2009). Further studies are needed to clarify the different mechanisms for interaction of these two molecules with megalin.

4) Chloride channel ClC-5

ClC-5 is a 746-amino acid protein originally assumed to belong to the voltage-gated chloride channel family (Uchida, 2000), but more recent evidence suggests that it may function as a H^+/Cl^- exchanger (Plans, Rickheit, & Jentsch, 2009; Scheel, Zdebik, Lourdel, & Jentsch, 2005). In kidney, ClC-5 is highly expressed in PTC and α- and β-intercalated cells of collecting ducts (Jentsch, 2005). In PTC, ClC-5 is located at the apical endosomes together with electrogenic V-type H^+-ATPases (Jentsch, 2005), where it has a complementary function in endosomal acidification (Gunther, Luchow, Cluzeaud, Vandewalle, & Jentsch, 1998). The physiological relevance of ClC-5 in renal functions became evident when mutations in the *CLCN5* gene were identified in patients with Dent's disease, an X-linked renal tubular disorder (Jentsch, 2005). This disorder is characterized by low molecular weight proteinuria, hypercalciuria, nephrocalcinosis, nephrolithiasis, aminoaciduria, phosphaturia, glycosuria and renal failure (Wrong, Norden, & Feest, 1994). The precise mechanism of this abnormality is not entirely clear but possibly results from defective acidification and/or reduced expression of megalin and cubilin in PTC (Christensen et al., 2003; Tanuma et al., 2007).

Regulation of Megalin Expression

Cellular expression of megalin was found to be downregulated by the action of TGF-ß (Russo, del Re, Brown, & Lin, 2007). We also found that megalin expression is upregulated in opossum-derived cultured PTC by treatment with insulin or high-concentration glucose (17.5 mM), whereas it is downregulated by angiotnsin II (Hosojima et al., 2009). Furthermore, we demonstrated that there is competitive cross talk between anigotensin II type 1 receptor- and insulin-mediated signaling pathways in the regulation of megalin expression in the cells, suggesting a counter-balanced mechanism that regulates megalin expression and functions in PTC (Hosojima et al., 2009).

Decreased megalin expression in PTC has been found in the early diabetic stages in experimental animals (Russo et al., 2007; Tojo et al., 2001). It is also suggested that the functions of megalin are impaired in patients in the early stages of diabetic nephropathy, since low-molecular-weight proteinuria are frequently observed in patients at these stages (Hong, Hughes, Chia, Ng, & Ling, 2003; Pontuch, Jensen, Deckert, Ondrejka, & Mikulecky, 1992). Thus, the altered regulation of megalin expression and functions must be significantly responsible for the early development of proteinuria/albuminuria in diabetic patients. The mechanisms of the regulation remain to be further investigated.

Intracellular Adaptor Proteins and Motors that Interact with Megalin

Various intracellular adaptor proteins, such as JIP1 and JIP2, SEMCAP-1 (GIPC), ANKRA, Dab2, PDS-95, MegBP and ARH, bind to megalin's cytoplasmic tail (Gotthardt et al., 2000; Larsson et al., 2003; Lou, McQuistan, Orlando, & Farquhar, 2002; Nagai, Meerloo, Takeda, & Farquhar, 2003; Oleinikov, Zhao, & Makker, 2000; Petersen et al., 2003; Rader, Orlando, Lou, & Farquhar, 2000). ARH and Dab2 are components of the clathrin coat, and they bind to the first and third NPXY motifs of megalin, respectively, through their PTB domains (Nagai et al., 2003; Oleinikov et al., 2000). ARH and Dab2 are known to interact with motor proteins as described below. Dab2 is also known to mediate signal transduction (Hocevar, Smine, Xu, & Howe, 2001; Prunier, Hocevar, & Howe, 2004).

The mechanisms of intracellular transport of megalin are largely unknown. Reverse-direction molecular motor myosin VI was found to link to Dab2 and GIPC, which binds to the cytoplasmic tail of megalin, and is assumed to be involved in the endocytosis in PTC (Hasson, 2003). However, myosin VI knockout mice, used as an animal model for deafness, showed no apparent renal manifestation presenting proteinuria (Avraham et al., 1995).

We recently identified that another motor protein, nonmuscle myosin heavy chain IIA (NMHC IIA), binds to Dab2 and is involved in megalin-mediated endocytosis (Hosaka et al., 2009). Genetic alterations of NMHC-IIA are known to cause inherited human diseases, known as *MYH9* disorders, which are characterized by giant platelets, thrombocytopenia and granulocyte inclusions (Kelley, Jawien, Ortel, & Korczak, 2000; Seri et al., 2000). The spectrum of diseases due to mutations in the gene includes May-Hegglin anomaly, Sebastian syndrome, Fechtner syndrome and Epstein syndrome (Arrondel et al., 2002; Heath et al., 2001; Kelley et al., 2000; Seri et al., 2000). It has been also reported that all of these disorders are related to development of kidney disease (Heath et al., 2001; Seri et al., 2002). The manifestation of kidney disease in *MYH9* disorders indicates the importance of NMHC-IIA in maintaining normal kidney functions, which has been also verified by two recent genome wide scan analyses (Kao et al., 2008; Kopp et al., 2008).

Another megalin-binding adaptor protein ARH also associates with motor and centrosomal proteins and is involved in centrosome assembly and cytokinesis (Lehtonen et al., 2008). The relevance of ARH's association with such molecules in the regulation of megalin transport remains undetermined.

Handling of Albumin in PTC, Related to Mechanisms of Albuminuria

Albumin (~69 kDa) is the most abundant circulating protein, carrying a variety of substances in plasma. Glomerular albumin filtration is assumed to be 3-6 g/day in humans (Gekle, 2005). Only negligible amounts of albumin are detected in urine, and the substantial remaining of glomerular-filtered albumin is reabsorbed in PTC via endocytosis, mediated by megalin and CUBAM. Albuminuria is an important clinical sign of kidney disease such as diabetic nephropathy (Mogensen, 1984; Viberti et al., 1982), as well as it is a risk marker of cardiovascular disease (CVD) (Gerstein et al., 2001; Wachtell et al., 2003). Impaired endocytic functions of PTC for albumin are relevant to the mechanisms of albuminuria.

After endocytosis, albumin is considered to be transferred to lysosomes for degradation to amino acids (Maunsbach, 1966). On the contrary, Comper's group has claimed the presence of a retrieval or transcytic pathway of albumin in PTC (Comper, Hilliard, Nikolic-Paterson, & Russo, 2008). A recent analysis using neonatal Fc receptor (FcRn) knockout mice supports the retrieval pathway in PTC where the receptor appears to play a critical role to recollect albumin from the glomerular filtrates (Sarav et al., 2009). Application of *in vivo* visualization techniques of albumin degradation would be useful for further studies on this issue (Slattery et al., 2008).

The association of albuminuria with development of CVD may be related to the impairment of metabolic or synthetic functions of PTC that may contribute to systemic vascular damage. For instance, vitamin D deficiency, which is caused by megalin dysfunction, is independently associated with increased cardiovascular mortality (Dobnig et al., 2008; Pilz et al., 2008). Selenoprotein P, a major carrier of selenium, is taken up by megalin (Olson, Winfrey, Hill, & Burk, 2008) and provides selenium for synthesizing glutathione peroxidase 3 (GPx3) in PTC (Avissar et al., 1994; Maser, Magenheimer, & Calvet, 1994). GPx3 is secreted into the extracellular space from where it enters blood and acts as antioxidant (Whitin, Bhamre, Tham, & Cohen, 2002). Therefore, reduced uptake of selenoprotein P in PTC due to impaired megalin function may result in decreased GPx3 synthesis in the cells and may be associated with development of vascular diseases.

Overloaded Endocytosis-Induced PTC Injury

Overloaded endocytosis in PTC due to increased glomerular protein filtration has been postulated to be a cause of tubulointerstitial injury. Albumin overload to PTC induces oxidative stress and upregulated changes in stress-related gene expression (Shalamanova, McArdle, Amara, Jackson, & Rustom, 2007). Megalin is identified as the key molecule to initiate the pathogenic process (Motoyoshi et al., 2008). Overloaded cultured PTC with albumin induces decreased expression of megalin and its dissociation with PKB, leading the cells to apoptosis by reduced Bad phosphorylation (Caruso-Neves, Pinheiro, Cai, Souza-Menezes, & Guggino, 2006). In diabetes, advanced glycation end products (AGE) are generated in the circulation and involved in various types of cellular damage (Dronavalli, Duka, & Bakris, 2008). Megalin also mediates the endocytosis of glomerular-filtered AGE in PTC (Saito et al., 2003; Saito et al., 2005), which causes toxicity in the cells (Sebekova et al., 1998; Verbeke, Perichon, Friguet, & Bakala, 2000). In metabolic syndrome or dyslipidemia, free fatty acids are delivered to PTC with carrier proteins such as albumin or liver-type fatty acid binding protein (Oyama et al., 2005). Because of their hyper-metabolic functions with high oxygen demands, PTC are vulnerable to hypoxic injury. Endocytic machinery is also affected by hypoxic cellular stimuli (Wang et al., 2009). Metabolically overloaded PTC are activated to express proinflammatory cytokines, such as MCP1 and TNFα, and lead to apoptosis (Motoyoshi et al., 2008) or epithelial-mesenchymal transition (Burns, Kantharidis, & Thomas, 2007; Strutz, 2009).

Megalin-Mediated Signal Transduction

Membrane proteins such as those belonging to the Notch and amyloidal precursor protein families are involved in signal transduction pathways via regulated intramembrane proteolysis (RIP) (Brown, Ye, Rawson, & Goldstein, 2000; Louvi & Artavanis-Tsakonas, 2006). Biemesderfer and his colleagues identified that megalin also undergoes RIP. They showed 1) that high levels of γ-secretase are expressed in the brush border and involced in the endocytic pathway of PTC where the enzyme colocalizes with megalin, 2) that megalin is subjected to PKC-regulated, metalloprotease-mediated ectodomain shedding that produces a 35- to 40-kDa megalin COOH-terminal fragment (MCTF), and 3) that the MCTF is membrane bound and is constitutively processed by γ-secretase activity to produce a soluble megalin intracellular domain (MICD) (Zou et al., 2004). They also found evidence suggesting that the COOH-terminal domain of megalin regulates megalin and NHE3 gene expression (Li, Cong, & Biemesderfer, 2008). These findings strongly indicate that not only be megalin involved in scavenging functions, but also that it participates in signal transduction in PTC.

Clinical Application of Detection of Megalin Shed in Urine

Urinary megalin excretion in humans was first analyzed with immunoblotting procedures using urine samples that had been pretreated with dialyisis and lyophilization (Norden et al., 2002). In the urine samples from normal subjects, they identified that megalin is present as a soluble form detected only with anti-ectodomain antibodies but not with anti-CT antibodies. They also found that the soluble form of urinary megalin is decreased in 8 of 9 families with Dent's disease and in 2 families with Lowe's syndrome, which are both genetic disorders of endocytic functions in PTC. In contrast, increased urinary excretion of megalin, evaluated using gel-based liquid chromatography-mass spectrometry, was reported in microalbuminuric patients with type 1 diabetes (Thrailkill et al., 2009). We are establishing ELISA systems to quantitate megalin proteins in human urine, which will be useful for clinically evaluating PTC injury.

Conclusion

Megalin, an endocytic receptor, mediates the conservation of nutrients and carrier-bound vitamins and trace elements in glomerular filtrates via interaction with various molecules in PTC. Megalin also plays a critical role in uptake of pathological substances or overloaded endocytosis that may lead to cellular damage. Megalin-mediated signal transduction may be also involved in the process. Further studies are imperative to elucidate the molecular mechanisms, generate novel biomarkers and develop therapeutic strategies of PTC damage.

Acknowledgments

The authors thank Taeko Soma and Ryoko Niizuma for their help with preparing the manuscript.

References

Aminoff, M., Carter, J. E., Chadwick, R. B., Johnson, C., Grasbeck, R., Abdelaal, M. A., et al. (1999). Mutations in CUBN, encoding the intrinsic factor-vitamin B12 receptor, cubilin, cause hereditary megaloblastic anaemia 1. *Nat Genet, 21*(3), 309-313.

Arrondel, C., Vodovar, N., Knebelmann, B., Grunfeld, J. P., Gubler, M. C., Antignac, C., et al. (2002). Expression of the nonmuscle myosin heavy chain IIA in the human kidney and screening for MYH9 mutations in Epstein and Fechtner syndromes. *J Am Soc Nephrol, 13*(1), 65-74.

Avissar, N., Ornt, D. B., Yagil, Y., Horowitz, S., Watkins, R. H., Kerl, E. A., et al. (1994). Human kidney proximal tubules are the main source of plasma glutathione peroxidase. *Am J Physiol, 266*(2 Pt 1), C367-375.

Avraham, K. B., Hasson, T., Steel, K. P., Kingsley, D. M., Russell, L. B., Mooseker, M. S., et al. (1995). The mouse Snell's waltzer deafness gene encodes an unconventional myosin required for structural integrity of inner ear hair cells. *Nat Genet, 11*(4), 369-375.

Bachmann, S., Schlichting, U., Geist, B., Mutig, K., Petsch, T., Bacic, D., et al. (2004). Kidney-specific inactivation of the megalin gene impairs trafficking of renal inorganic sodium phosphate cotransporter (NaPi-IIa). *J Am Soc Nephrol, 15*(4), 892-900.

Biemesderfer, D., DeGray, B., & Aronson, P. S. (2001). Active (9.6 s) and inactive (21 s) oligomers of NHE3 in microdomains of the renal brush border. *J Biol Chem, 276*(13), 10161-10167.

Biemesderfer, D., Nagy, T., DeGray, B., & Aronson, P. S. (1999). Specific association of megalin and the Na+/H+ exchanger isoform NHE3 in the proximal tubule. *J Biol Chem, 274*(25), 17518-17524.

Bobulescu, I. A., & Moe, O. W. (2009). Luminal Na(+)/H (+) exchange in the proximal tubule. *Pflugers Arch, 458*(1), 5-21.

Brown, M. S., Ye, J., Rawson, R. B., & Goldstein, J. L. (2000). Regulated intramembrane proteolysis: a control mechanism conserved from bacteria to humans. *Cell, 100*(4), 391-398.

Burns, W. C., Kantharidis, P., & Thomas, M. C. (2007). The role of tubular epithelial-mesenchymal transition in progressive kidney disease. *Cells Tissues Organs, 185*(1-3), 222-231.

Caruso-Neves, C., Pinheiro, A. A., Cai, H., Souza-Menezes, J., & Guggino, W. B. (2006). PKB and megalin determine the survival or death of renal proximal tubule cells. *Proc Natl Acad Sci U S A, 103*(49), 18810-18815.

Christensen, E. I., Devuyst, O., Dom, G., Nielsen, R., Van der Smissen, P., Verroust, P., et al. (2003). Loss of chloride channel ClC-5 impairs endocytosis by defective trafficking of megalin and cubilin in kidney proximal tubules. *Proc Natl Acad Sci U S A, 100*(14), 8472-8477.

Christensen, E. I., Moskaug, J. O., Vorum, H., Jacobsen, C., Gundersen, T. E., Nykjaer, A., et al. (1999). Evidence for an essential role of megalin in transepithelial transport of retinol. *J Am Soc Nephrol, 10*(4), 685-695.

Christensen, E. I., Verroust, P. J., & Nielsen, R. (2009). Receptor-mediated endocytosis in renal proximal tubule. *Pflugers Arch.*

Comper, W. D., Hilliard, L. M., Nikolic-Paterson, D. J., & Russo, L. M. (2008). Disease-dependent mechanisms of albuminuria. *Am J Physiol Renal Physiol, 295*(6), F1589-1600.

Coudroy, G., Gburek, J., Kozyraki, R., Madsen, M., Trugnan, G., Moestrup, S. K., et al. (2005). Contribution of cubilin and amnionless to processing and membrane targeting of cubilin-amnionless complex. *J Am Soc Nephrol, 16*(8), 2330-2337.

Davis, C. G., Goldstein, J. L., Sudhof, T. C., Anderson, R. G., Russell, D. W., & Brown, M. S. (1987). Acid-dependent ligand dissociation and recycling of LDL receptor mediated by growth factor homology region. *Nature, 326*(6115), 760-765.

Dobnig, H., Pilz, S., Scharnagl, H., Renner, W., Seelhorst, U., Wellnitz, B., et al. (2008). Independent association of low serum 25-hydroxyvitamin d and 1,25-dihydroxyvitamin d levels with all-cause and cardiovascular mortality. *Arch Intern Med, 168*(12), 1340-1349.

Dronavalli, S., Duka, I., & Bakris, G. L. (2008). The pathogenesis of diabetic nephropathy. *Nat Clin Pract Endocrinol Metab, 4*(8), 444-452.

Fyfe, J. C., Madsen, M., Hojrup, P., Christensen, E. I., Tanner, S. M., de la Chapelle, A., et al. (2004). The functional cobalamin (vitamin B12)-intrinsic factor receptor is a novel complex of cubilin and amnionless. *Blood, 103*(5), 1573-1579.

Gekle, M. (2005). Renal tubule albumin transport. *Annu Rev Physiol, 67*, 573-594.

Gerstein, H. C., Mann, J. F., Yi, Q., Zinman, B., Dinneen, S. F., Hoogwerf, B., et al. (2001). Albuminuria and risk of cardiovascular events, death, and heart failure in diabetic and nondiabetic individuals. *JAMA, 286*(4), 421-426.

Gotthardt, M., Trommsdorff, M., Nevitt, M. F., Shelton, J., Richardson, J. A., Stockinger, W., et al. (2000). Interactions of the low density lipoprotein receptor gene family with cytosolic adaptor and scaffold proteins suggest diverse biological functions in cellular communication and signal transduction. *J Biol Chem, 275*(33), 25616-25624.

Gunther, W., Luchow, A., Cluzeaud, F., Vandewalle, A., & Jentsch, T. J. (1998). ClC-5, the chloride channel mutated in Dent's disease, colocalizes with the proton pump in endocytotically active kidney cells. *Proc Natl Acad Sci U S A, 95*(14), 8075-8080.

Hasson, T. (2003). Myosin VI: two distinct roles in endocytosis. *J Cell Sci, 116*(Pt 17), 3453-3461.

Heath, K. E., Campos-Barros, A., Toren, A., Rozenfeld-Granot, G., Carlsson, L. E., Savige, J., et al. (2001). Nonmuscle myosin heavy chain IIA mutations define a spectrum of autosomal dominant macrothrombocytopenias: May-Hegglin anomaly and Fechtner, Sebastian, Epstein, and Alport-like syndromes. *Am J Hum Genet, 69*(5), 1033-1045.

Hernando, N., Gisler, S. M., Pribanic, S., Deliot, N., Capuano, P., Wagner, C. A., et al. (2005). NaPi-IIa and interacting partners. *J Physiol, 567*(Pt 1), 21-26.

Hjalm, G., Murray, E., Crumley, G., Harazim, W., Lundgren, S., Onyango, I., et al. (1996). Cloning and sequencing of human gp330, a Ca(2+)-binding receptor with potential intracellular signaling properties. *Eur J Biochem, 239*(1), 132-137.

Hocevar, B. A., Smine, A., Xu, X. X., & Howe, P. H. (2001). The adaptor molecule Disabled-2 links the transforming growth factor beta receptors to the Smad pathway. *EMBO J, 20*(11), 2789-2801.

Hong, C. Y., Hughes, K., Chia, K. S., Ng, V., & Ling, S. L. (2003). Urinary alpha1-microglobulin as a marker of nephropathy in type 2 diabetic Asian subjects in Singapore. *Diabetes Care, 26*(2), 338-342.

Hosaka, K., Takeda, T., Iino, N., Hosojima, M., Sato, H., Kaseda, R., et al. (2009). Megalin and nonmuscle myosin heavy chain IIA interact with the adaptor protein Disabled-2 in proximal tubule cells. *Kidney Int, 75*(12), 1308-1315.

Hosojima, M., Sato, H., Yamamoto, K., Kaseda, R., Soma, T., Kobayashi, A., et al. (2009). Regulation of megalin expression in cultured proximal tubule cells by angiotensin II type 1A receptor- and insulin-mediated signaling cross talk. *Endocrinology, 150*(2), 871-878.

Jentsch, T. J. (2005). Chloride transport in the kidney: lessons from human disease and knockout mice. *J Am Soc Nephrol, 16*(6), 1549-1561.

Kalantry, S., Manning, S., Haub, O., Tomihara-Newberger, C., Lee, H. G., Fangman, J., et al. (2001). The amnionless gene, essential for mouse gastrulation, encodes a visceral-endoderm-specific protein with an extracellular cysteine-rich domain. *Nat Genet, 27*(4), 412-416.

Kantarci, S., Al-Gazali, L., Hill, R. S., Donnai, D., Black, G. C., Bieth, E., et al. (2007). Mutations in LRP2, which encodes the multiligand receptor megalin, cause Donnai-Barrow and facio-oculo-acoustico-renal syndromes. *Nat Genet, 39*(8), 957-959.

Kao, W. H., Klag, M. J., Meoni, L. A., Reich, D., Berthier-Schaad, Y., Li, M., et al. (2008). MYH9 is associated with nondiabetic end-stage renal disease in African Americans. *Nat Genet, 40*(10), 1185-1192.

Kelley, M. J., Jawien, W., Ortel, T. L., & Korczak, J. F. (2000). Mutation of MYH9, encoding non-muscle myosin heavy chain A, in May-Hegglin anomaly. *Nat Genet, 26*(1), 106-108.

Kopp, J. B., Smith, M. W., Nelson, G. W., Johnson, R. C., Freedman, B. I., Bowden, D. W., et al. (2008). MYH9 is a major-effect risk gene for focal segmental glomerulosclerosis. *Nat Genet, 40*(10), 1175-1184.

Kozyraki, R., Fyfe, J., Verroust, P. J., Jacobsen, C., Dautry-Varsat, A., Gburek, J., et al. (2001). Megalin-dependent cubilin-mediated endocytosis is a major pathway for the apical uptake of transferrin in polarized epithelia. *Proc Natl Acad Sci U S A, 98*(22), 12491-12496.

Larsson, M., Hjalm, G., Sakwe, A. M., Engstrom, A., Hoglund, A. S., Larsson, E., et al. (2003). Selective interaction of megalin with postsynaptic density-95 (PSD-95)-like membrane-associated guanylate kinase (MAGUK) proteins. *Biochem J, 373*(Pt 2), 381-391.

Leheste, J. R., Rolinski, B., Vorum, H., Hilpert, J., Nykjaer, A., Jacobsen, C., et al. (1999). Megalin knockout mice as an animal model of low molecular weight proteinuria. *Am J Pathol, 155*(4), 1361-1370.

Lehtonen, S., Shah, M., Nielsen, R., Iino, N., Ryan, J. J., Zhou, H., et al. (2008). The endocytic adaptor protein ARH associates with motor and centrosomal proteins and is involved in centrosome assembly and cytokinesis. *Mol Biol Cell, 19*(7), 2949-2961.

Li, Y., Cong, R., & Biemesderfer, D. (2008). The COOH terminus of megalin regulates gene expression in opossum kidney proximal tubule cells. *Am J Physiol Cell Physiol, 295*(2), C529-537.

Lou, X., McQuistan, T., Orlando, R. A., & Farquhar, M. G. (2002). GAIP, GIPC and Galphai3 are concentrated in endocytic compartments of proximal tubule cells: putative role in regulating megalin's function. *J Am Soc Nephrol, 13*(4), 918-927.

Louvi, A., & Artavanis-Tsakonas, S. (2006). Notch signalling in vertebrate neural development. *Nat Rev Neurosci, 7*(2), 93-102.

Maser, R. L., Magenheimer, B. S., & Calvet, J. P. (1994). Mouse plasma glutathione peroxidase. cDNA sequence analysis and renal proximal tubular expression and secretion. *J Biol Chem, 269*(43), 27066-27073.

Maunsbach, A. B. (1966). Absorption of I-125-labeled homologous albumin by rat kidney proximal tubule cells. A study of microperfused single proximal tubules by electron microscopic autoradiography and histochemistry. *J Ultrastruct Res, 15*(3), 197-241.

Moestrup, S. K., Birn, H., Fischer, P. B., Petersen, C. M., Verroust, P. J., Sim, R. B., et al. (1996). Megalin-mediated endocytosis of transcobalamin-vitamin-B12 complexes suggests a role of the receptor in vitamin-B12 homeostasis. *Proc Natl Acad Sci U S A, 93*(16), 8612-8617.

Mogensen, C. E. (1984). Microalbuminuria predicts clinical proteinuria and early mortality in maturity-onset diabetes. *N Engl J Med, 310*(6), 356-360.

Motoyoshi, Y., Matsusaka, T., Saito, A., Pastan, I., Willnow, T. E., Mizutani, S., et al. (2008). Megalin contributes to the early injury of proximal tubule cells during nonselective proteinuria. *Kidney Int, 74*(10), 1262-1269.

Nagai, M., Meerloo, T., Takeda, T., & Farquhar, M. G. (2003). The adaptor protein ARH escorts megalin to and through endosomes. *Mol Biol Cell, 14*(12), 4984-4996.

Norden, A. G., Lapsley, M., Igarashi, T., Kelleher, C. L., Lee, P. J., Matsuyama, T., et al. (2002). Urinary megalin deficiency implicates abnormal tubular endocytic function in Fanconi syndrome. *J Am Soc Nephrol, 13*(1), 125-133.

Nykjaer, A., Fyfe, J. C., Kozyraki, R., Leheste, J. R., Jacobsen, C., Nielsen, M. S., et al. (2001). Cubilin dysfunction causes abnormal metabolism of the steroid hormone 25(OH) vitamin D(3). *Proc Natl Acad Sci U S A, 98*(24), 13895-13900.

Oleinikov, A. V., Zhao, J., & Makker, S. P. (2000). Cytosolic adaptor protein Dab2 is an intracellular ligand of endocytic receptor gp600/megalin. *Biochem J, 347 Pt 3*, 613-621.

Olson, G. E., Winfrey, V. P., Hill, K. E., & Burk, R. F. (2008). Megalin mediates selenoprotein P uptake by kidney proximal tubule epithelial cells. *J Biol Chem, 283*(11), 6854-6860.

Oyama, Y., Takeda, T., Hama, H., Tanuma, A., Iino, N., Sato, K., et al. (2005). Evidence for megalin-mediated proximal tubular uptake of L-FABP, a carrier of potentially nephrotoxic molecules. *Lab Invest, 85*(4), 522-531.

Petersen, H. H., Hilpert, J., Militz, D., Zandler, V., Jacobsen, C., Roebroek, A. J., et al. (2003). Functional interaction of megalin with the megalinbinding protein (MegBP), a novel tetratrico peptide repeat-containing adaptor molecule. *J Cell Sci, 116*(Pt 3), 453-461.

Pilz, S., Marz, W., Wellnitz, B., Seelhorst, U., Fahrleitner-Pammer, A., Dimai, H. P., et al. (2008). Association of vitamin D deficiency with heart failure and sudden cardiac death in a large cross-sectional study of patients referred for coronary angiography. *J Clin Endocrinol Metab, 93*(10), 3927-3935.

Plans, V., Rickheit, G., & Jentsch, T. J. (2009). Physiological roles of CLC Cl(-)/H (+) exchangers in renal proximal tubules. *Pflugers Arch, 458*(1), 23-37.

Pontuch, P., Jensen, T., Deckert, T., Ondrejka, P., & Mikulecky, M. (1992). Urinary excretion of retinol-binding protein in type 1 (insulin-dependent) diabetic patients with microalbuminuria and clinical diabetic nephropathy. *Acta Diabetol, 28*(3-4), 206-210.

Prunier, C., Hocevar, B. A., & Howe, P. H. (2004). Wnt signaling: physiology and pathology. *Growth Factors, 22*(3), 141-150.

Rader, K., Orlando, R. A., Lou, X., & Farquhar, M. G. (2000). Characterization of ANKRA, a novel ankyrin repeat protein that interacts with the cytoplasmic domain of megalin. *J Am Soc Nephrol, 11*(12), 2167-2178.

Riquier, A. D., Lee, D. H., & McDonough, A. A. (2009). Renal NHE3 and NaPi2 partition into distinct membrane domains. *Am J Physiol Cell Physiol, 296*(4), C900-910.

Russo, L. M., del Re, E., Brown, D., & Lin, H. Y. (2007). Evidence for a role of transforming growth factor (TGF)-beta1 in the induction of postglomerular albuminuria in diabetic nephropathy: amelioration by soluble TGF-beta type II receptor. *Diabetes, 56*(2), 380-388.

Saito, A., Nagai, R., Tanuma, A., Hama, H., Cho, K., Takeda, T., et al. (2003). Role of megalin in endocytosis of advanced glycation end products: implications for a novel protein binding to both megalin and advanced glycation end products. *J Am Soc Nephrol, 14*(5), 1123-1131.

Saito, A., Pietromonaco, S., Loo, A. K., & Farquhar, M. G. (1994). Complete cloning and sequencing of rat gp330/"megalin," a distinctive member of the low density lipoprotein receptor gene family. *Proc Natl Acad Sci U S A, 91*(21), 9725-9729.

Saito, A., Takeda, T., Sato, K., Hama, H., Tanuma, A., Kaseda, R., et al. (2005). Significance of proximal tubular metabolism of advanced glycation end products in kidney diseases. *Ann N Y Acad Sci, 1043*, 637-643.

Sarav, M., Wang, Y., Hack, B. K., Chang, A., Jensen, M., Bao, L., et al. (2009). Renal FcRn reclaims albumin but facilitates elimination of IgG. *J Am Soc Nephrol, 20*(9), 1941-1952.

Scheel, O., Zdebik, A. A., Lourdel, S., & Jentsch, T. J. (2005). Voltage-dependent electrogenic chloride/proton exchange by endosomal CLC proteins. *Nature, 436*(7049), 424-427.

Sebekova, K., Schinzel, R., Ling, H., Simm, A., Xiang, G., Gekle, M., et al. (1998). Advanced glycated albumin impairs protein degradation in the kidney proximal tubules cell line LLC-PK1. *Cell Mol Biol (Noisy-le-grand), 44*(7), 1051-1060.

Seetharam, B., Christensen, E. I., Moestrup, S. K., Hammond, T. G., & Verroust, P. J. (1997). Identification of rat yolk sac target protein of teratogenic antibodies, gp280, as intrinsic factor-cobalamin receptor. *J Clin Invest, 99*(10), 2317-2322.

Seetharam, B., Levine, J. S., Ramasamy, M., & Alpers, D. H. (1988). Purification, properties, and immunochemical localization of a receptor for intrinsic factor-cobalamin complex in the rat kidney. *J Biol Chem, 263*(9), 4443-4449.

Seri, M., Cusano, R., Gangarossa, S., Caridi, G., Bordo, D., Lo Nigro, C., et al. (2000). Mutations in MYH9 result in the May-Hegglin anomaly, and Fechtner and Sebastian syndromes. The May-Heggllin/Fechtner Syndrome Consortium. *Nat Genet, 26*(1), 103-105.

Seri, M., Savino, M., Bordo, D., Cusano, R., Rocca, B., Meloni, I., et al. (2002). Epstein syndrome: another renal disorder with mutations in the nonmuscle myosin heavy chain 9 gene. *Hum Genet, 110*(2), 182-186.

Shalamanova, L., McArdle, F., Amara, A. B., Jackson, M. J., & Rustom, R. (2007). Albumin overload induces adaptive responses in human proximal tubular cells through oxidative stress but not via angiotensin II type 1 receptor. *Am J Physiol Renal Physiol, 292*(6), F1846-1857.

Slattery, C., Lee, A., Zhang, Y., Kelly, D. J., Thorn, P., Nikolic-Paterson, D. J., et al. (2008). In vivo visualization of albumin degradation in the proximal tubule. *Kidney Int, 74*(11), 1480-1486.

Strutz, F. M. (2009). EMT and proteinuria as progression factors. *Kidney Int, 75*(5), 475-481.

Takeda, T., Yamazaki, H., & Farquhar, M. G. (2003). Identification of an apical sorting determinant in the cytoplasmic tail of megalin. *Am J Physiol Cell Physiol, 284*(5), C1105-1113.

Tanner, S. M., Aminoff, M., Wright, F. A., Liyanarachchi, S., Kuronen, M., Saarinen, A., et al. (2003). Amnionless, essential for mouse gastrulation, is mutated in recessive hereditary megaloblastic anemia. *Nat Genet, 33*(3), 426-429.

Tanuma, A., Sato, H., Takeda, T., Hosojima, M., Obayashi, H., Hama, H., et al. (2007). Functional characterization of a novel missense CLCN5 mutation causing alterations in proximal tubular endocytic machinery in Dent's disease. *Nephron Physiol, 107*(4), p87-97.

Thrailkill, K. M., Nimmo, T., Bunn, R. C., Cockrell, G. E., Moreau, C. S., Mackintosh, S., et al. (2009). Microalbuminuria in type 1 diabetes is associated with enhanced excretion of the endocytic multiligand receptors megalin and cubilin. *Diabetes Care, 32*(7), 1266-1268.

Tojo, A., Onozato, M. L., Ha, H., Kurihara, H., Sakai, T., Goto, A., et al. (2001). Reduced albumin reabsorption in the proximal tubule of early-stage diabetic rats. *Histochem Cell Biol, 116*(3), 269-276.

Uchida, S. (2000). In vivo role of CLC chloride channels in the kidney. *Am J Physiol Renal Physiol, 279*(5), F802-808.

Verbeke, P., Perichon, M., Friguet, B., & Bakala, H. (2000). Inhibition of nitric oxide synthase activity by early and advanced glycation end products in cultured rabbit proximal tubular epithelial cells. *Biochim Biophys Acta, 1502*(3), 481-494.

Verroust, P. J., Kozyraki, R., Hammond, T. G., Moestrup, S. K., & Christensen, E. I. (2000). Physiopathologic role of cubilin and megalin. *Adv Nephrol Necker Hosp, 30*, 127-145.

Viberti, G. C., Hill, R. D., Jarrett, R. J., Argyropoulos, A., Mahmud, U., & Keen, H. (1982). Microalbuminuria as a predictor of clinical nephropathy in insulin-dependent diabetes mellitus. *Lancet, 1*(8287), 1430-1432.

Wachtell, K., Ibsen, H., Olsen, M. H., Borch-Johnsen, K., Lindholm, L. H., Mogensen, C. E., et al. (2003). Albuminuria and cardiovascular risk in hypertensive patients with left ventricular hypertrophy: the LIFE study. *Ann Intern Med, 139*(11), 901-906.

Wang, Y., Roche, O., Yan, M. S., Finak, G., Evans, A. J., Metcalf, J. L., et al. (2009). Regulation of endocytosis via the oxygen-sensing pathway. *Nat Med, 15*(3), 319-324.

Whitin, J. C., Bhamre, S., Tham, D. M., & Cohen, H. J. (2002). Extracellular glutathione peroxidase is secreted basolaterally by human renal proximal tubule cells. *Am J Physiol Renal Physiol, 283*(1), F20-28.

Wrong, O. M., Norden, A. G., & Feest, T. G. (1994). Dent's disease; a familial proximal renal tubular syndrome with low-molecular-weight proteinuria, hypercalciuria, nephrocalcinosis, metabolic bone disease, progressive renal failure and a marked male predominance. *Qjm, 87*(8), 473-493.

Yammani, R. R., Seetharam, S., & Seetharam, B. (2001). Identification and characterization of two distinct ligand binding regions of cubilin. *J Biol Chem, 276*(48), 44777-44784.

Zou, Z., Chung, B., Nguyen, T., Mentone, S., Thomson, B., & Biemesderfer, D. (2004). Linking receptor-mediated endocytosis and cell signaling: evidence for regulated intramembrane proteolysis of megalin in proximal tubule. *J Biol Chem, 279*(33), 34302-34310.

In: Advances in Medicine and Biology Research, Vol. 14 ISBN: 978-1-61761-930-4
Editor: Leon V. Berhardt

Chapter X

In Vitro Expanded CD8+ NKT-Like Cells Possess Potent Cytotoxic and Immunosuppressive Activities

Li Zhou[1,2,3], Xing Zhong[1], Qing-Sheng Mi[1,2,3] and Jin-Xiong She[1,2*]
Center for Biotechnology and Genomic Medicine, Medical College of Georgia, 1120 15th Street, Augusta, GA 30912, USA[1]
Department of Pathology, Medical College of Georgia, 1120 15th Street, Augusta, GA 30912[2]
Henry Ford Hospital Immunology Program, Department of Dermatology, Henry Ford Health system, One Ford Place, Detroit, MI 48202[3]

Abstract

CD8+ natural killer T (NKT) cells possess potent cytolytic abilities and may be implicated in immune tolerance; however, the phenotype and function of this minor cell population are poorly defined. We have recently developed a protocol to culture CD8+ NKT-like cells and demonstrated the immunosuppressive activity of these cells towards CD4+ responder T cells. Data in this study indicate that these CD8+ NKT-like cells are anergic to stimulation by anti-CD3 and antigen presenting cells (APC) and can also efficiently suppress the proliferation of CD8+ naïve T cells. Most of the suppressive activity is cell contact-dependent, although soluble factors may also play a role in the function of their suppressive activity. The CD8+ NKT-like cells can efficiently kill antigen presenting cells but not naive T cells. APC-free suppression assays further demonstrate that the CD8+ NKT-like cells suppress T cell proliferation using APC-independent mechanisms including direct blockade of IL-2 production by responder T cells. Gene expression profiling suggested that over 3000 genes differ between the cultured CD8+ NKT-like cells and the naïve CD8+ T cells by more than 2-fold with >100 genes upregulated by more than 10-fold in the CD8+ NKT-like cells. A large proportion

*Correspondance: Jin-Xiong She: Center for Biotechnology and Genomic Medicine, Medical College of Georgia, 1120 15th Street, Augusta, GA 30912, jshe@mcg.edu, 706-721-3400 (o), 706-721-3688 (fax)

of the highly upregulated genes are soluble factors or surface markers that have previously been implicated in immune suppression or are likely to possess immunosuppressive properties. Many of these genes are regulated by two key cytokines, IL-10 and IFN-γ, suggesting that these pathways play a critical role in the suppressive function of these cells.

Abbreviations

APC: Antigen presenting cell
DC: dendritic cell
EGTA: ethylene glycol tetraacetic acid
Fas-L: Fas-ligand
INF-γ: interferon-γ
NKT: natural killer T cell
OVA: ovalbumin
TCR: T cell receptor
Treg: Regulatory T cells
7-AAD: 7-Aminoactinomycin D

Introduction

Both $CD8^+$ T cells and natural killer (NK) cells are major lymphocyte subsets with potent cytotoxic activities. A minor lymphocyte population known as NKT cells have been identified in both mice and human [1–4]. The murine NKT cells express phenotypic markers that are typically found on T cells such as CD3 and the αβ T-cell receptor (TCR) and markers for NK cells such as NK1.1 and DX5[5]. Two major NKT cell populations have been recognized in mice [6,7]. The first population is well-characterized invariant NKT (iNKT) cells that express invariant Vα14-Jα18 TCR in mice [8–10] and the Vα24-Jα18 in humans [11,12]. These iNKT cells are restricted by the major histocompatibility complex (MHC) class I-like molecule Cd1d and recognize glycolipid antigen α-galactosylceramide, a synthetic variant of a murine sponge-derived glycolipid [8,13]. These iNKT cells produce large amounts of interleukin-4 (IL-4) and IFN-γ upon activation and they have been shown to play a critical role in regulating immune response [8,13]. The second population of NKT cells expresses a variable TCR repertoire and is not restricted by Cd1d. These NKT cells express mainly CD8 or are negative for both CD8 and CD4 [6]. The whole $\alpha\beta TCR^+NK1.1^+$ NKT population represents 1-2% of splenocytes in B6 mice, and, of these cells, approximately 20% are $CD8^+$ [6]. The phenotype and function of the natural $CD8^+$ NKT cells are still poorly defined because of the extremely low frequency of these cells in normal mice and humans. Some MHC class I-restricted TCR-transgenic mouse strains (OT-I, P14 and H-Y) contain higher frequencies of transgenic $CD8^+$ T cells that co-express NK cell marker NK1.1 [14]. These transgenic $CD8^+$ NKT cells are endowed with effector properties such as cytokine production and antigen-specific cytotoxicity. Tumor-bearing B6 mice were also shown to have a population of NKT cells that co-express CD8 and NK1.1 [15]. These cells can be maintained in long-term culture with IL-4 and produce large amounts of interferon-γ (IFN-γ) following activation. These CD8$^+$ NKT cells show a potent NK-like cytotoxic activity against multiple tumor targets and their cytotoxic activity is CD1d independent [15]. CD8$^+$ cells with NK phenotype are also expandable using a culture condition that includes IFN-γ, anti-CD3 and rIL-2[16]. Such expanded CD8$^+$ NKT cells can efficiently kill tumor cells both *in vitro* and *in vivo* with only limited capacity to cause graft-versus-host disease (GVHD)[16].

Although the potent cytotoxic activity for CD8$^+$ NKT cells is widely recognized, less well known is their potential role in immune tolerance. This role is initially suggested by the increased frequency of CD8$^+$ NKT cells in mice with neonatal tolerance [17], which is induced by injection of splenic lymphocytes from allogeneic donors into neonatal recipient mice [17–19]. Recent studies have identified a novel population of Qa-1-restricted, NK marker expressing CD8αα$^+$TCRαβ$^+$ T cells, which target activated T cells for immune regulation with multiple suppression mechanisms including cytotoxic activity [20,21]. In our recent study, we developed a protocol that allowed the generation of CD8$^+$ T cells that express NK markers (referred to as CD8$^+$ NKT-like cells) and demonstrated their potent function in suppressing the proliferation of CD4$^+$ responder T cells [22]. Our previous gene expression profiling studies suggested that CD8$^+$ NKT-like cells and naïve CD8$^+$ T cells differ by more than 2-fold for about 3000 genes, among which 113 genes are up-regulated by more than 10-fold in the CD8$^+$ NKT-like cells. Pathway analysis of the expression data identified IL-10 and IFN-g as two critical nodes linking many of the up regulated genes that may be implicated in immune suppression. In the current study, we further characterized these CD8$^+$ NKT-like cells and investigated the mechanisms underlying the suppression activities. Our data indicate that these CD8$^+$ NKT-like cells are anergic to stimulation and can efficiently suppress the proliferation of both CD4$^+$ and CD8$^+$ naïve T cells. The majority of the suppressive activity is cell contact-dependent possibly through the killing of antigen presenting cells, although soluble factors may also play a role in their suppressive activity. The APC-free suppression assays further demonstrate that the CD8$^+$ NKT-like cells suppress T cell proliferation using APC-independent mechanisms including soluble factors IL-10 and INF-g among other potential molecules.

Materials and Methods

Mice

C57BL/6, OT-II TCR transgenic and OT-I TCR transgenic mice were purchased from Jackson Lab and housed/bred under specific pathogen-free conditions at the Medical College of Georgia Animal Barrier Facility. This study was approved by the Medical College of Georgia IACUC committee.

In Vitro Culture of $CD8^+$ and $CD4^+$ T Cells

AutoMACS-purified naïve $CD8^+$ or Mo-Flow sorted $CD4^+$ T cells at 10^6 cells/ml were cultured with anti-CD3 and anti-CD28 coupled to 4.5 mm paramagnetic Dynal beads (Invitrogen, Carlsbad, California , USA) supplemented with 2,000 IU/ml rIL-2 (PeproTech, Rocky Hill, NJ, USA) in complete medium, which consisted of 10% heat-inactivated fetal bovine serum (Sigma-Aldrich, St. Louis, MO, USA), nonessential amino acids, 0.5 mM sodium pyruvate, 5 mM Hepes, 1 mM glutaMax I (all from Invitrogen, Carlsbad, California , USA), and 55 μM mercaptoethanol (Sigma-Aldrich) in DMEM base. The cultures were monitored daily and maintained at 1-1.5 × 10^6 cells/ml by diluting with IL-2-supplemented complete medium for 8-13 days. At the end of the culture, the anti-CD3 and anti-CD28 beads were removed using a Dynal MPC-L magnet (Dynal Biotech), and the cells were routinely assayed for CD8, CD4, CD62L, and CD25 expression by flow cytometry and for *in vitro* suppression assays.

In Vitro Suppression Assays

Three different types of suppression assays were performed in this study. Most of the studies were performed using the CFSE system. Briefly, responder $CD4^+$ or $CD8^+$ T cells were labeled with 2.5 μM CFSE. These labeled responder cells ($1x10^5$) were cocultured at 37°C with different numbers of suppressor cells ($CD8^+$ NKT cells or $CD4^+CD25^+$ Tregs) in the presence of $1.5x10^5$ irradiated splenic APC (2000 rads) and 1.5 μg/ml anti-CD3 in a U-bottomed 96- well plate. After 72 h of culture, responder T cell proliferation was assessed by determining the dilution of CFSE using flow cytometry. The second type of suppression assay has identical culture conditions but the responder T cells were not labeled. The cultures were pulsed with 1 μCi/ well [^{3}H]thymidine for the last 16 h of 72h culture and the level of proliferation was assessed by the [^{3}H]thymidine incorporation by scintillation counting after cell harvesting. The third type of suppression assay is identical to the second assay in the absence of APC.

Flow Cytometry

$CD8^+$ T cells were enriched from spleens by negative selection using an AutoMACS (Miltenyi Biotec). The resulting $CD8^+$ cell purity is around 90-95%. Flow cytometric analyses were performed on a FACScalibur™ flow cytometer with CELLQuest™ software (Becton Dickinson). Cultures for ELISA analyses were carried out in 96-well plates with 1 x 10^5 $CD4^+CD25^-$ cells/well, 1.5 x 10^5 /well irradiated splenic APC and 1.5 μg/ml anti-CD3 in the presence or absence of 1 x 10^5 /well $CD8^+$ NKT cells or fresh $CD4^+CD25^+$ Tregs. The level of IL-10, IFN-γ, IL-4 and IL-2 in the culture supernatant was determined by ELISA Kits purchased from R&D Systems.

RT-PCR Analysis

Total RNA was isolated from re-purified $CD4^+$ responder T cells in the suppression assay using QIAshredder and RNeasy Mini Kit spin columns (Qiagen). cDNA was synthesized from 1-2 μg of sample RNA using SuperScript III reverse transcriptase and oligo(dT) 12-18 primer (Invitrogen Life Technologies). The same amount of RNA and cDNA was used for each Treg and Non-Treg sample pairs.

Cytotoxicity Assay

A Flow-based killing assay (FloKA) was used to measure *in vitro* cellular cytotoxicity of effector cells as previously described [23]. Purified target cells were washed with PBS, resuspended at 1 x 10^6/ml, and then labeled at 37°C for 10 min with 125nM of CFSE as described previously. Labeled target cells (1.5 x 10^5) were added to 96-well plate or 24-well plate (5 x 10^5) along with indicated effector cells in complete DMEM medium with or without anti-CD3 stimulation. The effector:target (E:T) ratio was 1:1 to mimic the suppression assay condition. Immediately before analysis, 1 μg/ml (final concentration) of 7-amino-actinomycin D (7-AAD; Calbiochem) was added to each sample. 7-AAD incorporation is used as a surrogate marker for late cell death/apoptosis as it intercalates with DNA in cells that have lost membrane integrity.

Results

Phenotypes of $CD8^+$ NKT-Like Cells

The phenotypes of the $CD8^+$ NKT-like cells cultured with anti-CD3/anti-CD28-coated beads and IL-2 were further characterized during the culture. The vast majority (~93%) of the cells used in the culture was $CD8^+$ and ~7% of the cells were $CD4^-CD8^-$ while very few cells were $CD4^+$ or $CD4^+CD8^+$ (Table 1). At day 6 - 8, the $CD4^-CD8^-$ cells increased in frequency for a brief period of time. The vast majority of the cultured cells at the end of the culture (day 10 - 13) are $CD8^+$ T cells (Table 1). As expected, almost all cultured cells are activated and express high levels of CD25 starting at day 4 (first time point analyzed) till the end of the culture. The percentages of $CD62L^+$ cells gradually decrease during the culture and became very low near the end of the culture (< 10%). Cell surface CTLA-4 were initially increased and then decreased during the culture and less than 5% of the expanded cells are positive by Day 10. $CD103^+$ cells quickly decreased from ~20% to <1% during the culture. Consistent with our previous study, the $CD8^+$ cells cultured under this condition gradually gained expression of several NK markers (Table 1 and Fig. 1). Naïve $CD8^+$ T cells comprise a very small percentage of cells positive for NK markers. By day 4 of the culture, NK markers are still very low (<6%) in the cultured $CD8^+$ cells. As shown in Table 1, as many as 15% of the cultured $CD8^+$ cells became positive for these markers around day 6, while 97% of the cultured cells expressed at least one NK marker (CD244) and most likely multiple NK markers including NK1.1, NKG2A, CD94 and DX5 (Fig. 1).

Table 1. Cell surface markers during the culture for CD8$^+$ NKT-like cells.

Markers	Fresh	Day 4	Day 6	Day 8	Day 10	Day 13
CD4$^+$	0.2	0.4	0.9	0.1	0.3	0.8
CD8$^+$	92.8	91.2	79.2	86.7	92.9	96.0
CD4$^+$CD8$^+$	0	0.3	0.9	0.2	0.3	0.8
CD4$^-$CD8$^-$	7.0	8.8	20.9	12.3	6.2	4.0
CD25$^+$	0.05	99.5	99.9	99.8	99.6	98.2
CD62L$^+$	78.8	85.5	26.6	18.4	7.6	6.3
CTLA-4$^+$	1.1	17.1	15.8	3.6	3.9	nd
CD103	19.8	nd	2.3	0.4	0.6	nd
NK1.1$^+$	1.1	0.9	6.8	28.0	76.3	83.0
NKG2A$^+$	1.4	5.7	4.5	15.5	26.1	27.2
CD244$^+$	2.6	0.2	15.0	50.6	87.9	97.1
CD94$^+$	0.3	1.5	4.5	1.5	27.2	30.2
DX5$^+$	nd	0.4	nd	nd	14.7	27.5

nd: not done.

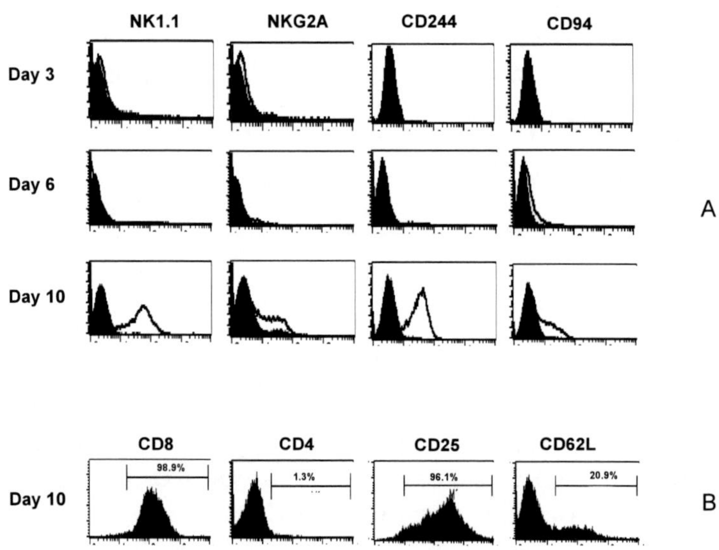

Figure 1. Phenotypic changes of CD8+ NKT-like cells during the culture. A. Four NK markers were shown for three different time points during the culture. The expression levels for the specific antibodies were overlaid with isotype control. B. Four T cell surface markers CD8, CD4, CD25 and CD62L expression level on day-10 cultured CD8+ NKT-like cells were tested with flow cytometry. The positive ratio of each marker was labeled on the histograms.

CD8$^+$ NKT-Like Cells Suppress Proliferation of Both CD4$^+$ and CD8$^+$ T Cells

Consistent with our previous results [22], the cultured CD8$^+$ NKT-like cells do not proliferate upon stimulation by APC (T cell-depleted spleen cells) and anti-CD3 (Fig. 2a) and that they can inhibit the proliferation of CD4$^+$CD25$^-$ naïve responder T cells in a dose-dependent manner (Fig. 2a). Here, we further show that the $CD8^+$ NKT-like cells also efficiently suppressed proliferation of $CD8^+$ naïve T cells in addition to $CD4^+CD25^-$ T cells (Fig. 2b). These $CD8^+$ NKT-like cells showed dose-dependent suppressive activity and more efficiently suppressed naïve $CD4^+CD25^-$ T cells than it did to naïve $CD8^+$ T cells (Fig. 2c) ($p < 10^{-6}$).

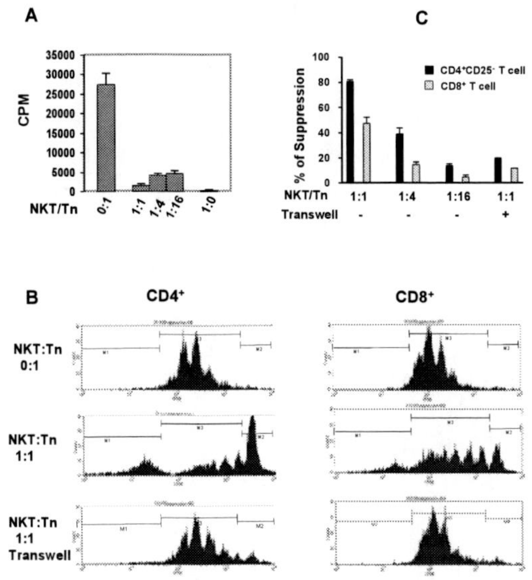

Figure 2. CD8+ NKT-like cells suppress the proliferation of naïve CD4+ and CD8+ T cells. (A) Dose-dependent suppression of CD4+CD25- responder T cells by CD8+ NKT-like cells. CD4+CD25- naïve T cells (Tn) isolated from B6 spleens were cultured with graded numbers of CD8+ NKT-like cells (NKT) in the presence of irradiated splenic APC and 1.5 ųg/ml anti-CD3. The cultures were pulsed with 1 μCi/well [3H]thymidine for the last 16 h of 72h culture and the level of proliferation was assessed by [3H]thymidine incorporation. (B) Representative suppression of CD4+ and CD8+ naïve T cell proliferation by CD8+ NKT-like cells at different NKT:Tn ratios. CFSE-labeled CD4+CD25- naïve T cells or naive CD8+ T cells isolated from B6 spleens were cultured with CD8+ NKT-like cells in the presence of irradiated splenic APC and 1.5 ųg/ml anti-CD3. After 72 h in the culture, CFSE dilution in the responder CD4+ or CD8+ T cells was analyzed by flow cytometry. In the transwell experiment, CD8+ NKT-like cells (1x105cells/well) are separated from APC and responder T cells. (C) Summary of suppression differences of CD8+ NKT-like cells. Data are presented as the mean and standard deviation of percentages of proliferated T cells among responder T cell precursors. ANOVA test suggest that CD8+ NKT-like cells suppress CD4+ naïve T cells significantly better than CD8+ naïve T cells (p < 10-6).

In Vitro Cultured $CD8^+$ NKT-Like Cells Possess Potent Cytotoxic Activity

Since $CD8^+$ T cells generally possess cytotoxic activity and the generated $CD8^+$ NKT-like cells have high levels of expression of molecule related to cytotoxic activities, we next tried to determine whether the $CD8^+$ NKT-like cells possess cytotoxic activity and whether such activity is responsible for their suppressive activities observed in our suppression assays. The cytotoxic activity was measured using a flow cytometry-based killing assay. Initially, freshly purified splenic T cells and APCs (T cell-depleted spleen cells) were used as killing targets for the $CD8^+$ NKT-like cells. CFSE-labeled target cells were co-cultured with $CD8^+$ NKT-like cells at a 1:1 ratio. Target cell death was measured by 7-AAD incorporation [23].

While naïve splenic T cells are resistant, splenic APCs are much more sensitive to killing by $CD8^+$ NKT-like cells in the killing assay (Fig. 3a). Furthermore, $CD8^+$ NKT-like cells have to be activated to possess killing function (Fig. 3a). Because perforin/granzymes are highly upregulated in the $CD8^+$ NKT-like cells, the perforin/granzyme pathway inhibitor EGTA was used to determine whether this pathway was important to the killing of splenic APCs by $CD8^+$ NKT-like cells. As shown in Fig. 3b, EGTA had no impact on the killing function and, similarly, blocking of the adhesion molecule CD18 did not inhibit the cytotoxicity of $CD8^+$ NKT-like cells.

We further tested the cytotoxic activities of $CD8^+$ NKT-like cells towards other cell types including the DC2.4 dendritic cell line. DC2.4 cells pulsed with OT-I OVA peptides were incubated with $CD8^+$ NKT-like cells cultured from OT-I TCR transgenic mice. As shown in Fig. 3c, $CD8^+$ NKT-like cells from OT-I mice exhibited minimal cytotoxicity to DC2.4 cells without peptide loading, while they kill OVA-loaded DC2.4 cells with high efficiency. Contrary to the killing of splenic APCs, EGTA abrogates almost completely the killing of DC2.4 cells by $CD8^+$ NKT-like cells. The killing of DC2.4 cells is also cell-contact dependent as suggested by the transwell experiment (Fig. 3c). Thus, $CD8^+$ NKT-like cells may possess multiple pathways involved in cytotoxicity and different pathways may be used depending on the target cell types.

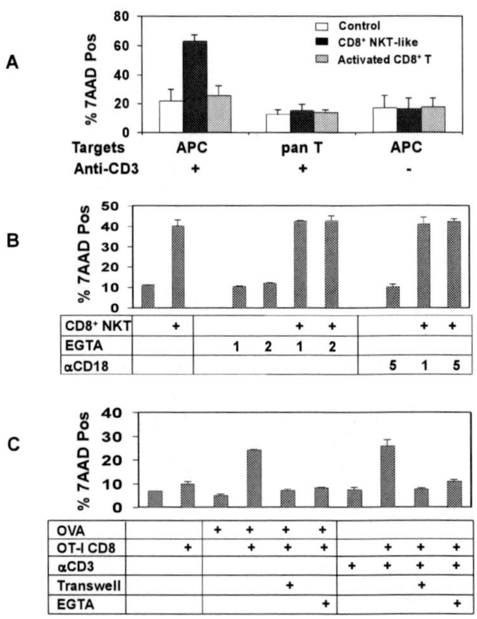

Figure 3. CD8+ NKT-like cells possess potent cytolytic abilities. (A) The killing ability of CD8+ NKT-like cells (1x105) was assessed by co-cultures with different target cells at a 1:1 ratio for 4 hr. Target cells positive for 7-AAD were measured by flow cytometry. The killing ability was compared between expanded CD8+ NKT-like cells and CD8+ T cells activated with plate bounded anti-CD3 for 48 hours. The controls did not contain CD8+ NKT-like or activated CD8+ T cells. (B) Killing of splenic APC by CD8+ NKT-like cells cannot be blocked by EGTA (1mM or 2mM) or neutralizing antibodies against CD18. Toxicity of EGTA (4 mM) and anti-CD18 (5μg/ml) was controlled by cultures of target cells without CD8+ NKT cells. (C) DC2.4 cells were used as target cells and co-cultured with CD8+ NKT-like cells from OT-I TCR transgenic mice at a 1:1 ratio for 4 hr. The killing function was tested with various combinations of experimental conditions (1.5□ g/ml of anti-CD3, OVA peptide, 4mM of EGTA Transwell).

The Suppression Activity of CD8+ NKT-Like Cells Involves Direct Interaction with Naïve T Cells

The cytotoxic activity towards APC used in the suppression assay may explain the suppressive activity of the CD8+ NKT-like cells. However, the molecular profile of these cells [22] suggested that they may possess other suppression mechanism(s). Therefore, we assessed the suppression activity of the CD8+ NKT-like cells using APC-free *in vitro* suppression assays. Briefly, CD4+CD25- T cell responders are cocultured with CD8+ NKT-like cells and anti-CD3 in the absence of APCs. As shown in Fig. 4, CD8+ NKT-like cells are highly efficient in suppressing responder T cells in this system. The suppression efficiency of CD8+ NKT-like cells is identical comparing the suppression assays with and without APC. These results suggest that CD8+ NKT-like cells can mediate suppression of T cell proliferation through APC-independent mechanisms.

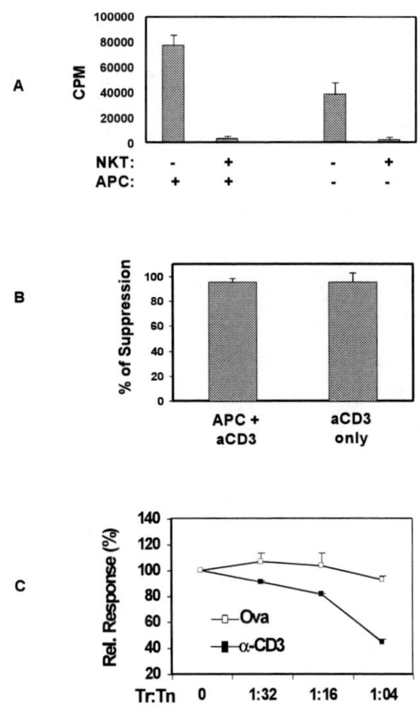

Figure 4. Suppression mechanisms for CD8+ NKT-like cells. (A) Proliferation of CD4+ naïve T cells (1x105/well) stimulated by soluble anti-CD3 (1.5μg/ml) in the presence or absence of APC (1x105/well) and CD8+ NKT-like cells (1x105/well). The cultures were pulsed with 0.5 μCi/well [3H]thymidine during the last 16h of 72h culture and [3H]thymidine incorporation was measured at the end of the culture. (B) The same data in (A) are presented as percentages of suppression for CD8+ NKT-like cells. Suppression % was calculated based on the proliferation of naïve CD4+ T cells in the absence of CD8+ NKT-like cells in each assay system. (C) CD8+ NKT-like suppression requires TCR-dependent activation. CD8+ NKT-like cells (Tr) from B6 mice, naïve T cells (Tn) from OT-II transgenic mice (1x105 cells/well) and APCs (1.5x105) from B6 mice were cocultured at various Tr/Tn ratios in the presence of OVA peptide or 1.5μg/ml of anti-CD3. Proliferation of OT-II transgenic T cells was measured by CFSE. The proliferation of OT-II T cells stimulated with OVA peptide or anti-CD3 in the absence of CD8+ NKT-like cells (Tr/Tn = 0) was considered as the reference (100% proliferation). The proliferation in other conditions was presented as relative response compared to the reference.

Suppression Function of $CD8^+$ NKT-Like Cells Requires Proper Activation

To determine whether the $CD8^+$ NKT-like cells are constitutively suppressive, $CD8^+$ NKT-like cells cultured from B6 mice were examined for their ability to suppress naïve T cells from OT-II TCR transgenic mice. $CD8^+$ NKT-like cells, OT-II transgenic naïve T cells and APCs were cocultured in the presence of OVA peptide (to activate only the OT-II T cells) or anti-CD3 (to activate both the OT-II naïve T cells and $CD8^+$ NKT cells from B6 mice). As shown in Fig. 5, the $CD8^+$ NKT-like cells expanded from B6 mice did not inhibit the proliferation of the OT-II T cells stimulated by the OVA peptide. However, the proliferation of OT-II naïve T cells was significantly inhibited by the expanded $CD8^+$ NKT-like cells in a dose-dependent manner when anti-CD3 is used in the suppression assay. These results suggest that the $CD8^+$ NKT-like cells require TCR-specific activation for effective suppression, an observation consistent with the functional features of *in vitro* expanded $CD4^+CD25^+$ Tregs [24].

Suppressive Activity of $CD8^+$ NKT-Like Cells is Mostly Cell Contact-Dependent

To test whether the suppression activity is cell contact-dependent, the suppressive activity of $CD8^+$ NKT-like cells was tested by trans-well experiments that separate $CD8^+$ NKT-like cells from the responder T cells and APCs with semi-permeable membranes. The suppressive activity of $CD8^+$ NKT-like cells was significantly reduced when they are placed in the upper chamber of the Trans-well (Fig. 2b). This is true irrespective of the types of responder T cells, $CD4^+CD25^-$ or $CD8^+$ T cells. However, the separation of $CD8^+$ NKT-like cells from responder cells and APC by semi-permeable membrane did not completely abolish the suppressive activity (Fig. 2b and 3c), suggesting that soluble factors also contribute to the suppressive activity of $CD8^+$ NKT-like cells.

$CD8^+$ NKT-Like Cells Inhibit Activation and Il-2 Production of Responder T Cells

It has been shown that $CD4^+CD25^+$ Tregs suppress polyclonal T cell activation *in vitro* by inhibiting IL-2 production and addition of IL-2 to the culture media almost completely abrogates the suppressive activities of $CD4^+CD25^+$ Tregs [25–27]. To investigate the role of IL-2 in the suppression function of $CD8^+$ NKT-like cells, we compared IL-2 in the culture media of anti-CD3 and APC stimulated $CD4^+CD25^-$ T cells, expanded $CD8^+$ NKT-like cells and the coculture of the two populations (Fig. 5a). IL-2 was readily detectable in the supernatants of $CD4^+CD25^-$ cell culture, undetectable in the culture of $CD8^+$ NKT-like cells, and reduced in the supernatants of the co-culture dose-dependently on the $CD8^+$ NKT-like cells. When exogenous IL-2 was added to the co-cultures, the suppressive function of $CD8^+$ NKT-like was inhibited in a dose-dependent manner (Fig. 5b). These results are consistent with the possibility that $CD8^+$ NKT-like cells may block the production of Il-2 of $CD4^+CD25^-$ cells. To test the possibility that $CD8^+$ NKT-like cells may inhibit proliferation of responder T cells by sequestering IL-2, we directly examined the production of IL-2 mRNA by RT-PCR. Responder $CD4^+$ T cells were isolated from the co-culture wells of suppression assays at the 48h time point. Similar to $CD4^+CD25^+$ T cells, $CD8^+$ NKT-like cells inhibited IL-2 transcription by responder T cells (Fig. 5c). These results strongly support the view that $CD8^+$ NKT-like cells exert their inhibitory effects by blocking the induction of IL-2 production of responder T cells at the RNA transcription level, which is consistent with the suppression mechanism of $CD4^+CD25^+$ Tregs [25].

IL-2 production and the up-regulation of CD25 surface expression are early events of naïve $CD4^+$ T cell activation in response to anti-CD3 stimulation [28]. Although $CD8^+$ NKT-like cells inhibited the induction of IL-2 mRNA synthesis by $CD4^+CD25^-$ responder T cells, it was still unclear at what stage of the T cell activation process the responder T cells were arrested by $CD8^+$ NKT-like cells. To answer this question, we examined the change of CD25 surface marker on $CD4^+CD25^-$ responder T cells. Similar to $CD4^+CD25^+$ regulatory T cells [29,30], $CD8^+$ NKT-like cells efficiently inhibited CD25 up-regulation on responder $CD4^+CD25^-$ T cells (Fig 5d). Furthermore, the surface expression of CD25 on responder cells correlates with their proliferation (Fig. 5d).

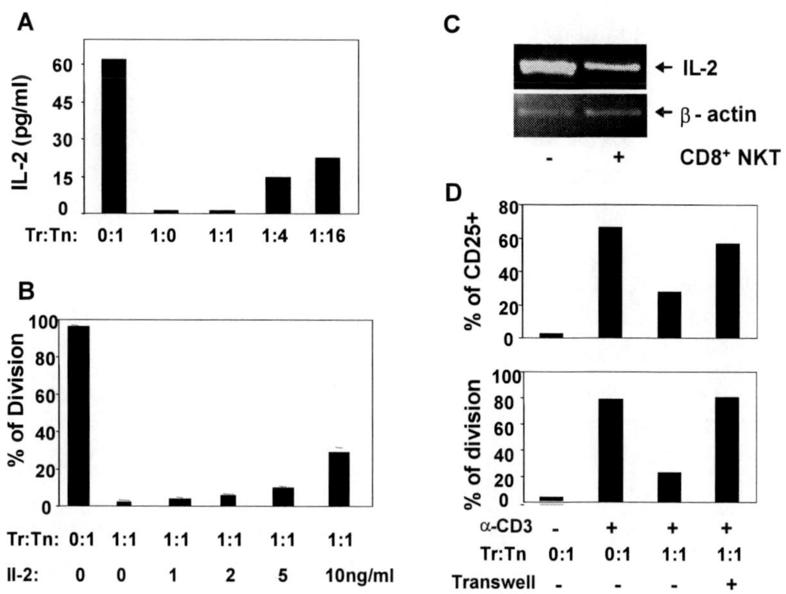

Figure 5. CD8+ NKT-like cells inhibit IL-2 production in responder T cells. (A) CD4+CD25- cells (1x105) and splenic APC (1.5x105) were cultured with 1.5μg/ml of anti-CD3 and different numbers of CD8+ NKT-like cells (Tr). Supernatants were taken at the 72 h time point and IL-2 was quantified by ELISA. (B) Various amounts of exogenous rIL-2 were added to the culture of CD4+CD25- with 1:1 ratio of CD8+ NKT-like cells in the same condition as in (A). At the 72 h culture time point, cell proliferation was assessed by the dilution of CFSE. Data are presented as the ratio of proliferated precursor cells in the total CD4+ responder T cells based on the CFSE dilution. (C) CD4+ responder T cells in the suppression culture were purified at the 48h time point. IL-2 mRNA levels were determined by RT-PCR using β-actin normalization. (D) CD4+CD25- T cells labeled with CFSE were cultured with APC alone or together with expanded CD8+ NKT cells at a 1:1 ratio in the presence or absence of soluble anti-CD3. 72 h after the initiation of the culture, cell proliferation and CD25 expression of responder CD4+ T cells were analyzed in parallel by flow cytometry. Suppressive effects are presented as % of proliferated precursor cells in the total CD4+ responder T cells based on the CFSE dilution. Data are representative of two independent experiments.

Discussion

Cell culture with anti-CD3/anti-CD28-coated beads and high dose IL-2 was previously shown to expand $CD4^{+}Foxp3^{+}$ Treg cells that can suppress the proliferation of responder T cells and prevent the development of autoimmune diseases [31]. Previous studies from our laboratory demonstrated that culture of naïve $CD8^{+}$ T cells with similar condition produces large numbers of $CD8^{+}$ NKT-like cells that co-express CD8 and several NK cell markers including NK1.1, DX5, CD244, NKG2A, and CD94. These cells were shown to have potent immunosuppressive activity and the molecular profiles of these immunosuppressive cells were extensively investigated [20]. However, the immunological mechanisms underlying the observed suppression activity were not addressed in our previous studies. Similar to our studies, $CD8^{+}$ NKT cells can also be cultured from murine spleen, bone marrow or thymus with INF-γ followed by culture with anti-CD3-coated flask in the presence of IL-2 [16]. The $CD8^{+}$ NKT cells expanded with the INF-γ protocol were shown to have antitumor effect. The cytolytic ability of the INF-γ-expanded cells seem to be due to the production of INF-γ (6,000pg/mL) [16]. More recently, Tasken's group from Norway showed the generation of highly suppressive $CD8^{+}CD25^{+}Foxp3^{+}$ cells from human $CD8^{+}CD25^{-}$ PBMCs using continuous antigen stimulation procedure. During the course of stimulation, those cells acquired immunosuppressive properties in parallel with up-regulation and expression of cytotoxic molecules including perforin and Granzymes and they inhibit $CD4^{+}$ and $CD8^{+}$ T cell proliferation and cytokine production mainly through a contact-dependent mechanism [32]. In line with these reports, the $CD8^{+}$ NKT-like cells generated using our system also express high levels of INF-γ (>8,000pg/mL) and other cell death molecules including perforin, Fas-L and multiple granzymes [16]. Therefore, the $CD8^{+}$ NKT-like cells may exert their immune suppressive function through their cytotoxic abilities. This possibility was extensively tested in this study using *in vitro* killing assays. We demonstrated that the *in vitro* generated $CD8^{+}$ NKT-like cells can efficiently kill APCs but not naive T cells. Therefore, cytotoxicity to APC may contribute to the reduced proliferation of responder T cells in the suppression assays that contain APC. We further demonstrated that the suppression of $CD8^{+}$ NKT-like cells was related to the inhibition of IL-2 production in responder cells and exogenous IL-2 was able to partially reverse the suppressive activity of the $CD8^{+}$ NKT-like cells. The contact between $CD8^{+}$ NKT-like cells and APCs as well as between $CD8^{+}$ NKT-like cells and responder T cells are implicated in suppression. These characteristics of the $CD8^{+}$ NKT-like cells are similar to that of the $CD4^{+}CD25^{+}$ regulatory T cells.

Previous studies on $CD8^{+}$ T cell mediated immune suppression indicate that multiple mechanisms are involved in their regulation mechanisms including direct killing of the target cell, negative signaling directly on the target or on APCs and secretion of soluble factors, such as immunosuppressive cytokines [33]. Consistent with these studies, our APC-free suppression assays clearly indicated that the $CD8^{+}$ NKT-like cells can suppress responder T cell proliferation in the absence of APC, suggesting that other than cytotoxicity, the $CD8^{+}$ NKT-like cells possess other suppression mechanisms. Using $IL\text{-}10^{-/-}$ and $IFN\text{-}\gamma^{-/-}$ mice, our previous studies indicate that IL-10 and IFN-γ play an important role in the in vitro generation and suppression functions of $CD8^{+}$ NKT-like cells, which is consistent with the current transwell experimental results, suggesting the involvement of soluble factors in the *in vitro* suppression activities of $CD8^{+}$ NKT-like cells. The detailed mechanisms of the $CD8^{+}$ NKT-like cells suppress T cell proliferation in an APC-independent manner has not been fully investigated. Our microarray data [22] revealed several candidate molecules that are up-regulated in the $CD8^+$ NKT cells compared to fresh $CD8^+$ T cells. These include Tim3, OPG, 4-1BB, GITR, Ox40, Icos, Trail, RANKL, Pdcd1, Pdcd1 ligand 2. Many of these molecules have been previously identified to be involved in the suppression function of $CD4^+CD25^+$ T cells and may be involved in $CD8^+$ NKT suppression function as well, based on the fact they share many functional characteristics. In addition to the cell surface molecules, many soluble molecules such as Esm1, Spp1, Fgl2, Tnfrsf11b, Lgals3, Lgals1, and IL-24, which are known or have the potential to be implicated in the immune suppression function, may also contribute to the potent suppression activities of these $CD8^+$ NKT-like cells. Future studies on both surface and soluble molecules should help illuminate the molecular mechanism underlying the suppressive activity of suppressor T cells.

Acknowledgment

The authors wish to thank Drs. Leszek Ignatowiz and Raffel Pacholczyk for their critical comments on the manuscript. The study was partially supported by a grant from the National Institutes of health (2PO1 AI-42288) to JXS.

References

Baker J, Verneris MR, Ito M, Shizuru JA, Negrin RS (2001) Expansion of cytolytic CD8(+) natural killer T cells with limited capacity for graft-versus-host disease induction due to interferon gamma production. *Blood* 97: 2923-2931.

Ballas ZK, Rasmussen W (1990) NK1.1+ thymocytes. Adult murine CD4-, CD8- thymocytes contain an NK1.1+, CD3+, CD5hi, CD44hi, TCR-V beta 8+ subset. *J Immunol* 145: 1039-1045.

Bendelac A, Rivera MN, Park SH, Roark JH (1997) Mouse CD1-specific NK1 T cells: development, specificity, and function. *Annu Rev Immunol* 15: 535-562.

Bienvenu B, Martin B, Auffray C, Cordier C, Becourt C, Lucas B (2005) Peripheral CD8+CD25+ T lymphocytes from MHC class II-deficient mice exhibit regulatory activity. *J Immunol* 175: 246-253.

BILLINGHAM RE, BRENT L, MEDAWAR PB (1953) Actively acquired tolerance of foreign cells. *Nature* 172: 603-606.

Dellabona P, Padovan E, Casorati G, Brockhaus M, Lanzavecchia A (1994) An invariant V alpha 24-J alpha Q/V beta 11 T cell receptor is expressed in all individuals by clonally expanded CD4-8- T cells. *J Exp Med* 180. 1171-1176.

Earle KE, Tang Q, Zhou X, Liu W, Zhu S, Bonyhadi ML, Bluestone JA (2005) In vitro expanded human CD4+CD25+ regulatory T cells suppress effector T cell proliferation. *Clin Immunol* 115: 3-9.

Eberl G, Lees R, Smiley ST, Taniguchi M, Grusby MJ, MacDonald HR (1999) Tissue-specific segregation of CD1d-dependent and CD1d-independent NK T cells. *J Immunol* 162: 6410-6419.

Godfrey DI, MacDonald HR, Kronenberg M, Smyth MJ, Van Kaer L (2004) NKT cells: what's in a name? *Nat Rev Immunol* 4: 231-237.

Grossman WJ, Verbsky JW, Barchet W, Colonna M, Atkinson JP, Ley TJ (2004) Human T regulatory cells can use the perforin pathway to cause autologous target cell death. *Immunity* 21: 589-601.

Itoh M, Takahashi T, Sakaguchi N, Kuniyasu Y, Shimizu J, Otsuka F, Sakaguchi S (1999) Thymus and autoimmunity: production of CD25+CD4+ naturally anergic and suppressive T cells as a key function of the thymus in maintaining immunologic self-tolerance. *J Immunol* 162: 5317-5326.

Kawamura H, Kameyama H, Kosaka T, Kuwahara O, Bannai M, Kawamura T, Watanabe H, Abo T (2002) Association of CD8+ natural killer T cells in the liver with neonatal tolerance phenomenon. *Transplantation* 73: 978-992.

Kronenberg M, Gapin L (2002) The unconventional lifestyle of NKT cells. *Nat Rev Immunol* 2: 557-568.

Lanier LL, Le AM, Civin CI, Loken MR, Phillips JH (1986) The relationship of CD16 (Leu-11) and Leu-19 (NKH-1) antigen expression on human peripheral blood NK cells and cytotoxic T lymphocytes. *J Immunol* 136: 4480-4486.

MacDonald HR (1995) NK1.1+ T cell receptor-alpha/beta+ cells: new clues to their origin, specificity, and function. *J Exp Med* 182: 633-638.

Mahic M, Henjum K, Yaqub S, Bjornbeth BA, Torgersen KM, Tasken K, Aandahl EM (2008) Generation of highly suppressive adaptive CD8(+)CD25(+)FOXP3(+) regulatory T cells by continuous antigen stimulation. *Eur J Immunol* 38: 640-646. 10.1002/eji.200737529 [doi].

Ohteki T, MacDonald HR (1994) Major histocompatibility complex class I related molecules control the development of CD4+8- and CD4-8- subsets of natural killer 1.1+ T cell receptor-alpha/beta+ cells in the liver of mice. *J Exp Med* 180: 699-704.

Piccirillo CA, Shevach EM (2001) Cutting edge: control of CD8+ T cell activation by CD4+CD25+ immunoregulatory cells. *J Immunol* 167: 1137-1140.

Porcelli S, Yockey CE, Brenner MB, Balk SP (1993) Analysis of T cell antigen receptor (TCR) expression by human peripheral blood CD4-8- alpha/beta T cells demonstrates preferential use of several V beta genes and an invariant TCR alpha chain. *J Exp Med* 178: 1-16.

Ridge JP, Fuchs EJ, Matzinger P (1996) Neonatal tolerance revisited: turning on newborn T cells with dendritic cells. *Science* 271: 1723-1726.

Sarantopoulos S, Lu L, Cantor H (2004) Qa-1 restriction of CD8+ suppressor T cells. *J Clin Invest* 114: 1218-1221. 10.1172/JCI23152 [doi].

Schmidt RE, Murray C, Daley JF, Schlossman SF, Ritz J (1986) A subset of natural killer cells in peripheral blood displays a mature T cell phenotype. *J Exp Med* 164: 351-356.

Shevach EM, McHugh RS, Piccirillo CA, Thornton AM (2001) Control of T-cell activation by CD4+ CD25+ suppressor T cells. *Immunol Rev* 182: 58-67.

Smith TR, Kumar V (2008) Revival of CD8+ Treg-mediated suppression. *Trends Immunol* 29: 337-342. S1471-4906(08)00128-2 [pii];10.1016/j.it.2008.04.002 [doi].

Smith TR, Kumar V (2008) Revival of CD8+ Treg-mediated suppression. *Trends Immunol* 29: 337-342. S1471-4906(08)00128-2 [pii];10.1016/j.it.2008.04.002 [doi].

Stremmel C, Exley M, Balk S, Hohenberger W, Kuchroo VK (2001) Characterization of the phenotype and function of CD8(+), alpha / beta(+) NKT cells from tumor-bearing mice that show a natural killer cell activity and lyse multiple tumor targets. *Eur J Immunol* 31: 2818-2828.

Sykes M (1990) Unusual T cell populations in adult murine bone marrow. Prevalence of CD3+CD4-CD8- and alpha beta TCR+NK1.1+ cells. *J Immunol* 145: 3209-3215.

Takahashi T, Kuniyasu Y, Toda M, Sakaguchi N, Itoh M, Iwata M, Shimizu J, Sakaguchi S (1998) Immunologic self-tolerance maintained by CD25+CD4+ naturally anergic and suppressive T cells: induction of autoimmune disease by breaking their anergic/suppressive state. *Int Immunol* 10: 1969-1980.

Tang Q, Henriksen KJ, Bi M, Finger EB, Szot G, Ye J, Masteller EL, McDevitt H, Bonyhadi M, Bluestone JA (2004) In vitro-expanded antigen-specific regulatory T cells suppress autoimmune diabetes. *J Exp Med* 199: 1455-1465.

Thornton AM, Shevach EM (1998) CD4+CD25+ immunoregulatory T cells suppress polyclonal T cell activation in vitro by inhibiting interleukin 2 production. *J Exp Med* 188: 287-296.

Wingender G, Berg M, Jungerkes F, Diehl L, Sullivan BA, Kronenberg M, Limmer A, Knolle PA (2006) Immediate antigen-specific effector functions by TCR-transgenic CD8+ NKT cells. *Eur J Immunol* 36: 570-582.

Yoshimoto T, Paul WE (1994) CD4pos, NK1.1pos T cells promptly produce interleukin 4 in response to in vivo challenge with anti-CD3. *J Exp Med* 179: 1285-1295.

Zhou L, Wang H, Zhong X, Jin Y, Mi QS, Sharma A, McIndoe R, Garge N, Podolsky R, She JX (2008) The IL-10 and IFN-gamma pathways are essential to the potent immunosuppressive activity of cultured CD8+ NKT-like cells. *Genome Biology* 9: R119.

In: Advances in Medicine and Biology Research, Vol. 14
Editor: Leon V. Berhardt
ISBN: 978-1-61761-930-4

Chapter XI

Severe Late Onset Pompe Disease: Management and Treatments

David Orlikowski*, Hélène Prigent, Nadine Pellegrini and Djillali Annane

Home ventilation unit and intensive care, Functional testing department, Raymond Poincaré teaching hospital, 104 boulevard Raymond Poincaré 92380 Garches, France.

Abstract

Pompe disease is a rare genetic metabolic myopathy caused by deficiency of lysosomal acid α-glucosidase (GAA). Age of onset and rate of progression cover a broad clinical spectrum. Presentation of disease ranges from rapidly progressive infantile form with cardiac, skeletal and respiratory muscles always fatal without specific treatment to juvenile forms with muscular, respiratory and sometimes cardiac impairment and late onset forms consisting in progressive limb girdle myopathy beginning in adulthood, usually around the third decade, involving skeletal muscles but sometimes more specifically respiratory muscles.

Severe forms of Late Onset Pompe Disease (LOPD) are characterized by a progressive respiratory involvement that can lead to severe respiratory failure, which can require ventilatory support by mechanical ventilation, and/or muscular involvement that can lead to loss of ambulation. The management of these patients usually requires a multidisciplinary approach involving the neurologist, the myology specialist, the pulmonologist and the physical medicine specialist.

In this book chapter, we describe the particularity of these severe forms: clinical presentation with the description and relationship of the muscular and respiratory impairments. Muscular and functional explorations, respiratory and sleep explorations and criteria for starting ventilation are also discussed focusing on Pompe disease particularities. Recent advances in therapeutics, like enzyme replacement therapy (ERT) have mainly focused on infantile forms; however, late onset forms due to their frequency

*Corresponding author: David Orlikowski, Email: david.orlikowski@rpc.aphp.fr, Tel: 0033147107776, Fax: 0033147104446

and to the heterogeneous presentation and progression they are susceptible to present, might also benefit from those advances.

Introduction

Epidemiology and Clinical Presentation of Late onset Forms

The spectrum of Pompe Disease (LOPD) is wide and varies from a rapidly progressive infantile disease with cardiorespiratory insufficiency, which is usually fatal before 1 yr of age (1) in absence of treatment, to a slowly progressive adult onset disease with muscle weakness, respiratory insufficiency (1, 2) and sleep-disordered breathing (3) defining the Late Onset Pompe Disease (LOPD) forms. Adults typically present with progressive myopathy that may resemble limb-girdle muscular dystrophy (4). In our experience this is a myopathy usually evolving during the second part of life; on a series of 29 adult patients with confirmed LOPD, the age at symptom onset ranged from 16 to 70 yrs (41±11) (5). In contrast to other neuromuscular diseases, where respiratory insufficiency occurs after loss of ambulation, respiratory symptoms in LOPD can be one of the first manifestations and respiratory impairment can be major while the ability to walk still remains (6).

However, both the presenting symptoms and the clinical outcomes in adults seem to vary widely. Although respiratory muscle involvement is rarely inaugural (7, 8), it may occur early in the course of the disease and may cause death in the absence of specific follow-up and treatment by nocturnal non invasive ventilation, requiring even in some cases invasive ventilation (9). The onset of respiratory impairment may be very subtle and about two-third of the patients experience a progressive respiratory muscles weakness, with a decrease in vital capacity (VC) and cough ability, and the development of sleep disorder breathing leading to chronic respiratory insufficiency and even to acute respiratory failure in the absence of intervention such as ventilation (9).

The best definition of severe LOPD seems to be patients requiring a walking assistance and a respiratory assistance (invasive or not) for part of or for the entire day. In a recent questionnaire study of 54 LOPD patients of the Dutch Neuromuscular Diseases Association, Hagemans *et al.* (10) found that 15 patients used both ventilatory assistance and a wheelchair, 6 patients started ventilatory assistance during the same year that they were diagnosed and 3 patients started both assistance modalities during the same year.

Clinical Presentation and Prevalence of Patients with Severe Motor Deficit

Usually, muscle weakness in LOPD is symmetrical but differs between the upper and the lower part of limbs with a higher muscular impairment in the proximal part. These data were recently confirmed by a whole body MRI study (11) where severe muscular destruction was observed in LOPD with a gradual progression related to the duration of disease (figure 1).

Muscular impairment and weakness prevails on limb girdles, usually involving first the pelvic girdle through the following muscles: glutei, adductors and flexors of hip, then also of quadriceps and harmstrings; however, the scapular girdle may also be affected, often secondarily, mainly involving the deltoid muscle but also the fixative muscles of the scapula, with a decrease of the active abduction of shoulders. In distal part of the limb, muscular strength is preserved for a very long time. The trunk is also affected with a deficit of abdominal muscles and of flexors of the neck. The walking gait is rather characteristic, with hyperlordosis and external rotation of hips.

At the time of completion of the questionnaire in the Hageman's study (10), 87% of the interviewed patients experienced walking problems, ranging from waddling gait to a complete inability to walk. Fifteen percent used walking aids, but did not need a wheelchair. Twenty six patients (84%) used a wheelchair. Half of them (12) always needed their wheelchair for mobility. Half alternated the use of a wheelchair with the use of walking aids such as a walking frame or a cane, depending on the distance to be covered. The mean age at which patients started to use a wheelchair was 46.1±12.4 years, ranging from 22 to 71 years (10). Among the 29 LOPD patients studied by Pellegrini *et al* (5), 23 patients used a wheelchair, starting at a mean age of 46 yrs. 7 patients (24%) had a Walton score higher or equal to 6 (Gardner, Medwin and Walton score ranges from 0 for normal to 10 for bed bound) (12).

Muscular Assessment in LOPD

Because specific movements are impaired in LOPD a regular functional evaluation is therefore necessary: for example in Hageman's study, 16 % were unable to raise arms above the head, 35 % were unable to rise from a chair and 41 % were unable to climb a staircase (10).

Proposed Motor and Functional Evaluation in LOPD

Articular assessment, in search of limitation of articular amplitudes at lower and upper limbs level (even if there is generally no contracture), can be performed by:

- Interrogation, searching for pain, which can observed most frequently at the spine, the shoulders or the thighs
- Analytical muscular assessment including:
    - Muscular testing, made by muscular groups according to the Modified Medical Research Concil (Medical Research Council. Aids to the investigation peripheral nerve scold. London: Majesty's Her Stationery Office; 1976)
    - Active shoulders abduction assessment
    - A quantified force measurement if it is possible, on quadriceps and the flexors of the knee, as well as on the triceps brachial and the flexors of the elbow
    - A measurement of the termino–lateral force the thumb/index grip, always using the same dynamometer and a similar position of the upper limb for the measurement

- Functional assessment including:
  - Brooke score for upper limbs and Vignos score for lower limbs (13, 14).
  - Motor function measurement scale (MFM), which must be realized by a specifically trained user.
  - The 6 minutes walking test , which must be performed between 2 20-meters distant marks
  - Barré test for upper limbs and (over 150 seconds) and Mingazzini test for lower limbs (over 75 seconds)
  - Timed tests:
    - 10 meters walking test (clarify if technical help is needed).
    - Climbing up of 8 steps (clarify if technical help is needed).
    - Passage from lying to sited position

Functional urinary signs must be searched and explored by ultrasound exam of urinary tract and urodynamic assessment.

BROOKE's SCALE of the upper limbs : attribute a rank / 6 according to the possibilities of the patient with its 2 upper limbs[1]:

1) Can make a complete abduction the tense arms, the physiological position (arm along the body) until join hands above the head
2) Can put hands on the head only by folding elbows
3) Cannot raise hands above the head, but can carry a full glass (180 ml) in the mouth (possibly by using 2 hands)
4) can raise hands to the mouth, but cannot carry a full glass (180 ml) in the mouth
5) Cannot carry the hands in his(her,its) mouth, but can hold a pencil or take a coin on the table
6) Cannot carry the hands in the mouth, and does not have no useful usage of the hands

VIGNOS's SCALE : attribute a rank / 10 according to the capacity of movement[2]

1) Walking, and climbs stairs without help
2) Walking, and climbs stairs by means of a banister
3) Walking, and climbs stairs slowly by means of a banister (+ of 25 seconds for 8 standard stairs)
4) Walking without help and gets up from a chair, but cannot climb stairs
5) Walking without help), but cannot get up from a chair nor climb stairs
6) Walk only with help, or walk only with a big equipment of lower limbs
7) Walking with a big equipment of lower limbs, but needs help to remain in balance

---

[1]Brooke MH, Griggs RC, Mendell JR, Fenichel GM, Shumate JB, Pellegrino RJ : Clinical trial in Duchenne dystrophy : I. The design of the protocol. Muscle Nerve 1981 ; 4 : 186 - 97

[2]Vignos PJ, Spencer GE, Archibald KC : Management of Progressive Muscular Dystrophy of Childhood — JAMA 184 : 89 – 96, 1963

8) Is standing up with a big equipment of lower limbs, but is incapable to walkeven with help
9) Is in wheelchair. The flexors of the elbow are at least 3 (MRC scale, movement against gravity)
10) Is in wheelchair or in the bed. The flexors of the elbow are unless 3 (MRC scale, movement impossible against gravity)

Respiratory Management

LOPD is characterized by a progressive respiratory involvement that can lead to severe respiratory failure which can require ventilatory support by mechanical ventilation.

The mechanism of respiratory failure is predominantly progressive diaphragmatic weakness which can lead to intense orthopnea for some patients. Respiratory involvement is frequent in LOPD and was reported in a third of the patients in early descriptions of the disease (2). However more recent and well documented studies described respiratory muscle weakness in up to 89.7% of the patients (6). Respiratory failure can reveal LOPD (2, Laforet, 2000 #97, 7, 15, 16). Respiratory muscle weakness is not related to the patients' age (5, 15). While Laforet *et al* (16) did not find any relation between limb and respiratory muscle weakness in 21 patients, mixing LOPD and juvenile forms, Pellegrini *et al* (5) found a week correlation between respiratory and locomotor function in 29 LOPD patients. Duration since symptoms onset seems to be the best predictor of disease progression and respiratory involvement (3, 6, 17). Progression of the disease varies substantially between patients (15, 17, 18). During the long-term follow-up of 16 patients mixing LOPD and juvenile forms, Van der Beek *et al* (18)observed a simultaneous decline in respiratory and locomotor function in only two thirds of the patients. Progression of respiratory muscle weakness can precede locomotor decline, some patients developing severe respiratory failure requiring mechanical ventilation while they are still able to walk (5, 18, 19). Documentation of the evolution of respiratory function showed a mean decline of predicted VC ranging from 1.6% to 4.6% per year (5, 6, 20, 21)

Respiratory and Sleep Exploration

Respiratory Exploration

Although the routine respiratory follow-up of LOPD is not standardized, the progressive respiratory muscle weakness, the variability of patient's disease progression and the absence of strong correlation between respiratory and locomotor function incite to perform a systematic evaluation upon diagnosis and a regular follow-up of respiratory function.

Routine respiratory evaluation should include VC measurement with spirometry, as a decrease in VC is an early sign of respiratory involvement. Measurements of CV in the upright and supine positions are helpful as a > 20% drop of CV in the supine strongly points to major diaphragmatic weakness (22, 23); it is often observed in LOPD patients (3, 6).

Respiratory muscle strength should also be assessed, at least by non-invasive techniques with the measurement of maximal inspiratory (MIP) and expiratory (MEP) pressures. MIP reflects inspiratory muscle strength and can be measured by sniff inspiratory nasal pressure or maximum inspiratory mouth pressure which are complementary non-invasive volitional techniques (24, 25). MEP allows evaluating expiratory muscle weakness and should be associated to cough evaluation by peak cough flow. A peak cough flow <200 L/min has been considered a sign of impaired cough (26), although a recent study suggests that values <255 L/min are already predictive of a risk of ineffective cough during respiratory tract infections (27).

Blood gases allow the assessment of gas exchange and acid-base status and the evaluation of the functional consequence of respiratory muscle weakness and the development of respiratory failure (28).

The rate of respiratory assessment in LOPD is not standardized; however the observed rate of decline of respiratory muscle function has prompted some teams to recommend a respiratory evaluation every 6 to 12 months in Pompe disease patients (21).

Sleep Exploration

According to the severity of diaphragmatic weakness and the importance of CV decline, respiration is usually first compromised during sleep with occurrence of hypoventilation during REM sleep. Only later with the progression of the disease will diurnal hypoventilation appear. Therefore, polysomnography should be performed regularly to detect sleep disordered breathing(29).

Mellies *et al.* (3) systematically studied disease course, respiratory function and nocturnal respiration in 20 patients with LOPD and 7 with juvenile forms. Ventilatory capacity, respiratory muscle function and polysomnography were systematically assessed. Sleep disorder breathing (SDB) was defined by a respiratory disturbance index (RDI) associating apneic, hyponeic and desaturation events higher than 10/H; SDB was subclassified according to the presence or not of nocturnal hypoventilation (defined as elevated PtCO2 > 50 mmHg or SaO2< 90% of either REM sleep alone of REM and non-REM sleep). Nocturnal hypoventilation associated to a normal RDI was also considered SDB. Nine patients needed help for ambulation or were unable to walk alone. Lung and respiratory function were well correlated. Thirteen patients (48%) presented diaphragmatic weakness (defined as a postural drop on vital capacity > 20%); 10 of them presented hypercapnic respiratory failure (defined as daytime hypocia and hypercapnia > 45 mmHg) which was associated to a longer disease duration. Sleep disorder breathing was present in 13 patients (48%), 11 of which exhibited diaphragmatic weakness . It was due to obstructive sleep apnea syndrome in only one patient, to REM sleep hypopneas in 3 patients, REM sleep hypopneas with hypoventilation in 4 patients and continuous hypoventilation whatever the sleep stage in the 5 remaining patients. Introduction of mechanical ventilation reversed the blood gazes exchanges abnormalities. A VC below 40% was a strong predictor of continuous nocturnal hypoventilation with a sensitivity of 80% and specificity of 93%.

In more severe patients, Pellegrini *et al* (5) studied 11 ventilated patients included: 9 had diurnal hypercapnia 5.9 kPa (45 mmHg) and 2 patients had nocturnal oxyhaemoglobin desaturation with morning headaches. Five patients underwent overnight polysomnography before and after MV initiation. Prior to MV initiation, 3 had significant apnoea–hypopnoea index per hour of sleep and percentage of total sleep time spent with arterial oxygen saturation (SaO2) < 90%. When polysomnography was repeated in the same patients after MV initiation both apnoea–hypopnoea index and percentage of total sleep time with SaO2 < 90% improved.

Prevalence of Patients Needing Invasive or Non-Invasive Ventilation

Among the 29 LOPD patients of Pellegrini study, 16 patients (55%) using MV and 5 patients (17%) had a tracheostomy. MV was initiated at a mean age of 50±12 yrs (range 20–70 yrs). Most of the patients were ventilated with an assist-control volumetric mode (5).

In Hagemans's study, 20 patients (37%) used artificial ventilation: non-invasive by nose hood or facemask (n =16) or invasive by trachea canulla (n = 4). The median duration of artificial ventilation was 11.5 h per day (range 8–24 h). All 20 patients needed ventilation during the night. Twelve patients used it during daytime. The mean age of the ventilator dependent patients was 56.7 ± 13.2 years (range 32–81 years) and the mean age at the artificial ventilation initiation was 48.6 ± 16.3 years (range 15–78 years). Six patients started using artificial ventilation in the same year that they were diagnosed. The time between first complaints and diagnosis for these patients was 24.8 ± 15.5 years, ranging from 1 to 39 years (10).

Criteria for Starting Ventilation

Neuromuscular respiratory failure usually evolves insidiously, this is especially true in Pompe disease where a discrepancy between muscular and respiratory weakness can be observed. Progressive and unspecific symptoms such as headache, mild and late dyspnea and orthopnea (in case of diaphragmatic paralysis) or sleep disorders are major indicators to perform respiratory explorations. A pluridisciplinary follow up, including clinical exam, measurements of blood gases, vital capacity, maximal expiratory or inspiratory pressures and sleep polysomnographic exam when necessary are useful to detect criteria for long term mechanical ventilation (hypercapnia >45 mmHg, nocturnal desaturations < 88%, VC <60% , PImax < 60 cm H_2O)(28, 30).

The start of home ventilation is a major decision in patients with neuromuscular disease as it entails physical, psychological, social and sometimes financial consequences. The launch of home ventilation should be anticipated as often as possible in order to avoid its initiation during an acute respiratory failure episode. Patients with Pompe disease and their family must be followed and informed regarding this risk, even if substitutive treatment is started because in late onset forms there are actually no proof that enzymotherapy can prevent the need for nocturnal ventilation. Home ventilation is usually begun during a sufficient hospital stay to optimise both the device and the ventilator setting. Use of cough assistance techniques must be proposed for Pompe patients requiring mechanical ventilation (9).

In the Pellegrini's study (5),11 patients met international criteria for nocturnal noninvasive MV (30). When comparing MV users and non-users, no difference in age was found but VC was significantly different between the two groups, as expected lower in MV users, but differences were also noted for locomotor function higher Walton score, lower MMRC score for lower limb and lower shoulder abduction.

MV was initiated during an episode of acute respiratory failure in five patients, with two of these patients needing to undergo tracheostomy during the same episode. Tracheostomy was also performed in two patients who failed to respond adequately to noninvasive MV and in one patient after a stroke responsible for impaired swallowing. Most of the patients were ventilated with an assist-control volumetric mode and a back-up rate set two to three breaths below the awake spontaneous breathing rate, according to consensus conference guidelines (31).

Relationship between Respiratory Impairment and Muscular Impairment (Figures 1 And 2)

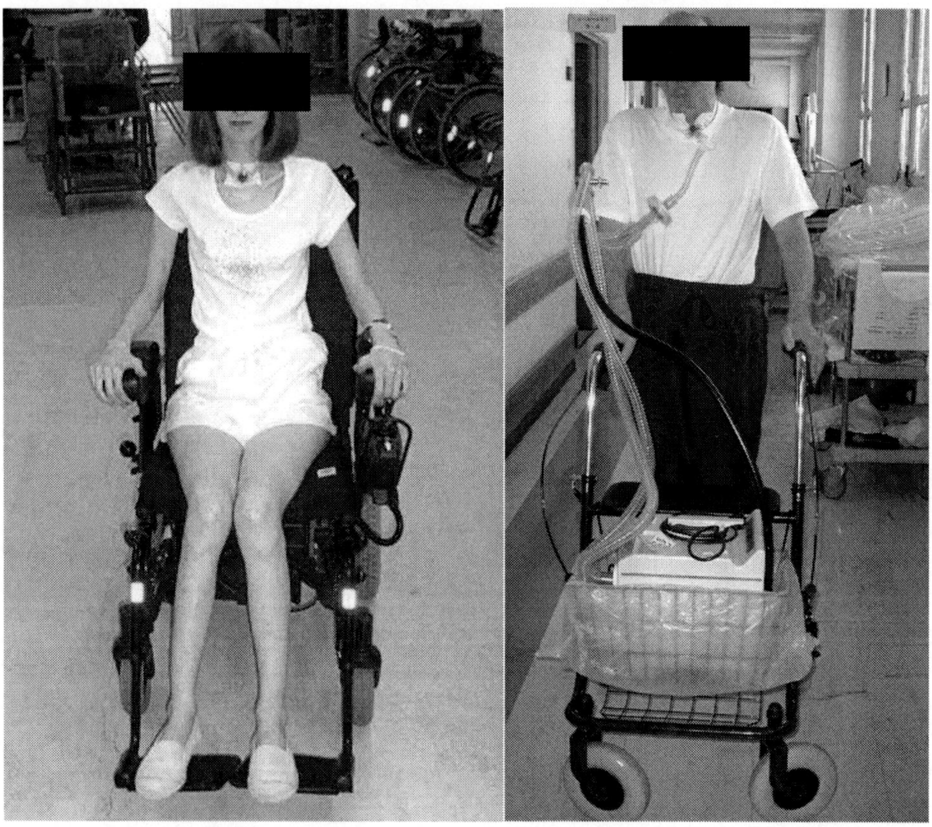

Figure 1 and 2: two patients with severe late onset forms requiring mechanical ventilation (invasive) and a wheel chair in one case (a 20 years old young women with late onset juvenile form) or walking device (a 50 years old patient with adult late onset form).

Pellegrini *et al* (5) observed that locomotor function was more severely impaired in the patients with rather than without MV, although age was similar in the two groups. Two hypotheses were advanced by the authors to explain this finding. First, both respiratory function and locomotor function may reflect muscle weakness, as previously suggested by Hagemans *et al*.(10). For the authors this hypothesis was consistent with the weak correlation between VC and locomotor function parameters. The second hypothesis was that MV may adversely affect locomotor function. However, the present study showed a weak relationship between respiratory function and locomotor function in adults with late-onset Pompe's disease. Disease severity was not related to age. Retrospective data indicated a gradual decline in respiratory and locomotor function over time, as previously described by Wokke *et al.* (6).

Effect of Enzymotherapy

Enzyme replacement therapy was first introduced in infantile forms with severe muscular, respiratory and cardiac impairment and improved dramatically the prognosis of these patients (32). Nevertheless infantile forms are less frequent than late onset forms where only muscular and respiratory impairment are present.

Experience with enzyme therapy in LOD is sill limited. However, a recently published multicenter, randomized, double-blind, placebo-controlled trial on the efficacy and safety of intravenous recombinant alphaglucosidase (Myozyme ®), 20 mg/kg every 15 days, conducted in 90 patients older than 8 years, ambulatory and not invasively ventilated, showed that, after 78 weeks of alphaglucosidase alfa treatment, 6 minutes walking test distance improved and pulmonary function stabilized (with a significantly improved VC compared to the placebo group). On the secondary outcomes, both maximal inspiratory pressures (not significantly) and expiratory pressures (significantly) also improved in the treated group compared to the placebo group. Thirty one patients used ventilatory support and 39 patients needed a walking device but no data were provided regarding improvement in use of these technical helps (33).

In severely affected patients, Van der Ploeg recently reported the outcomes of 18 patients requiring wheelchair and mechanical ventilation in 17. Ten improved respiratory and 13 motor function and the 5 remaining seemed to stabilize. Most of the patient reported improvement in their quality of life and an earlier intervention in the disease course was associated with a greater clinical improvement (34). In a study on 5 severe forms wheelchair and ventilatory dependent (1 invasively ventilated, 3 ventilated 24/24h) patients with a mean age of 48±14 years, at 52 weeks of treatment 2 were able to be disconnected from ventilator during 1 an 2 hours respectively. Two showed mild improvements in slow VC, sitting and supine, Pi max and Pe max and transdiaphragmatic pressures (35).

Conclusion

Muscular and respiratory impairments are often present in muscular disease. In late onset form of Pompe disease, a weak correlation between these 2 aspects has been described. Its explained that loss of ambulation may occur sometimes after initiation of nocturnal ventilation or that tracheostomy is needed despite mild limb muscular impairment. Both regularly respiratory and muscular evaluations are then required for the follow up of these patients. Enzyme replacement therapy, recently introduced, seems to stabilize respiratory and/or muscular function, even in severe forms, but it is too early to state on effects on mechanical ventilation.

References

[1] Engel AG, Gomez MR, Seybold ME, Lambert EH. The spectrum and diagnosis of acid maltase deficiency. *Neurology* 1973;23(1):95-106.

[2] Rosenow EC, 3rd, Engel AG. Acid maltase deficiency in adults presenting as respiratory failure. *Am J Med* 1978;64(3):485-91.

[3] Mellies U, Ragette R, Schwake C, Baethmann M, Voit T, Teschler H. Sleep-disordered breathing and respiratory failure in acid maltase deficiency. *Neurology* 2001;57(7):1290-5.

[4] Engel AG. Acid maltase deficiency in adults: studies in four cases of a syndrome which may mimic muscular dystrophy or other myopathies. *Brain* 1970;93(3):599-616.

[5] Pellegrini N, Laforet P, Orlikowski D, Pellegrini M, Caillaud C, Eymard B, et al. Respiratory insufficiency and limb muscle weakness in adults with Pompe's disease. *Eur Respir J* 2005;26(6):1024-31.

[6] Wokke JH, Escolar DM, Pestronk A, Jaffe KM, Carter GT, van den Berg LH, et al. Clinical features of late-onset Pompe disease: a prospective cohort study. *Muscle Nerve* 2008;38(4):1236-45.

[7] Keunen RW, Lambregts PC, Op de Coul AA, Joosten EM. Respiratory failure as initial symptom of acid maltase deficiency. *J Neurol Neurosurg Psychiatry* 1984;47(5):549-52.

[8] Sivak ED, Salanga VD, Wilbourn AJ, Mitsumoto H, Golish J. Adult-onset acid maltase deficiency presenting as diaphragmatic paralysis. *Ann Neurol* 1981;9(6):613-5.

[9] Mellies U, Lofaso F. Pompe disease: a neuromuscular disease with respiratory muscle involvement. *Respir Med* 2009;103(4):477-84.

[10] Hagemans ML, Hop WJ, Van Doorn PA, Reuser AJ, Van der Ploeg AT. Course of disability and respiratory function in untreated late-onset Pompe disease. *Neurology* 2006;66(4):581-3.

[11] Dlamini N, Jan W, Norwood F, Sheehan J, Spahr R, Al-Sarraj S, et al. Muscle MRI findings in siblings with juvenile-onset acid maltase deficiency (Pompe disease). *Neuromuscul Disord* 2008;18(5):408-9.

[12] Walton JR, L. . Clinical examination, differential diagnosis and classification. 6th Edn ed. New York: *Churchill Livingstone;* 1994.

[13] Brooke MH, Griggs RC, Mendell JR, Fenichel GM, Shumate JB, Pellegrino RJ. Clinical trial in Duchenne dystrophy. I. The design of the protocol. *Muscle Nerve* 1981;4(3):186-97.

[14] Vignos PJ, Jr., Spencer GE, Jr., Archibald KC. Management of progressive muscular dystrophy in childhood. *Jama* 1963;184:89-96.

[15] Muller-Felber W, Horvath R, Gempel K, Podskarbi T, Shin Y, Pongratz D, et al. Late onset Pompe disease: clinical and neurophysiological spectrum of 38 patients including long-term follow-up in 18 patients. *Neuromuscul Disord* 2007;17(9-10):698-706.

[16] Laforet P, Nicolino M, Eymard PB, Puech JP, Caillaud C, Poenaru L, et al. Juvenile and adult-onset acid maltase deficiency in France: genotype-phenotype correlation. *Neurology* 2000;55(8):1122-8.

[17] Hagemans ML, Winkel LP, Hop WC, Reuser AJ, Van Doorn PA, Van der Ploeg AT. Disease severity in children and adults with Pompe disease related to age and disease duration. *Neurology* 2005;64(12):2139-41.

[18] van der Beek NA, Hagemans ML, van der Ploeg AT, Reuser AJ, van Doorn PA. Pompe disease (glycogen storage disease type II): clinical features and enzyme replacement therapy. *Acta Neurol Belg* 2006;106(2):82-6.

[19] Hagemans ML, Winkel LP, Van Doorn PA, Hop WJ, Loonen MC, Reuser AJ, et al. Clinical manifestation and natural course of late-onset Pompe's disease in 54 Dutch patients. *Brain* 2005;128(Pt 3):671-7.

[20] Mellies U, Stehling F, Dohna-Schwake C, Ragette R, Teschler H, Voit T. Respiratory failure in Pompe disease: treatment with noninvasive ventilation. *Neurology* 2005;64(8):1465-7.

[21] Van der Beek NA, Hagemans ML, Reuser AJ, Hop WC, Van der Ploeg AT, Van Doorn PA, et al. Rate of disease progression during long-term follow-up of patients with late-onset Pompe disease. *Neuromuscul Disord* 2009;19(2):113-7.

[22] Allen SM, Hunt B, Green M. Fall in vital capacity with posture. *Br J Dis Chest* 1985;79(3):267-71.

[23] Fromageot C, Lofaso F, Annane D, Falaize L, Lejaille M, Clair B, et al. Supine fall in lung volumes in the assessment of diaphragmatic weakness in neuromuscular disorders. *Arch Phys Med Rehabil* 2001;82(1):123-8.

[24] Steier J, Kaul S, Seymour J, Jolley C, Rafferty G, Man W, et al. The value of multiple tests of respiratory muscle strength. *Thorax* 2007;62(11):975-80.

[25] Terzi N, Orlikowski D, Fermanian C, Lejaille M, Falaize L, Louis A, et al. Measuring inspiratory muscle strength in neuromuscular disease: one test or two? *Eur Respir J* 2008;31(1):93-8.

[26] Bach JR, Saporito LR. Criteria for extubation and tracheostomy tube removal for patients with ventilatory failure. A different approach to weaning. *Chest* 1996;110(6):1566-71.

[27] Sancho J, Servera E, Diaz J, Marin J. Predictors of ineffective cough during a chest infection in patients with stable amyotrophic lateral sclerosis. *Am J Respir Crit Care Med* 2007;175(12):1266-71.

[28] ATS/ERS Statement on respiratory muscle testing. *Am J Respir Crit Care Med* 2002;166(4):518-624.

[29] Case LE, Kishnani PS. Physical therapy management of Pompe disease. *Genet Med* 2006;8(5):318-27.

[30] Clinical indications for noninvasive positive pressure ventilation in chronic respiratory failure due to restrictive lung disease, COPD, and nocturnal hypoventilation--a consensus conference report. *Chest* 1999;116(2):521-34.

[31] Make BJ, Hill NS, Goldberg AI, Bach JR, Criner GJ, Dunne PE, et al. Mechanical ventilation beyond the intensive care unit. Report of a consensus conference of the American College of Chest Physicians. *Chest* 1998;113(5 Suppl):289S-344S.

[32] Kishnani PS, Corzo D, Nicolino M, Byrne B, Mandel H, Hwu WL, et al. Recombinant human acid [alpha]-glucosidase: major clinical benefits in infantile-onset Pompe disease. *Neurology* 2007;68(2):99-109.

[33] van der Ploeg AT, Clemens PR, Corzo D, Escolar DM, Florence J, Groeneveld GJ, et al. A randomized study of alglucosidase alfa in late-onset Pompe's disease. *N Engl J Med* 2010;362(15):1396-406.

[34] van der Ploeg AT, Marsden DL. Response to Enzyme Replacement Therapy in 18 Juvenile and Adult Patients with Pompe disease. *In: 59th annual meeting of American Academy of Neurology*. Boston; 2007. p. 257.

[35] Orlikowski DL, P. Pellegrini, N. Prigent, H. Carlier, P. Carlier, R. Eymard, B. Lofaso, F. Annane, D. Effects of one year enzyme replacement therapy in severe form of late onset Pompe disease. Severe Late Onset Treatment Study (french SLOTS). *In: 60th annual meeting of American Academy of Neurology.* Chicago; 2008.

In: Advances in Medicine and Biology Research, Vol. 14
Editor: Leon V. Berhardt
ISBN: 978-1-61761-930-4

Chapter XII

Biofilm Development Kinetics

Masashi Nakano

Professor Emeritus, the University of Tokyo, Graduate school of Agriculture and Life Science, nakano27@vesta.ocn.ne.jp

Abstract

The microbial biofilm development is theoretically analyzed in combination of Monod-based microbial kinetic equation with a biofilm architecture model. It is assumed that the development of biofilm is primarily dominated by the net decay rates of microorganisms while an increase in the population of microorganisms are expressed by the sum of the net growth and decay rates. The biofilms are modeled so that they consist of the stack of micro-colonies and a density function of micro-colonies within biofilms is proposed in order to express the biofilm architecture in the direction perpendicular to the surface of substratum. By modeling a micro-colony as disk-shaped, it has been demonstrated that the maximum thickness of biofilms is in a range of 20 - 55 μm at the mean density of $7x10^{-6}$ g/cm^3 of microorganisms in biofilms, while it has been suggested that an increase in the population of microorganisms significantly causes an increase in the density of microorganisms in micro-colonies besides an increase in both the maximum thickness of biofilms and the number of the stacks of micro-colony in biofilms.

Keywords: Biofilm, Micro-colony, Microorganisms, Microbial kinetics, Modeling.

1. Introduction

The formation and development of microbial biofilms have been studied using such tools as confocal scanning laser microscope, nuclear magnetic resonance imaging, scanning vibrating electrode technique and exo-polysaccharide analysis since the first description by Zobel and Anderson [1, 2]. In addition, a number of modeling studies have been performed to predict the biofilm development since the process was extremely complicate and affected by a number of factors. The models presented for biofilm development have been built based on the insights that biofilms will develop owing to the combined effects of the growth of microorganisms with the transport dynamics of nutrient or substrate as well as biomass and biofilm itself [3-9].

Although biofilms consist of various components such as inert residues, biogenic extracellular products, substances peeling away from the interface of substratum and substances migrating into biofilms from the outside, the development of biofilm is considered to be essentially due to an increase in the population of microorganisms and the habitat architecture built by living microorganisms within biofilms. Therefore, in order to analyze the formation and development of the microbial biofilms, it is important to evaluate the growth and decay kinetics of microorganisms and the ecological behavior of microorganisms in biofilms as well as especially the structure of biofilms that changes with time owing to an increases in population of microorganisms.

The growth and decay kinetics of microorganisms can be expressed using the net growth and decay parameters introduced in Monod-based kinetic equation [10, 11]. The primary ecological behavior of microorganisms might be expressed in the facts that microorganisms form micro-colonies and biofilms consist of the heaps of micro-colony [2, 12]. The structure of biofilms could be identified by the distribution function of micro-colonies in biofilms in the horizontal and vertical directions.

This chapter demonstrates that the biofilm development can be evaluated using the mathematical models for both the growth kinetics of microorganisms and the physical biofilm structure represented as the stacks of micro-colonies by assuming that the density function of micro-colonies might be related to the mean density of microorganisms in biofilms [13].

2. Growth Kinetics of Microorganisms and Biofilm

The changes of population of microorganisms with time could be generally expressed by the sum of the net growth and decay rates using Monod-based kinetic equation.

$$\frac{dM}{dt} = \mu M - \kappa M \tag{1}$$

where M is the population of microorganisms, μ and κ are the biological parameters referring to the net growth and decay, respectively, and t is time.

Both the biological parameters μ and κ could be expressed as the function of time since the growth and decay rates of microorganisms is considered to change with both the ageing and the density of the microorganisms living in micro-colonies and biofilms. The function could be given as follows;

$$\mu = \mu_{\max}\left(1 - a_{\mu} \cdot \exp\left(-b_{\mu} t^2\right)\right) \tag{2}$$

and

$$\kappa = \kappa_{\max}\left(1 - a_{\kappa} \cdot \exp\left(-b_{\kappa} t^2\right)\right) \tag{3}$$

where μ_{max} and κ_{max} are the maximum of μ and κ, respectively, and a and b are the biological constants peculiar to the relevant microorganisms. The subscripts μ and κ refer to the constants for the growth and decay, respectively. The restrained conditions, $a_\kappa > a_\mu$ and $b_\kappa < b_\mu$, could be given since the population of microorganisms increases gradually with time at an initial stage. In addition, the condition that κ_{max} is equal to μ_{max} could be given since the population of microorganisms will be in a stable state by balancing the net decay rate with the net growth rate at the final stage.

While the population of microorganisms increases as time progresses, the thickness of biofilms increases with time at an initial stage owing to the reflecting of the growth and decay kinetics and it reaches a certain steady state with the maximum values at the final stage owing to the influence of an increases in population and the increasing environmental restriction factors against the growth of microorganisms. These influences might be considered to be mainly expressed by the net decay rate of microorganisms. Therefore, the development of the biofilm thickness could be assumed to be followed by the changes of the decay coefficient κ with time.

Hence, we assume that the biofilm thickness χ could be expressed as follows:

$$\chi = \chi_{max}\left(1.1 - a_\chi \cdot \exp\left(-b_\chi t^2\right)\right) \tag{4}$$

where χ_{max} is the maximum thickness of biofilms, a_χ and b_χ are the biological constants peculiar to the relevant microorganisms and the conditions that $a_\chi = a_\kappa$ and $b_\chi = b_\kappa$ might be given by assuming that the changes of χ with time are significantly followed by the changes of decay coefficient κ with time. A constant of 1.1 in equation is introduced by assuming that the initial thickness of bio-film will be around 10 % of the maximum thickness in order to avoid the divergence of the integral in computing the changes of the density of microorganism in biofilms with time.

3. Population of Microorganisms

The population of microorganisms in the biofilms with the surface area A could be given by $M = \int_0^{\chi} A\rho dx$, where ρ is the density of microorganisms at a certain location in biofilms and x is the spatial coordinate in biofilms. Hence, Equation (1) leads to the following equation

$$\frac{\partial}{\partial t}\left(\int_0^{\chi} A\rho dx\right) = (\mu - \kappa)\int_0^{\chi} A\rho dx \tag{5}$$

Let us consider the mean density $\bar{\rho}$ of microorganisms in biofilms since the population will be different in the direction perpendicular to the surface of materials that biofilms are formed.

$$\bar{\rho} = \frac{1}{\chi}\int_0^{\chi} \rho dx \tag{6}$$

Substituting of Equation (6) in Equation (5) leads to:

$$\frac{\partial}{\partial t}\left(\chi\bar{\rho}\right)=\left(\mu-\kappa\right)\chi\bar{\rho} \tag{7}$$

Rearranging Equation (7) (see Appendix A), we obtains:

$$\frac{\partial\bar{\rho}}{\partial t}=\left\{\left(\mu-\kappa\right)-\frac{\partial}{\partial t}\left(\ln\chi\right)\right\}\bar{\rho} \tag{8}$$

Equation (8) indicates that the thickness χ of biofilms plays a role restraining the increases in the population of microorganisms. Integrating Equation (8) with respect time t, we obtain:

$$\frac{\bar{\rho}}{\rho_0}=\exp\left[\int_0^t\left\{\left(\mu-\kappa\right)-\frac{\partial}{\partial t}\left(\ln\chi\right)\right\}dt\right] \tag{9}$$

where ρ_0 is the initial density of microorganisms in biofilms.

Now, we can define the non-dimensional mean density $\bar{\rho}_{non}$ as $\bar{\rho}_{non}=\bar{\rho}/\rho_0$. Hence, the mean density $\bar{\rho}$ of microorganisms in biofilms is expressed as $\bar{\rho}=\rho_0\bar{\rho}_{non}$ and the population M of microorganisms in biofilms might be expressed by $M=A\chi\rho_0\bar{\rho}_{non}$ using this non-dimensional mean density $\bar{\rho}_{non}$.

4. Structure of Biofilm

Let us consider that microorganisms inhabit micro-colonies and micro-colonies are heaped up. Consequently, biofilms are formed on the surface of materials while the population of microorganisms increases with time. The micro-colonies may be initially scattered on the surface of materials and they gradually reduce the distances between adjacent micro-colonies while constructing a stack and forming biofilms. In this process, the micro-colonies are considered to be distributed in the direction perpendicular to the surface of materials while they horizontally keep the space between adjacent micro-colonies, as shown in Figure 1. In the space formed between micro-colonies, there are various substances such as inert residues, biogenic extracellular products and other chemical matters. In addition, the transport of gasses and nutrient elements are performed for the survival of microorganisms [2, 14-16].

In order to describe this process of biofilm development, let us introduce a distribution function of micro-colonies in biofilms in the direction perpendicular to the surface of materials in such a form that the biofilms could keep space between adjacent micro-colonies, as follows:

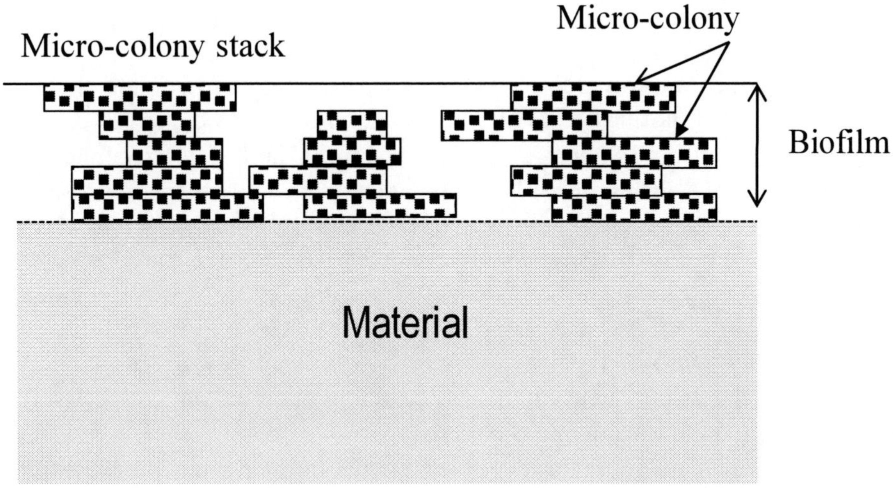

Figure 1. Schematic structure of biofilm. Micro-colonies are scattered and construct stacks on the surface of materials.

$$n_k = n_{suf} \frac{\delta}{1+\alpha}\left(1+\alpha e^{-\gamma k^2}\right) \tag{10}$$

and horizontally, let us introduce the following:

$$n_{suf} = P \cdot n_{suf}^{\max} \tag{11}$$

where k is the number of layers in the stacks of micro-colonies and n_k is the number of micro-colonies in the kth layer of stacks, while the subscript *suf* and the superscript *max* refer to the number of micro-colonies in the first layer and their maximum value, respectively. α, γ and δ are the biological constants. The constants α and δ must be determined so that n_k is equal to n_{suf} when k=1 and so that a sufficient space between adjacent micro-colonies is kept for the diffusion of gas and nutrients in biofilms. Equation (10) refers to a probability density function of micro-colonies in biofilms.

We have introduced a coefficient P into Equation (11) in order to assess the number of micro-colonies in the first layer of the micro-colony stacks. Physically speaking, P could define the mean distances between two adjacent micro-colonies in the first layer (See Appendix B). In this study, P was considered to be expressed by the linear function of the mean density of microorganisms $\bar{\rho} = \rho_0 \bar{\rho}_{non}$ in biofilms as an acceptable approximation so that the number of micro-colonies in the first layer increases with an increase in the population of microorganisms in biofilms. Consequently, Equation (11) could indirectly express the horizontal distribution of micro-colonies in biofilms.

$$P = G \cdot \rho_0 \bar{\rho}_{non} + F \tag{12}$$

When the number of the micro-colony stacks reaches k at a certain time, the population of microorganisms in the biofilms with surface area A also could be expressed using both the density of microorganisms in micro-colonies and the probability density function (Equation (10)) of micro-colonies in the direction perpendicular to the surface. We then obtain the following.

$$M = sh\rho_{bc} A \int^{k} n_k dk \tag{13}$$

where s and h are the surface area and the thickness of a micro-colony, respectively, and ρ_{bc} is the density of microorganisms in micro-colonies.

This population of microorganisms must be equal to the population expressed using the non-dimensional mean density $\bar{\rho}_{non}$ owing to the principle of mass conservation. Hence, we could describe the following.

$$sh\rho_{bc} A \int^{k} n_k dk = A\chi \rho_0 \bar{\rho}_{non} \tag{14}$$

The biofilms will gradually develop and the thickness will reach a specific steady state values at a time when no increases of the mean density of microorganisms are observed. The biohilm thickness in this specific steady state refers to the maximum thickness of biofilms.

Equation (14) also could hold for the biofilms with the maximum thickness as follows:

$$sh\rho_{bc} A \int^{k_{\max}} n_k dk = A\chi_{\max} \left[\rho_0 \bar{\rho}_{non} \right]_{\max} \tag{15}$$

where the subscript *max* refers to the maximum values of the relevant factors.

Introducing a relation of $\chi_{\max} = hk_{\max}$, Equation (15) leads to:

$$\frac{1}{k_{\max}} \int^{k_{\max}} n_k dk = \frac{1}{s\rho_{bc}} \left[\rho_0 \bar{\rho}_{non} \right]_{\max} \tag{16}$$

By solving Equation (16) using Equation (10) and the maximum mean density $\left[\rho_0 \bar{\rho}_{non} \right]_{\max}$ of microorganisms that is given by Equation (9), we can obtain $k_{\max}$ and consequently know the maximum thickness $\chi_{\max}$ of biofilms using a relation of $\chi_{\max} = hk_{\max}$.

As a result, we could evaluate the biofilm development from Equation (4) using this $\chi_{\max}$ computed, provided both the biological constants a_χ and b_χ peculiar to the relevant microorganisms.

5. Case Studies

5.1 Mean Density of Microorganisms

The acceptable biological constants used for case studies are shown in Table 1. The initial density ρ_0 of microorganisms adhered was given in two different values of $4\text{x}10^{-8}$ and $4\text{x}10^{-9}$ g/cm^3 on the basis of biomass abundance observed in fresh groundwater by Harvey et al. [17].The constant $\mu_{\max}$ has been estimated from the values shown in literatures [6, 18-23]. The constants a_μ and b_μ used for expressing the biological parameter μ were given so that all of microorganisms keep the same values for the growth regardless of their age. The constant a_κ used for κ was given as unity in order to demonstrate the significant changes of κ with b_κ.

Table 1. Biological constants selected for computing.

Case No.	ρ_0 g/cm^3	μ_{max} day^{-1}	a_μ [-]	b_μ [-]	κ_{max} day^{-1}	a_κ [-]	b_κ [-]
(i)	$4\text{x}10^{-9}$	2.4	0.5	4.0	2.4	1.0	0.043
(ii)	$4\text{x}10^{-8}$	2.0	0.5	4.0	2.0	1.0	0.050
(iii)	$4\text{x}10^{-8}$	2.4	0.5	4.0	2.4	1.0	0.083
(iv)	$4\text{x}10^{-9}$	2.0	0.5	4.0	2.0	1.0	0.034

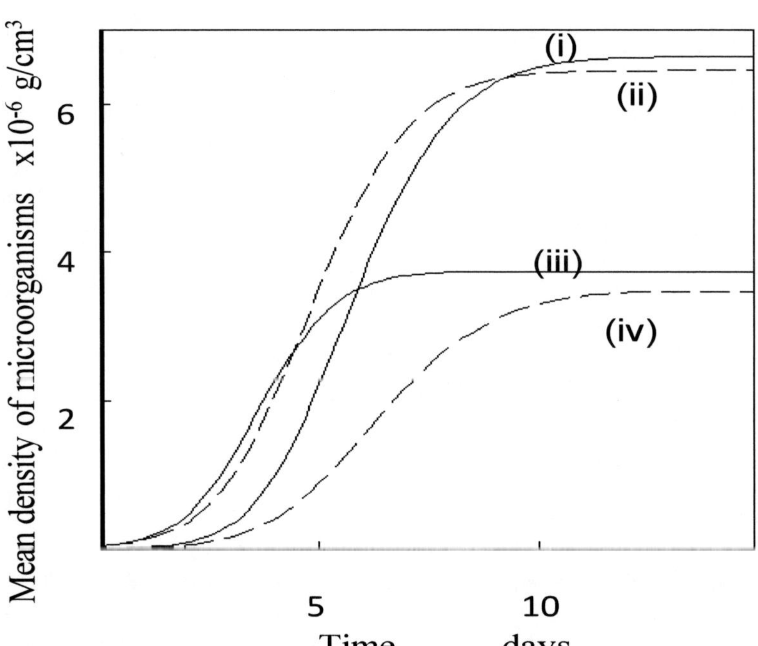

Figure 2. Changes of mean density of microorganisms in biofilms with time. The figures indicate the case number shown in Table 1.

Figure 2 illustrates the changes of the mean density $\bar{\rho}$ of microorganisms in biofilms with time after a few amounts of microorganisms adhere to the surface of materials. The mean density increased abruptly at the beginning of the growth and reached a specific steady state at the final stage in any cases. In this final state, the net growth rate was equilibrated with the net decay rate of microorganisms. The mean density was larger as b_κ was smaller, in comparison of events (ii) and (iii) with the same initial density, even though μ_{max} was smaller in event (ii). In addition, there was a certain specific time at when the mean density reached a steady state and they was larger in events (i) and (iv), provided the smaller value of b_κ. At this specific time, the net decay rates was analyzed to be equal to the net growth rates of microorganisms

It was noted that the almost same maximum mean density of microorganisms was computed in events (i) and (ii) as well as in events (iii) and (iv), although the different combinations of the constant μ_{max}, a_μ, b_μ, κ_{max}, a_κ, b_κ and the initial density ρ_0 were used.

5.2 Maximum Thickness of Biofilms

We need the information on the shape and size of micro-colonies in order to compute the maximum thickness of biofilms. Now, let us consider a disk-shaped micro-colony since it is not difficult to suppose that a heap of the disk-shaped micro-colonies will be corn-shaped or mushroom-shaped [2, 12]. In addition, let a disk-shaped micro-colony keep a diameter of 10 μm and a thickness of 5 μm as an elementary size easy to solve Equation (16) on k_{max} [14, 24].

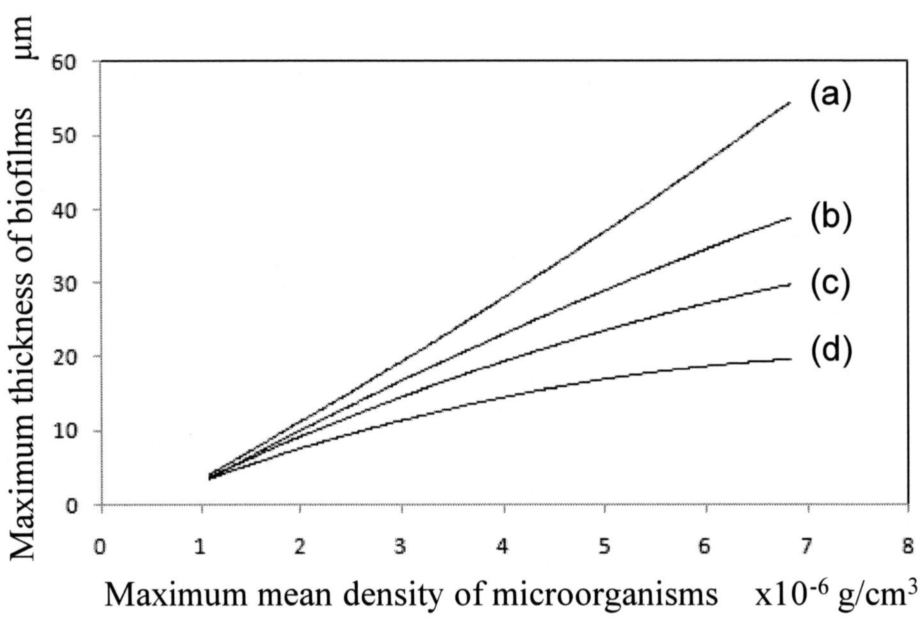

Figure 3. Maximum thickness of biofilms plotted against the various maximum mean densities of microorganisms in biofilm that were computed for a number of different combinations of μ_{max}, κ_b and ρ_0 at ρ_{bc} = 100 (a), 97.5 (b), 95.0 (c), and 90.0 x10^{-3} g/cm^3 (d).

Supposing δ =1.25, α =1.02, and γ = 0.5 for Equation (10) as a acceptable hypothesis on the basis of the computer simulation results shown by Costeron and his group [2, 12, 15] and assuming G = 0.208×10^6 and F = – 0.109 for Equation (12) as possible and useful values [13], we obtained the maximum thickness using the maximum mean density $\left[\rho_0 \bar{\rho}_{non}\right]_{max}$ shown in Figure 2.

Figure 3 illustrates the maximum thickness of biofilms plotted against the various maximum mean densities of microorganisms that were computed for a number of different combinations of μ_{max}, κ_b and ρ_0 at the several density ρ_{bc} of microorganisms in micro-colonies. The value of ρ_{bc} was estimated on the basis of the value of 90 mg/cm^3 that was used for simulating microbial growth dynamics by Benefield and Molz [19] and Molz et al. [24]. The maximum thickness increased with an increase in the maximum mean density of microorganisms in biofilms and the mode of increases were considerably affected by the density ρ_{bc} of microorganisms in micro-colonies. The maximum thickness of biofilms was in range 20-55 μm at the maximum mean density of 7x10^{-6} g/cm^3, provided the density of microorganisms in micro-colonies in the range of 90 - 100 x10^{-3} g/cm^3.

5.3 Development of Biofilm

Figure 4 shows the mode of the biofilm development computed for the events shown in Table 1 using the maximum thickness obtained in section 5.2. The constants used to compute them were given as $a_\chi = a_\kappa$, and $b_\chi = b_\kappa$ since the development of biofilms was considered to be followed by rather a net decay rate than a growth rate of microorganisms. The thickness of biofilms was followed by an increase in the mean density of microodganisms that was demonstrated in Figure 2 and reached a certain steady maximum value at a certain time elapsed.

Figure 5 shows the effects of the density ρ_{bc} of microorganisms in micro-colonies on the maximum thickness of biofilms. The maximum thickness increased with an increase in the density ρ_{bc} of microorganisms in micro-colonies. This observation has suggested that an increase in the population of microorganisms significantly causes an increase in the density ρ_{bc} of microorganisms in micro-colonies besides an increase in both the maximum thickness of biofilms and the number of the stacks of micro-colony in biofilms

In the numerous studies performed in the past, the thickness of biofilms has been observed with a large variability since the measurements were performed under different environmental conditions for microorganisms. The average thickness has been reported to be 33 to 42 μm or about 50 μm with a variability of about 13 to 90 μm in pure-culture biofilm systems [25-29], while the average thickness of 123 to 295 μm has been observed in the natural stream [30]. We might suppose that in nature the biofilm thickness was observed extremely larger because biofilms contained various substance included in flow water besides biomass and the relevant matters.

The maximum biofilm thickness predicted by our model was in harmony to the experimental results observed in the pure-culture system. Therefore, our model will express very well the mode of essential biofilm development due to the growth and decay of microorganisms in the events that we could ignore another external cause except an increase in the population of microorganisms.

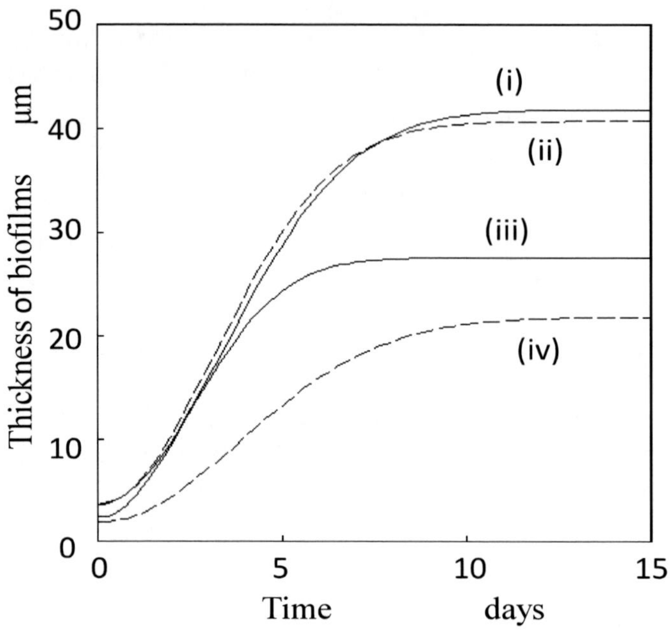

Figure 4. Changes of biofilm thickness with time. The thickness was followed by an increase in the mean density of microodganisms that was shown in Figure 2. The figures indicate the case number shown in Table 1. The density ρ_{bc} of microorganisms in micro-colonies was given as 97.5 x10^{-3} g/cm^3.

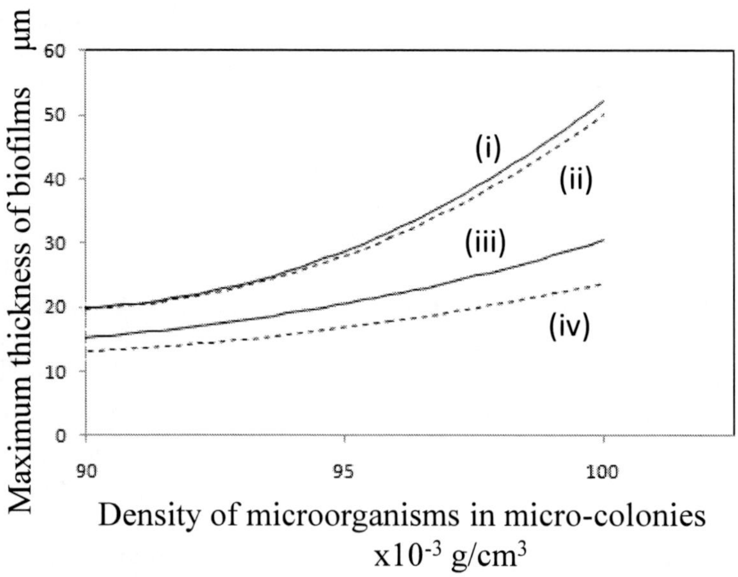

Figure 5. Effects of the density ρ_{bc} of microorganisms in micro-colonies on the maximum thickness of biofilms. The figures indicate the case number shown in Table 1. The mean densities of microorganisms in biofilms were 6.69 x10^{-6} g/cm^3 for case (i), 6.48 x10^{-6} g/cm^3 for (ii), 4.32 x10^{-6} g/cm^3 for (iii) and 3.50 x10^{-6} g/cm^3 for (iv).

Conclusion

The biofilms develop with an increase in the population of microorganisms. By modeling both the net growth and decay rate of microorganisms and the micro-colony system in biofilms, we conclude that the thickness of biofilms can be explained in terms of both the mean density of microorganisms in biofilms and the density of microorganisms in micro-colonies. We have reached a preliminary conclusion that the maximum thickness of biofilms is in range of 20 - 55 μm at the mean density of 7×10^{-6} g/cm^3 of microorganisms in biofilms. An increase in the population of microorganisms significantly causes an increase in the density ρ_{bc} of microorganisms in micro-colonies besides an increase in both the maximum thickness of biofilms and the number of the stacks of micro-colony in biofilms.

Appendix A: Integral of Equation (7)

$$\frac{\partial}{\partial t}(\chi\bar{\rho})=(\mu-\kappa)\chi\bar{\rho} \tag{i}$$

The partial derivative of the left hand side leads to the following:

$$\chi\frac{\partial\bar{\rho}}{\partial t}+\bar{\rho}\frac{\partial\chi}{\partial t}=(\mu-\kappa)\chi\bar{\rho} \tag{ii}$$

Rearranging of Equation (ii) leads to the following equation:

$$\frac{\partial\bar{\rho}}{\partial t}=\left\{(\mu-\kappa)-\frac{1}{\chi}\frac{\partial\chi}{\partial t}\right\}\bar{\rho}=\left\{(\mu-\kappa)-\frac{\partial}{\partial t}(\ln\chi)\right\}\bar{\rho} \tag{iii}$$

The integral form is:

$$\int_0^{\bar{\rho}}\frac{1}{\bar{\rho}}d\bar{\rho}=\int_0^t\left\{(\mu-\kappa)-\frac{\partial}{\partial t}(\ln\chi)\right\}dt \tag{iv}$$

The integral of the left hand side yields the following:

$$\int_0^{\bar{\rho}}\frac{1}{\bar{\rho}}d\bar{\rho}=\left[\ln\bar{\rho}\right]_{\rho_0}^{\bar{\rho}}=\left[\ln\bar{\rho}-\ln\rho_0\right]=\ln\frac{\bar{\rho}}{\rho_0} \tag{v}$$

Hence, equation (iv) leads to:

$$\ln\frac{\bar{\rho}}{\rho_0}=\int_0^t\left\{(\mu-\kappa)-\frac{\partial}{\partial t}(\ln\chi)\right\}dt \tag{vi}$$

Consequently, we obtain:

$$\frac{\bar{\rho}}{\rho_0} = \exp\int_0^t \left\{(\mu - \kappa) - \frac{\partial}{\partial t}(\ln \chi)\right\} dt \tag{vii}$$

Appendix B: Properties of Coefficient *P*

Let us consider the number of micro-colonies arranged on the straight line of length l. The number i of micro-colonies arranged is given by $i = l/D$, provided that the mean distances between the adjacent micro-colonies is D, where the mean distances are defined as a distance between the centers of gravity. The mean distances between the adjacent micro-colonies then can be given by $D = l/i$. The condition $D \geq d$ is given, provided the mean size of micro-colonies is d. Hence, the maximum number $i_{\max}$ of micro-colonies arranged on the line of length l is given by $i_{\max} = l/d$. Assuming that the number of micro-colonies arranged on a line of the unit length l could be given by $i = \beta(l/d)$ at a certain time, where β is a coefficient ($\beta \leq 1$), the mean distances D between the adjacent micro-colonies can be given by $D = l/[\beta(l/d)] = (1/\beta)d$. Hence, we obtain $\beta = d/D$.

According to this way, the number n_{suf} of micro-colonies arranged on the surface of the unit area l^2 is expressed by:

$$n_{suf} = \left(\beta \frac{l}{d}\right)^2 \tag{i}$$

The maximum number $n_{suf}^{\max}$ is given at $\beta = 1$ as follows:

$$n_{suf}^{\max} = \left(\frac{l}{d}\right)^2 \tag{ii}$$

By inserting equation (ii) into (i), we obtain the following

$$n_{suf} = \beta^2 \cdot n_{suf}^{\max} \tag{iii}$$

Inserting $\beta = d/D$ into equation (iii), we obtain:

$$n_{suf} = \left(\frac{d}{D}\right)^2 \cdot n_{suf}^{\max} \tag{iv}$$

In comparison of Equation (iv) with Equation (11) in the text, P is given as

$$P = \left(\frac{d}{D}\right)^2 \quad \text{(v)}$$

The coefficient *P* then is considered to be in inverse proportion to the square of a distance between two adjacent micro-colonies.

References

[1] Zobell, C.E., Anderson, D.Q. 1936. Observations on the multiplication of bacteria in different volumes of stored seawater and the influence of oxygen tension and solid surfaces. *Biol. Bull.* Woods Hole 71: 324-342.

[2] Costerton, J.W., Lewandowski, Z., Caldwell, D.E., Korber, D.R., Lappin-Scott, H.M. 1995. Microbial biofilms. *Annu. Rev. Microbiol.* 49: 711-745.

[3] Rittmann, B.E., McCarty, P.L. 1980a. Model of steady-state-biofilm kinetics. *Biotech. Bioeng.* 22: 2343-2357.

[4] Rittmann, B.E., McCarty, P.L. 1980b. Evaluation of steady-state-biofilm kinetics. *Biotech. Bioeng.* 22: 2359-2373.

[5] Rittmann, B.E., Manem, J.A. 1992. Development and experimental evaluation of a steady-state, multispecies biofilm model. *Biotech. Bioeng.* 39: 914-922.

[6] Wanner, O., Gujer, W. 1986. A multispecies biofilm model. *Biotech. Bioeng.* 28: 314-328.

[7] Wanner, O., Reichert, P. 1996. Mathematical modeling of mixed-culture biofilms. *Biotech. Bioeng.* 49: 172-184.

[8] Lee, M.W., Park, J.M. 2007. One-dimensional mixed-culture biofilm model considering different space occupancies of particulat components. *Water Res.* 41: 4317-4328.

[9] Shafahi, M., Vafai, K. 2009. Biofilm affected characteristics of porous structures. *Int. J. Heat and Mass Transfer* 52: 574-581.

[10] Monod, J. 1949. The growth of microorganismsl culture. *Ann. Rev. Microbiology* 3: 371-394.

[11] Murphy, E.M., Ginn, T.R. 2000. Modeling microbial processes in porous media. *Hydrogeology J.* 8: 142-158.

[12] Wimpenny, J.W.T., Colasanti, R. 1997. A unifying hypothesis for the structure of microbial bio-films based on cellular automaton models. *FEMS Microbiol.* Ecol. 22: 1-16.

[13] Nakano, M., Kawamura, K. 2010. Estimating the degradation of compacted bentonite by a conceptual model based on microbial growth dynamics. *Appl. Clay Sci.* 47: 43-50.

[14] Keevil, C.W., Walker, J.T. 1992. Nomarsky DIC microscopy and image analysis of biofilm. *Binary Comput. Microbiol.* 4: 93-95.

[15] Costerton, J.W., Lewandowski, Z., Caldwell, D.E., Korber, D.R., De Beer, D., James, G. 1994. Biofilms: The customized microniche. *J. Bacteriol.* 176: 2137-2142.

[16] Walker, J.T., Mackerness, C.W., Rogers, J., Keevil, C.W. (1995). Heteroheneous masaic-a haven for waterborne pathogens. In: Lappin-Scott, H.M. and Costerton, J.W., (Eds.), *Microbial Biofilms* (pp. 196-204). Cambridge, Cambridge University Press.

[17] Harvey, W.H., Smith, R.L., George, L. 1984. Effect of organic contamination upon microbial distributions and eterotrophic uptake in a Cape Cod, Mass, aquifer. *App. Environ. Microbiol.* 48: 1197-1201.

[18] MacDonald, D.G., Clark, R.H. 1970. The oxidation of aqueous ferrous sulphate by Thiobacillus ferrooxidans. *Can. J. Chem. Engineering* 48: 669-676.

[19] Benefield, L., Molz, F. 1985. Mathematical simulation of a bio-film process. *Biotechnol. Bioeng.* 27: 921-931.

[20] Konishi, Y., Asai, S. Kato, H. 1990. Microorganismsl dissolution of pyrite by *Thiobacillus ferrooxidans*. *Bioprocess Eng.* 5: 231-237.

[21] Konishi, Y., Asai, S., Yoshida, N. 1995. Growth kinetics of *Thiobacillus thiooxidans* on the surface of elemental sulfur. *Amer. Soc. for Microbiology* 61: 3617-3622.

[22] Atkinson, B., Mavituna, F. Biochemical Engineering and Biotechnology Handbook. 2nd ed. New York: *Macmillan, Stockton Press*; 1991.

[23] Nica, D., Davis, J.L., Kirby, L., Zuo, G., Roberts, D.J. 2000. Isolation and characterization of microorganisms involved in the biodeterioration of concrete in sewers. *Int. Biodeterioration and Biodegradation* 46: 61-68.

[24] Molz, F.J., Widdowson, M.A., Benefield, L.D. 1986. Simulation of microbial growth dynamics coupled to nutrient and oxygen transport in porous media. *Water Resour. Res.* 22: 1207-1216.

[25] Cunningham, A.B., Characklis, W.G., Abedeen, F., Crawford, D. 1991. Influence of biofilm accumulation on porous media hydrodynamics. *Environ. Sci. Technol.* 25: 1305-1311.

[26] Korber, D.R., Lawrence, J.R., Hendry, M.J., Caldwell, D.E. 1993. Analysis of spatial variability within mot^+ and mot^- *Pseudomonas fluorescens* biofilms using representative elements. *Biofouling* 7: 339-358.

[27] Korber, D.R., James, G.A., Costerton, J.W. 1994. Evaluation of fleroxacin activity against established *Pseudomonas fluorescens* biofilma. *Appl. Environ. Microbiol.* 60: 1663-1669.

[28] Stewart, P.S., Peyton, B.M., Drury, W.J., Murga, R. 1993. Quantitative observations of heterogeneities in *Pseudomonas aeruginosa* biofilms. *Appl. Environ. Microbiol.* 59: 327-329.

[29] Wanner, O., Cunningham, A.B., Lundman, R. 1995. Modeling biofilm accumulation and mass transport in a porous medium under high substrate loading. *Biotechnol. Bioeng.* 47: 703-712.

[30] Battin, T.J., Kaplan, L.A., Newbold, J.D., Cheng, X., Hansen, C. 2003. Effects of current velocity on the nascent architecture of stream microbial biofilms. *Appl. Environ. Microbiol.* 69: 5443-5452.

In: Advances in Medicine and Biology Research, Vol. 14 ISBN: 978-1-61761-930-4
Editor: Leon V. Berhardt

Chapter XIII

Definition of Empirical Physical Parameter of Food Texture for Infants and that for Dysphagic Patients and People Having Difficulty in Swallowing Using Official Puncture Tests in Japan

Yukinori Sato*
Faculty of Life and Environmental Sciences,
refectural University of Hiroshima, Japan

Abstract

The puncture test is an empirical test that uses a texturometer. The maximum force on a plunger is analyzed as it penetrates a sample. In Japan, the physical properties of liquid, gel-like, and semi-solid foods for babies and people with dysphagia are measured using the puncture test. This study analyzed the forces on the base and the curved surfaces of the plunger during the puncture test. It was foundthat the contribution of the curved surface to the overall force on the plunger could not be ignored. It was concluded that a simple calculation involving the combined forces on both surfaces of the plunger and the area of its base is inappropriate. Based on these findings, the compressive forces reported in the literature should be recalculated if they were estimated without considering the contribution of the curved surface to the compressive force.

Keywords: puncture test, baby food, dysphagia, texture, mastication

* **Correspondence:** Yukinori SATO, Ph.D., Faculty of Life and Environmental Sciences, Prefectural University of Hiroshima, Nanatsuka 562, Hiroshima 727-0023, Japan. E-mail: yukisato@pu-hiroshima.ac.jp

Introduction

Chewing behavior affects human physical development and the acceptance of solid food consistency (Willson and Green, 2009). The physical properties of food affect the saliva flow rate and the number of chewing cycles, which influence the ability to swallow food (Engelen *et al.*, 2005). Indeed, increasing bolus viscosity improves many swallowing disorders, including dysphagia in the elderly and in patients with neurological diseases (Adeleye and Rachal, 2007; Logemann, 2007; Clavé *et al.*, 2008). Therefore, the physical properties of food may be intimately involved in the development and restoration of oral function at various ages. Solid-food diets are thought to increase the risk of eczema (Fergusson *et al.*, 1982) and asthma (Wilson *et al.*, 1998) in infants younger than 4 months old, although negative results have also been reported (Taylor *et al.*, 1983; Fergusson *et al.*, 1983). The effect of food consistency on the expression of insulin-like growth factors has been investigated in rats (Saito *et al.*, 2004). Food consistency may influence various physiological functions, including human growth.

Low dietary protein intake causes malnutrition (Wadud *et al.*, 2004). In Japan, the physical parameters of foods for infants and people with dysphagia have been defined officially (Ministry of Health, Labor and Welfare in Japan, 1996, 2009). These properties are measured using the puncture test (PT). Briefly, the food sample to be tested is poured into a non-rotational double-cylinder cell (diameter 40 mm) and its upper surface is penetrated by a cylindrical plunger (diameter 20 mm) moving at a fixed speed (10 mm/s). The physical property analyzed is the relationship between the force detected on the plunger sensor, F_t, and time. For liquid or paste foods, the PT used is called the back-extrusion method, since no puncture hole remains after the food is penetrated. It is a simple, inexpensive test of the physical properties of food (Steffe and Osorio, 1987).

It is important to evaluate the properties of physically distinct staple baby foods, milk, and gel-formed foods that are consumed as the ability to masticate develops. For the estimated 15 million people with dysphagia in the USA (Adeleye and Rachal, 2007) and 3 million in Japan (Sumi *et al.*, 2009), the food consumed needs to be more liquid when the condition is more severe. Consequently, to evaluate the physical parameters of foods for developing infants and people with dysphagia in Japan correctly, official PTs for liquid and paste foods must be developed.

Materials and Methods

Materials

1) 1% apple pectin solution

Pectin powder purified from apple pomace was purchased from Sigma (St. Louis, MO, USA). The galacturonic acid and methoxyl contents of the apple pectin were measured to be 79.0% and 8.7% (62.1% the degree of esterification), respectively, the latter by the method of Walter *et al.* (1993). All the reagents used were of analytical grade and were not purified further. The pectin powder was suspended and dissolved in warm water (50°C) using a test tube mixer. The sample was incubated overnight at 50°C and then used in the PTs.

2) Pumpkin paste

A commercial pumpkin paste for infants older than 10 months old (Ajisen Shioji; water content 84.6%) was used immediately after opening.

Methods

The liquid and paste samples were poured to within 5 mm of the tops of cylindrical 20-mm (H15), 35-mm (H30), or 70-mm (H65) cells (diameter 40 mm). Back-extrusion (Steffe and Osorio, 1987; Osorio and Steffe, 1987) was performed at 20°C and a steady speed of 10 mm/s using a polyacetal plunger (diameter 20 mm, depth 24 mm). F_t was measured using an RS 33005S texturometer (Yamaden, Japan) to analyze the physical properties on the F_t–L curve. The official Japanese method for infants uses the maximum value of F_t below L = 13.50 mm to calculate the apparent compressive force, τ_b (N/m^2), when the plunger travels L = 13.53 mm. For people with dysphagia, the maximum F_t value below L = 10.00 mm was adopted, although the plunger used was longer than the 8 mm depth prescribed in the official Japanese method (Ministry of Health, Labor and Welfare in Japan, 2009). The minimum value of F_t detected was 0.001 N.

Results and Discussion

1. F_t–L curve

Figure 1 shows the F_t–L curves for water and an aqueous 1% apple pectin solution. At a 5-mm clearance in sample container H15, the regression of the F_t–L relationship for water was almost linear (R^2 = 0.999), as was that for aqueous 1% apple pectin solution (R^2 = 0.991), with F_t increasing slightly above L = 13 mm. This slight increment may be responsible for the clearance between the bottom of the plunger and the sample cell. To clarify this, using the H30 cell, no increase was seen at L = 13.0 – 13.5 mm with a higher R^2 value of 0.998.

The Ft–L curves of the commercial pumpkin paste are shown in Figure 2. The curve is sigmoid for the H15 cell, and is similar to that for yuzu paste (Sato and Miyawaki, 2007). For the H30 or H65 cells, the F_t–L curves were linear, not sigmoid, indicating that there was little contribution of Ft at the base surface. The slopes of the lines for H30 or H65 were steeper than those for water, due to the viscosity of the paste. Therefore, the contribution of F_t at the curved surface was analyzed theoretically, in more detail.

Ignoring the buoyancy of the plunger, F_t for liquid foods is the sum of F_t at the curved and base surfaces.

$$F_T = 2\pi a L \tau_w + \pi a^2 \Delta P \qquad (1)$$

Eq. (1) can be rewritten as follows (Sato and Miyawaki, 2007):

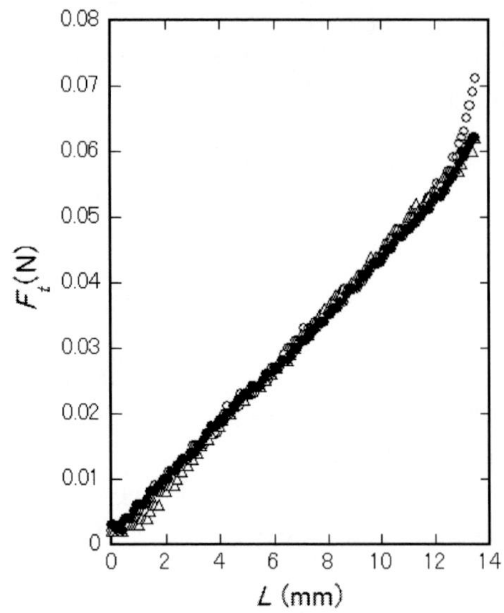

Figure 1. Extrusion forces exerted with 1% apple pectin solution and water according to the clearance between the base surface of the plunger and the outer cylinder cell wall.
○, 1% apple pectin solution extruded to a clearance of 1.47 mm after filling the outer cylinder cell to a depth of 15 mm; △, 1% apple pectin solution extruded at a clearance of 16.47 mm after filling the outer cylinder cell to a depth of 30 mm; ●, Water extruded to a clearance 1.47 mm after filling the outer cylinder cell to a depth of 15 mm. The apparent compressive force (N/m^2), τ_b, was calculated using F_t (N) x π^{-1} x 10^4.

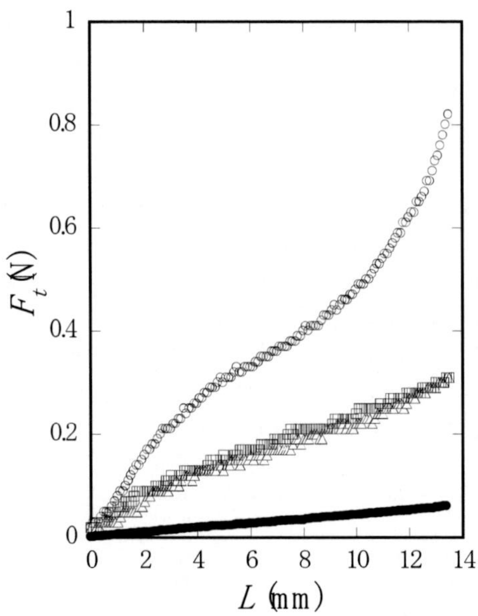

Figure 2. Extrusion force for a commercial pumpkin paste according to the clearance between the base surface of the plunger and the outer cylinder cell wall. ○, Clearance 1.47 mm; △, Clearance 16.47 mm; □, Clearance 51.47 mm; ●, water extruded to a clearance 1.47 mm. The apparent compressive force (N/m^2) was calculated using F_t (N) x π^{-1} x 10^4.

$$F_T = \frac{-\mu\ Vp\,[1+(\kappa^2-1)R^2/\{R^2(1+\kappa^2)\ln\kappa+(1-\kappa^2)\}}{\kappa\ R\ln\kappa} + 4\pi\ aL\mu\ Vp\left[\frac{\kappa}{R\{(1+\kappa^2)\ln\kappa+(1-\kappa^2)\}} + \frac{a}{R^2\{(1+\kappa^2)\ln\kappa+(1-\kappa^2)\}}\right] \quad (2)$$

Considering the coefficients a, V_p, R, and κ for the device used, Eq. (2) can be simplified as:

$$F_T = \alpha\mu + \beta\mu L \quad (3)$$

where α and β are constants specific to the device.

From the inconsistency with the linear equation described in Eq. (3), μ must change during plunger penetration, suggesting a difference in sample viscosity depending on its position: at the curved surface or the base surface of the plunger. The apparent increase in F_t at L = 13 mm for 1% apple pectin solution and at L = 5 mm for pumpkin paste in the H15 cell, compared with F_t for the H30 or H65 cells, may reflect the physical properties at the base surface of the plunger.

2. Contributions of F_t at the curved and base surfaces to the total of F_t on the plunger

When the plunger comes in contact with the surface of the sample liquid or paste, the area of the base surface is involved in calculating the apparent compressive force. During plunger penetration, only the area of the curved surface increases with plunger penetration, while the area of the base surface is fixed. Imagine a hypothetical plunger without a base surface. Figure 3 shows the relationship between F_t and the area of the curved surface. For water, the relationship is linear with R^2 = 0.998, so only the increasing area of the curved surface contributes to the increase in F_t. For a hypothetical plunger without a base surface, the apparent compressive force, τ_c, for 1% apple pectin solution or pumpkin paste can be expressed using the following equation.

$$\tau_c = F_T / 2\pi aL \quad (4)$$

In this model, $\tau_c = \tau_w$ or $\tau_c > \tau_w$ (the subscript w indicates water).

Figure 4 shows the τ_c–L curves for 1% apple pectin solution and water. The curves for water and 1% apple pectin are almost identical using the H15 cell. At L = 0, τ_c is infinity. With penetration, τ_c is a fixed line independent of L, implying that the area of the curved surface is predominately involved in the increase in F_t. At L > 12 mm, there is a marked change for 1% aqueous apple pectin solution using the H15 cell compared with the H30 cell, indicating that the change in μ at L > 12 mm is due to the increase in F_t at the base surface, which depends on the clearance between the base surface of the plunger and the outer cylinder. Although the contributions of the base and curved surfaces to F_t are not clear in Figure 1, describing the τ_c–L curve, it is easy to evaluate the contributions of each plunger surface to F_t. Therefore, the main contribution to F_t detected using the H30 cell is the curved surface. Since the physical parameters measured with both Japanese official methods should reflect the parameters of mastication due to comparing the compressive force, the results using the H15 and H30 cells should be comparable. However, owing to the contribution of the curved surface in the H30 cell, they are currently not comparable.

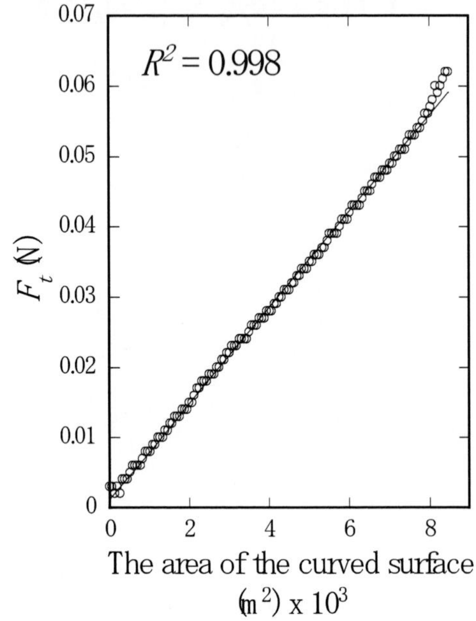

Figure 3. Relationship between the area of the curved surface of the plunger and the extrusion force for water.

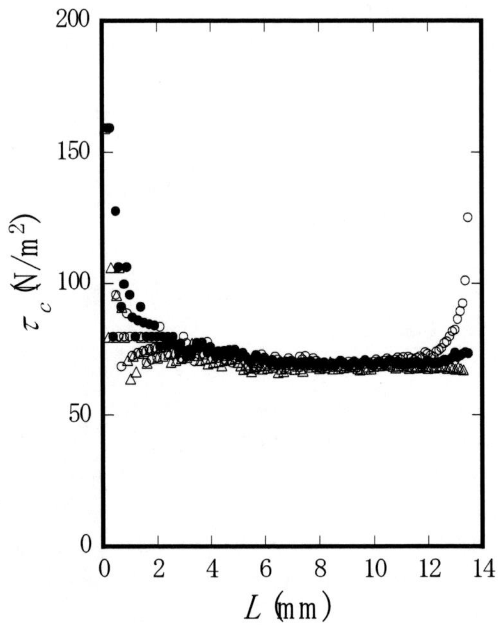

Figure 4. Apparent compressive force, τ_c, calculated from F_t using the area of the curved surface for 1% apple pectin solution and water depending on the clearance between the base surface of the plunger and the outer cylinder cell. ○, Clearance 1.47 mm after filling the outer cylinder cell to a depth of 15 mm; △, Clearance 16.47 mm after filling the outer cylinder cell to a depth of 30 mm; ●, Water at a clearance of 1.47 mm after filling the outer cylinder cell to a depth of 15 mm.

Table 1 shows the apparent and supplemented compressive forces of several food models and a paste using the Japanese official method for infant development (JOI) and the derived method for people with dysphagia (JODP). For the pumpkin paste, 54% of the JOI parameter estimated from F_t at $L = 13.50$ mm is from F_t at the base surface. For the JODP parameter at L = 10.00 mm, the base surface contributes 37%. Based on these findings, the apparent compressive forces reported in the literature should be recalculated due to the large contribution of F_t at the curved surface, which is not related to the compressive force that mimics mastication.

Table 1. Comparison of compressive force of several food models and pastes using Japanese official methods for infant development and for dysphagic patients and people having difficulty in swallowing. (N/m^2)

	For infant development		For dysphagic patients and people having difficulty in swallowing	
	Apparent	After supplemented	Apparent	After supplemented
Distilled water	186±15	3± 0 (2%)*	139± 2	3±15 (2%)
1% apple pectin solution	263±64	79±63 (30%)	140± 0	2± 2 (1%)
Pumpkin paste	2667±606	1452±626 (54%)	1461±278	537±230 (37%)

This assay was carried out in triplicate and n = 10 for a pumpkin paste.

* Average of the ratio between the supplemented compressive force and the apparent compressive force.

An epoch-making measurement of the physical properties of both solid and liquid food is based on the human bite force, which is several hundred Newtons (Bakke, 2006; Kawaguchi *et al.*, 2007). Regardless, the parameters in both JODP and JOI are both named *Katasa*, which is used elsewhere, making it an ambiguous term although their Kanji characters differ. The parameters calculated without including the contributions of each surface to F_t are difficult to relate to compressive forces, so they should be estimated after calculating the surface contributions.

Recently, the Japanese method for evaluating food for people with dysphagia (Ministry of Health, Labor and Welfare in Japan, 2009) was revised, but it is now more complicated owing to the inappropriate change in F_t induced by the shorter length of the plunger; precise evaluation of the physical properties of food is a crucial issue for people with dysphagia.

Conclusion

The empirical physical parameter used to quantify the texture of baby food using the puncture test in Japan was defined and the physical properties observed from the puncture penetration curve were elucidated and compared with the official Japanese method for quantifying food for people with dysphagia.

Both the curved and base surfaces of the plunger contribute to F_t and neither method should ignore the contribution of the curved surface. The parameter used to quantify the physical properties of baby food and food for people with dysphagia should not be calculated only from the force detected on the plunger and the area of the base surface.

Nomenclature

L : depth of the top of the plunger that immersed into the sample liquid or paste

a : radius of the cylinder plunger

τ_w : shear stress (Pa) at the curve surface for water without F_t at the base surface.

τ_c : shear stress (Pa) at the curve surface for sample food or models without F_t at the base surface.

ΔP : change in compressive force (Pa) when the plunger has reached to L (mm).

R : radius of the cylinder cell

κ : ratio of a to R $(= a/R)$

μ : theoretical viscosity of sample liquid or paste

V_p : velocity of the plunger

References

Adeleye B.; Rachal C. (2007). Comparison of the rheological properties of ready-to-serve and powdered instant food-thickened beverages at different temperatures for dysphagic patients. J. Am. Diet. Assoc. 107, 1176-1182.

Bakke M. (2006). Bite Force and Occlusion. Seminars Orthodontics, 12, 120-126.

Clavé P.; Arreola V.; Romea M.; Medina L.; Palomera E.; Serra-Prat M. (2008). Accuracy of the volume-viscosity swallow test for clinical screening of oropharyngeal dysphagia and aspiration. Clin. Nutr., 27, 806-815.

Engeleden L.; Fontijin-Tekamp A.; van der Bilt A. (2005).The influence of product and oral characteristics on swallowing. Arch. Oral Biol.,50, 739-746.

Fergusson D.M.; Horwood L.J.; Shannon F.T. (1982). Risk factors in childhood eczema. J. Epidemiol. Commu. Health., 36, 118-122.

Fergusson D.M.; Horwood L.J.; Shannon F.T. (1983). Asthma and infant diet. Arch. Disease Childhood, 58, 48-51.

Kawaguchi T.; Kawata T.; Kuriyagawa T.; Sasaki K. (2007). In Vivo 3-dimenshional measurement of the force exerted on a tooth during clenching. J. Biomech., 40, 244-251.

Logemann J. A. (2007). Swallowing disorder, Best Practice & Research Clinical Gastroenterology, 21, 563-573.

Ministry of Health, Labour and Welfare in Japan. (1996). Baby Food Shishinn nitsuite., Eisin 56.

Ministry of Health, Labour and Welfare in Japan. (2009). Tokubetsu Youto Shokuhin no Hyoujikyoka touni Tsuite, Shokuannhatsu Dai 0212001.

Osorio F.A. ; Steffe J.F. (1987). Back extrusion of power law fluids, J. Text. Std., 18, 43-63.

Saito T.; Fukui K.; Akutsu S.; Nakagawa Y.; Ishibashi K.; Nagata J.; Shuler C.F.; Yamane A. (2004). Effects of diet consistency on the expression of insulin-like growth factors (IGFs), IGF receptors and IGF binding proteins during the development of rat masseter muscle soon after weaning. Arch. Oral. Biol., 49, 777-782.

Sato Y.; Miyawaki O. (2007). Physical meaning of the parameters used for evaluating food for the elderly, J. Home Econ. Jpn., 58, 261-270 (In Japanese).

Steffe J.F.; Osorio F.A. (1987). Back-extrusion of non-Newtonian fluids, Food Technol., 41, 72-77.

Sumi Y.; Miura H.; Nagaya M.; Nagaosa S.; Umemura O. (2009). Relationship between oral function and general condition among Japanese nursing home residents. Arch. Gerontol. Geriatr., 48, 100-105.

Taylor B.; Wadsworth, J.; Golding, J.; Butler, N. (1983). Breast feeding, eczema, asthma, and hayfever. J. Epidemiol. Commu. Health., 37, 95-99.

Wadud S.; Abid H.; Ara H.; Kosar S.; Shah W.H. (2004). Production, quality evaluation and storage at ability of vegetable protein-based baby food. Food Chem., 85, 175-179.

Walter W. M.; Jr. Fleming H. P.; McFeeters R. F. (1993). Base-mediated firmness retention of sweet potato products. J. Food Sci., 58, 813-816.

Wilson A.C.; Forsyth J.S.; Greene S.A.; Irvine L.; Hau C.; Howie P.W. (1998). Relation of infant diet to childhood health: seven year follow up of cohort of children in Dundee infant feeding study. BMJ, 316, 21-25.

Wilson E.M; Green J.R. (2009). The development of jaw for mastication. Early Human Development, 85, 303-311.

Index

A

B

C

D

F

G

H

I

N

T

U

V